Honda CX/GL 500 & 650 V-Twins Owners Workshop Manual

by Jeremy Churchill

Models covered
CX500.   497cc.   UK 1978 to 1982, US 1978 to 1979
CX500 C.   497cc.   UK 1981 to 1982, US 1979 to 1982
CX500 D.   497cc.   US only 1979 to 1981
CX500 E.   497cc.   UK only 1982 to 1983
GL500.   497cc.   US only 1981 to 1982
GL500 D.   497cc.   UK only 1982 to 1983
GL500 I.   497cc.   US only 1981 to 1982
CX650 C.   673cc.   US only 1983
CX650 E.   673cc.   UK only 1983 to 1986
GL650 673cc.   US only 1983
GL650 D2.   673cc.   UK only 1984 to 1986
GL650 I.   673cc.   US only 1983

(442 - 3AS6)   ABCD

2

Haynes Publishing Group
Sparkford Nr Yeovil
Somerset BA22 7JJ England

Haynes Publications, Inc
859 Lawrence Drive
Newbury Park
California 91320 USA

Acknowledgements

Our grateful thanks are due to Richard Loder, of Loder's former Honda Centre, Dorchester who supplied the original machine and to Paul Branson Motorcycles of Yeovil who supplied the more recent models featured. We also thank Wally Cheshire, now of C.W. Motorcycles, Dorchester, for his help in supplying technical advice.

We would also like to thank the Avon Rubber Company who kindly supplied information and technical assistance on tyre fitting, and NGK Spark Plugs (UK) Ltd for information on spark plug maintenance and electrode conditions.

© Haynes Publishing Group 2019

A book in the **Haynes Owners Workshop Manual Series**

Printed in Malaysia

ISBN 978 1 85010 157 4

Library of Congress Catalog Card Number 86-80546

British Library Cataloguing in Publication Data
Churchill, Jeremy
    Honda CX/GL500 & 650 V-twins owners
    workshop manual. – 2nd ed. – (Owners
    workshop manual)
    1. Honda motorcycles
    I. Title II. Series
    629.28'775   TL448.H6
    ISBN 1 85010-157-4

Printed using NORBRITE BOOK 48.8gsm (CODE: 40N6533) from NORPAC; procurement system certified under Sustainable Forestry Initiative standard. Paper produced is certified to the SFI Certified Fiber Sourcing Standard (CERT - 0094271)

Contents

The CX500 (UK) model

The CX500 C-B model

About this manual

The purpose of this manual is to present the owner with a concise and graphic guide which will enable him to tackle any operation from basic routine maintenance to a major overhaul. It has been assumed that any work would be undertaken without the luxury of a well-equipped workshop and a range of manufacturer's service tools.

To this end, the machine featured in the manual was stripped and rebuilt in our own workshop, by a team comprising a mechanic, a photographer and the author. The resulting photographic sequence depicts events as they took place, the hands shown being those of the author and the mechanic.

The use of specialised, and expensive, service tools was avoided unless their use was considered to be essential due to risk of breakage or injury. There is usually some way of improvising a method of removing a stubborn component, providing that a suitable degree of care is exercised.

The author learnt his motorcycle mechanics over a number of years, faced with the same difficulties and using similar facilities to those encountered by most owners. It is hoped that this practical experience can be passed on through the pages of this manual.

Where possible, a well-used example of the machine is chosen for the workshop project, as this highlights any areas which might be particularly prone to giving rise to problems. In this way, any such difficulties are encountered and resolved before the text is written, and the techniques used to deal with them can be incorporated in the relevant section. Armed with a working knowledge of the machine, the author undertakes a considerable amount of research in order that the maximum amount of data can be included in the manual.

A comprehensive section, preceding the main part of the manual, describes procedures for carrying out the routine maintenance of the machine at intervals of time and mileage. This section is included particularly for those owners who wish to ensure the efficient day-to-day running of their motorcycle, but who choose not to undertake overhaul or renovation work.

Each Chapter is divided into numbered sections. Within these sections are numbered paragraphs. Cross reference throughout the manual is quite straightforward and logical. When reference is made 'See Section 6.10' it means Section 6, paragraph 10 in the same Chapter. If another Chapter were intended, the reference would read, for example, 'See Chapter 2, Section 6.10'. All the photographs are captioned with a section/paragraph number to which they refer and are relevant to the Chapter text adjacent.

Figures (usually line illustrations) appear in a logical but numerical order, within a given Chapter. Fig. 1.1 therefore refers to the first figure in Chapter 1.

Left-hand and right-hand descriptions of the machines and their components refer to the left and right of a given machine when the rider is seated normally.

Motorcycle manufacturers continually make changes to specifications and recommendations, and these, when notified, are incorporated into our manuals at the earliest opportunity.

The CX500 E-C model

The CX650 E-D model

Introduction to the Honda CX/GL 500 and 650 models

When the CX500 was introduced in early 1978 it came as a surprise, being a complete departure from the range of parallel fours and twins which then made up the bulk of Honda's medium to large capacity models. It was evident from the start that the machine had been designed with the interests of the ordinary motorcyclist in mind.

Honda had chosen the transverse V-twin to keep the wheelbase compact and had given it a narrow included angle of 80° as the best compromise between balance and engine width. By 'twisting' the cylinder heads to avoid the usual disadvantage of this layout, that of the carburettors interfering with the rider's knees, and by neatly incorporating the gearbox in the crankcase, they produced an engine unit of reasonable size. The characteristic V-twin power found favour with many riders and heralded a new host of engine layouts which soon began to supplant the parallel four which had been so common before this time.

The machine featured shaft drive and full electronic ignition for ease of maintenance, water-cooling to reduce engine noise and to provide longer life and many other features which combined to make it extremely attractive to many motorcyclists.

Honda soon realised this and launched two new versions in the US in 1979; the Custom featured reversed, highlighted Comstar wheels, the rear being smaller in diameter, a smaller fuel tank and conventional instruments to replace the original headlamp nacelle. The Deluxe model was similar to the original but featured the new wheels and instruments of the Custom model. These two became so successful that the standard model was not continued in 1980, being replaced by the Deluxe version. In the UK the basic model became available in two further versions which had only minor styling differences. Although a Custom model was imported, it was not continued after the initial batch had been sold.

As so many owners had fitted after-market touring equipment,

Honda responded by producing the GL500 Silver Wing model in 1981, which featured air-assisted suspension front and rear, with the Pro-Link system being fitted at the rear instead of twin suspension units. The GL500 Interstate model was fitted as standard with the touring fairing and panniers developed from the Gold Wing models; both GL500 models were fitted with a rear stowage box which could be replaced by an optional pillion seat when required. The GL500 D was imported into the UK in the following year; this was fitted with the Interstate fairing, and a conventional dual seat but had no panniers.

The more sporting inclinations of most UK owners were catered for with the introduction of the CX500 E Sports model in 1982. This was fitted with anti-dive front forks, Pro-Link rear suspension and disc brakes at front and rear; its more European styling included a headlamp fairing and the earlier cam chain tensioner problems were laid to rest by the fitting of an automatic tensioner (also on the GL500 D).

The need for increased performance led to the 500cc engine being increased in capacity to 673cc. This proved so successful that the much modified engine unit was incorporated in all models for 1983, the 500 versions being discontinued. The new models were the Custom CX650 C, the basic GL650 and the touring GL650 Interstate in the US, and the sports CX650 E and the touring GL650 D2 (in 1984) in the UK. All models are very similar to their 500-engined predecessors.

To assist the owner in identifying exactly the machine being worked on (essential for some working procedures and when ordering spare parts), each machine's full title is given below with the approximate dates of import, the frame number with which its production run commenced (and ended, if available) and any distinguishing features, where relevant: **Note:** *UK models are identified at all times in this Manual by the suffix letter indicating the production year; US models are identified by the year of production.*

Model	Frame number	Date of import	Notes
CX500	CX500 – 2003736 to 2051719	Feb 1978 to 1980	The original UK model – no suffix letter used
CX500 1978	CX500 – 2000001 on	1978	Round master cylinder
CX500 1979	CX500 – 2100006 on	1979	Square master cylinder
CX500-A	CX500 – 2200013 on	Feb 1980 to 1981	Square master cylinder, alloy radiator guard, flyscreen
CX500-B	CX500 – 2300004 on	Feb 1981 to 1982	Reversed, highlighted, Comstar wheels
CX500 C 1979	PC01 – 2000017 on	1979	Custom
CX500 C 1980	PC01 – 2100018 to 2125022	1980	Custom
CX500 C-B	PC01 – 2200006 on	Feb 1981 to 1983	The only UK Custom model
CX500 C 1981	PC010 *BM200006 to BM212752	1981	Custom
CX500 C 1982	PC010 *CM300001 to CM307389	1982	Custom
CX500 D 1979	PC01 – 4000005 on	1979	Deluxe
CX500 D 1980	PC01 – 4100002 to 4107865	1980	Deluxe
CX500 D 1981	PC011 *BM200011 to BM205030	1981	Deluxe
CX500 E-C	PC06 – 2000029 on	Feb 1982 to 1983	The UK 'Sports' model
GL500 1981	PC020 *BM000022 to BM007550	1981	Silver Wing
GL500 I 1981	PC021 *BM000007 to BM004396	1981	Silver Wing Interstate
GL500 D-C	N/Av	Jan 1982 to 1983	The UK touring model
GL500 1982	PC020 *CM100001 on	1982	Silver Wing
GL500 I 1982	PC021 *CM100001 on	1982	Silver Wing Interstate
CX650 C 1983	RC110 *DM000018 on	1983	Custom
CX650 E-D	N/Av	March 1983 to 1986	'Sports'
GL650 1983	RC100 *DM000004 on	1983	Silver Wing
GL650 D2-E	RC10E – 410001 on	Feb 1984 to 1986	Touring
GL650 I 1983	RC101 *DM000018 on	1983	Silver Wing Interstate

Note: *The digit indicated by the asterisk (*) in the new VIN numbers of US models varies from machine to machine.*

Model dimensions and weights

Overall length

CX500 C 1979, 1980	2150 mm (84.7 in)
CX500 C 1981, 1982	2160 mm (85.0 in)
CX650 C	2180 mm (85.8 in)
CX500 (UK), CX500 1978, 1979, all CX500 D models	2185 mm (86.0 in)
CX500-A, CX500-B, GL500 1981, 1982	2205 mm (86.8 in)
GL650	2215 mm (87.2 in)
GL500 D-C	2235 mm (88.0 in)
CX500 C-B, CX500 E-C	2240 mm (88.2 in)
CX650 E-D	2250 mm (88.6 in)
GL650 D2-E	2260 mm (89.0 in)
GL500 I 1981, 1982, GL650 I	2305 mm (90.8 in)

Overall width

CX500 (UK)	740 mm (29.1 in)
CX500 E-C	755 mm (29.7 in)
CX650 E-D	760 mm (29.9 in)
CX500-A, CX500-B	770 mm (30.3 in)
CX650 C	790 mm (31.1 in)
CX500 C 1982	855 mm (33.7 in)
CX500 1978, 1979, all CX500 D models	865 mm (34.1 in)
CX500 C-B, CX500 C 1979, 1980, all GL500 models	875 mm (34.5 in)
CX500 C 1981, GL650 D2-E, GL650 I	885 mm (34.8 in)
GL650	890 mm (35.1 in)

Overall height

CX500 (UK)	1125 mm (44.3 in)
All CX500 D models, CX650 C	1165 mm (45.9 in)
CX500 C-B, CX500 C 1979, 1980, 1981	1170 mm (46.1 in)
CX500 1978, 1979	1175 mm (46.3 in)
GL500 1981, 1982	1178 mm (46.4 in)
GL650	1184 mm (46.6 in)
CX500 E-C, CX650 E-D	1190 mm (46.9 in)
CX500 C 1982	1195 mm (47.1 in)
CX500-A, CX500-B	1205 mm (47.4 in)
GL650 D2-E, GL650 I	1480 mm (58.3 in)
GL500 D-C, GL500 I 1981, 1982	1505 mm (59.3 in)

Wheelbase

CX500 C 1981, 1982	1465 mm (57.7 in)
All Pro-Link models	1495 mm (58.9 in)
CX650 C	1515 mm (59.6 in)
All other models	1455 mm (57.3 in)

Seat height

GL650 D2-E, GL650 I	770 mm (30.3 in)
All GL500 models, GL650, CX650 C	780 mm (30.7 in)
CX500 C-B, CX500 C 1979, 1980, 1982	790 mm (31.1 in)
CX500-B, CX500-C 1981, CX500 E-C, CX650 E-D	795 mm (31.3 in)
All CX500 D models	800 mm (31.5 in)
CX500 (UK), CX500-A, CX500 1978, 1979	810 mm (31.9 in)

Ground clearance

GL500 I 1981	127 mm (5.0 in)
GL500 1981	132 mm (5.2 in)
All CX500 C and CX500 D models, GL650 D2-E, GL650 I	145 mm (5.7 in)
GL500 D-C, GL500 I 1982	148 mm (5.8 in)
All CX500 models, CX650 E-D, GL650	150 mm (5.9 in)
GL500 1982	152 mm (6.0 in)
CX650 C	155 mm (6.1 in)
CX500 E-C	165 mm (6.5 in)

Dry weight

CX650 C	196 kg (432 lb)
All CX500 models	200 kg (441 lb)
All CX500 C (US) models	202 kg (445 lb)
CX500 C-B, all CX500 D models	205 kg (452 lb)
GL500 1981, 1982	207 kg (456 lb)
CX500 E-C	208 kg (459 lb)
CX650 E-D	210 kg (463 lb)
GL650	217 kg (478 lb)
GL500 D-C	222 kg (489 lb)
GL650 D2-E	230 kg (507 lb)
GL650 I	240 kg (529 lb)
GL500 I 1981, 1982	247 kg (545 lb)

Ordering spare parts

When ordering spare parts for any Honda, it is advisable to deal direct with an official Honda agent, who should be able to supply most items ex-stock.

Always quote the engine and frame numbers in full, and colour when painted parts are required.

The frame number is located on the side of the steering head, and the engine number is stamped on the crankcase below the left-hand carburettor.

Use only parts of genuine Honda manufacture. Pattern parts are available, some of which originate from Japan, but in many instances they may have an adverse effect on performance and/or reliability. Honda do not operate a 'service exchange' scheme.

Some of the more expendable parts such as spark plugs, bulbs, tyres, oils and greases etc., can be obtained from accessory shops and motor factors, who have convenient opening hours, and can often be found not far from home. It is also possible to obtain parts on a Mail Order basis from a number of specialists who advertise regularly in the motorcycle magazines.

Location of frame number

Location of engine number

Safety first!

Professional motor mechanics are trained in safe working procedures. However enthusiastic you may be about getting on with the job in hand, do take the time to ensure that your safety is not put at risk. A moment's lack of attention can result in an accident, as can failure to observe certain elementary precautions.

There will always be new ways of having accidents, and the following points do not pretend to be a comprehensive list of all dangers; they are intended rather to make you aware of the risks and to encourage a safety-conscious approach to all work you carry out on your vehicle.

Essential DOs and DON'Ts

DON'T start the engine without first ascertaining that the transmission is in neutral.

DON'T suddenly remove the filler cap from a hot cooling system – cover it with a cloth and release the pressure gradually first, or you may get scalded by escaping coolant.

DON'T attempt to drain oil until you are sure it has cooled sufficiently to avoid scalding you.

DON'T grasp any part of the engine, exhaust or silencer without first ascertaining that it is sufficiently cool to avoid burning you.

DON'T allow brake fluid or antifreeze to contact the machine's paintwork or plastic components.

DON'T syphon toxic liquids such as fuel, brake fluid or antifreeze by mouth, or allow them to remain on your skin.

DON'T inhale dust – it may be injurious to health (see *Asbestos* heading).

DON'T allow any spilt oil or grease to remain on the floor – wipe it up straight away, before someone slips on it.

DON'T use ill-fitting spanners or other tools which may slip and cause injury.

DON'T attempt to lift a heavy component which may be beyond your capability – get assistance.

DON'T rush to finish a job, or take unverified short cuts.

DON'T allow children or animals in or around an unattended vehicle.

DON'T inflate a tyre to a pressure above the recommended maximum. Apart from overstressing the carcase and wheel rim, in extreme cases the tyre may blow off forcibly.

DO ensure that the machine is supported securely at all times. This is especially important when the machine is blocked up to aid wheel or fork removal.

DO take care when attempting to slacken a stubborn nut or bolt. It is generally better to pull on a spanner, rather than push, so that if slippage occurs you fall away from the machine rather than on to it.

DO wear eye protection when using power tools such as drill, sander, bench grinder etc.

DO use a barrier cream on your hands prior to undertaking dirty jobs – it will protect your skin from infection as well as making the dirt easier to remove afterwards; but make sure your hands aren't left slippery. Note that long-term contact with used engine oil can be a health hazard.

DO keep loose clothing (cuffs, tie etc) and long hair well out of the way of moving mechanical parts.

DO remove rings, wristwatch etc, before working on the vehicle – especially the electrical system.

DO keep your work area tidy – it is only too easy to fall over articles left lying around.

DO exercise caution when compressing springs for removal or installation. Ensure that the tension is applied and released in a controlled manner, using suitable tools which preclude the possibility of the spring escaping violently.

DO ensure that any lifting tackle used has a safe working load rating adequate for the job.

DO get someone to check periodically that all is well, when working alone on the vehicle.

DO carry out work in a logical sequence and check that everything is correctly assembled and tightened afterwards.

DO remember that your vehicle's safety affects that of yourself and others. If in doubt on any point, get specialist advice.

IF, in spite of following these precautions, you are unfortunate enough to injure yourself, seek medical attention as soon as possible.

Asbestos

Certain friction, insulating, sealing, and other products – such as brake linings, clutch linings, gaskets, etc – contain asbestos. *Extreme care must be taken to avoid inhalation of dust from such products since it is hazardous to health.* If in doubt, assume that they *do* contain asbestos.

Fire

Remember at all times that petrol (gasoline) is highly flammable. Never smoke, or have any kind of naked flame around, when working on the vehicle. But the risk does not end there – a spark caused by an electrical short-circuit, by two metal surfaces contacting each other, by careless use of tools, or even by static electricity built up in your body under certain conditions, can ignite petrol vapour, which in a confined space is highly explosive.

Always disconnect the battery earth (ground) terminal before working on any part of the fuel or electrical system, and never risk spilling fuel on to a hot engine or exhaust.

It is recommended that a fire extinguisher of a type suitable for fuel and electrical fires is kept handy in the garage or workplace at all times. Never try to extinguish a fuel or electrical fire with water.

Note: *Any reference to a 'torch' appearing in this manual should always be taken to mean a hand-held battery-operated electric lamp or flashlight. It does* **not** *mean a welding/gas torch or blowlamp.*

Fumes

Certain fumes are highly toxic and can quickly cause unconsciousness and even death if inhaled to any extent. Petrol (gasoline) vapour comes into this category, as do the vapours from certain solvents such as trichloroethylene. Any draining or pouring of such volatile fluids should be done in a well ventilated area.

When using cleaning fluids and solvents, read the instructions carefully. Never use materials from unmarked containers – they may give off poisonous vapours.

Never run the engine of a motor vehicle in an enclosed space such as a garage. Exhaust fumes contain carbon monoxide which is extremely poisonous; if you need to run the engine, always do so in the open air or at least have the rear of the vehicle outside the workplace.

The battery

Never cause a spark, or allow a naked light, near the vehicle's battery. It will normally be giving off a certain amount of hydrogen gas, which is highly explosive.

Always disconnect the battery earth (ground) terminal before working on the fuel or electrical systems.

If possible, loosen the filler plugs or cover when charging the battery from an external source. Do not charge at an excessive rate or the battery may burst.

Take care when topping up and when carrying the battery. The acid electrolyte, even when diluted, is very corrosive and should not be allowed to contact the eyes or skin.

If you ever need to prepare electrolyte yourself, always add the acid slowly to the water, and never the other way round. Protect against splashes by wearing rubber gloves and goggles.

Mains electricity and electrical equipment

When using an electric power tool, inspection light etc, always ensure that the appliance is correctly connected to its plug and that, where necessary, it is properly earthed (grounded). Do not use such appliances in damp conditions and, again, beware of creating a spark or applying excessive heat in the vicinity of fuel or fuel vapour. Also ensure that the appliances meet the relevant national safety standards.

Ignition HT voltage

A severe electric shock can result from touching certain parts of the ignition system, such as the HT leads, when the engine is running or being cranked, particularly if components are damp or the insulation is defective. Where an electronic ignition system is fitted, the HT voltage is much higher and could prove fatal.

Tools and working facilities

The first priority when undertaking maintenance or repair work of any sort on a motorcycle is to have a clean, dry, well-lit working area. Work carried out in peace and quiet in the well-ordered atmosphere of a good workshop will give more satisfaction and much better results than can usually be achieved in poor working conditions. A good workshop must have a clean flat workbench or a solidly constructed table of convenient working height. The workbench or table should be equipped with a vice which has a jaw opening of at least 4 in (100 mm). A set of jaw covers should be made from soft metal such as aluminium alloy or copper, or from wood. These covers will minimise the marking or damaging of soft or delicate components which may be clamped in the vice. Some clean, dry, storage space will be required for tools, lubricants and dismantled components. It will be necessary during a major overhaul to lay out engine/gearbox components for examination and to keep them where they will remain undisturbed for as long as is necessary. To this end it is recommended that a supply of metal or plastic containers of suitable size is collected. A supply of clean, lint-free, rags for cleaning purposes and some newspapers, other rags, or paper towels for mopping up spillages should also be kept. If working on a hard concrete floor note that both the floor and one's knees can be protected from oil spillages and wear by cutting open a large cardboard box and spreading it flat on the floor under the machine or workbench. This also helps to provide some warmth in winter and to prevent the loss of nuts, washers, and other tiny components which have a tendency to disappear when dropped on anything other than a perfectly clean, flat, surface.

Unfortunately, such working conditions are not always available to the home mechanic. When working in poor conditions it is essential to take extra time and care to ensure that the components being worked on are kept scrupulously clean and to ensure that no components or tools are lost or damaged.

A selection of good tools is a fundamental requirement for anyone contemplating the maintenance and repair of a motor vehicle. For the owner who does not possess any, their purchase will prove a considerable expense, offsetting some of the savings made by doing-it-yourself. However, provided that the tools purchased meet the relevant national safety standards and are of good quality, they will last for many years and prove an extremely worthwhile investment.

To help the average owner to decide which tools are needed to carry out the various tasks detailed in this manual, we have compiled three lists of tools under the following headings: *Maintenance and minor repair*, *Repair and overhaul*, and *Specialized*. The newcomer to practical mechanics should start off with the simpler jobs around the vehicle. Then, as his confidence and experience grow, he can undertake more difficult tasks, buying extra tools as and when they are needed. In this way, a *Maintenance and minor repair* tool kit can be built-up into a *Repair and overhaul* tool kit over a considerable period of time without any major cash outlays. The experienced home mechanic will have a tool kit good enough for most repair and overhaul procedures and will add tools from the specialized category when he feels the expense is justified by the amount of use these tools will be put to.

It is obviously not possible to cover the subject of tools fully here. For those who wish to learn more about tools and their use there is a book entitled *Motorcycle Workshop Practice Manual* (Bk no 1454) available from the publishers of this manual.

As a general rule, it is better to buy the more expensive, good quality tools. Given reasonable use, such tools will last for a very long time, whereas the cheaper, poor quality, item will wear out faster and need to be renewed more often, thus nullifying the original saving. There is also the risk of a poor quality tool breaking while in use, causing personal injury or expensive damage to the component being worked on.

For practically all tools, a tool factor is the best source since he will have a very comprehensive range compared with the average garage or accessory shop. Having said that, accessory shops often offer excellent quality tools at discount prices, so it pays to shop around. There are plenty of tools around at reasonable prices, but always aim to purchase items which meet the relevant national safety standards. If in doubt, seek the advice of the shop proprietor or manager before making a purchase.

The basis of any toolkit is a set of spanners. While open-ended spanners with their slim jaws, are useful for working on awkwardly-positioned nuts, ring spanners have advantages in that they grip the nut far more positively. There is less risk of the spanner slipping off the nut and damaging it, for this reason alone ring spanners are to be preferred. Ideally, the home mechanic should acquire a set of each, but if expense rules this out a set of combination spanners (open-ended at one end and with a ring of the same size at the other) will provide a good compromise. Another item which is so useful it should be considered an essential requirement for any home mechanic is a set of socket spanners. These are available in a variety of drive sizes. It is recommended that the ½-inch drive type is purchased to begin with as although bulkier and more expensive than the ⅜-inch type, the larger size is far more common and will accept a greater variety of torque wrenches, extension pieces and socket sizes. The socket set should comprise sockets of sizes between 8 and 24 mm, a reversible ratchet drive, an extension bar of about 10 inches in length, a spark plug socket with a rubber insert, and a universal joint. Other attachments can be added to the set at a later date.

Maintenance and minor repair tool kit

Set of spanners 8 – 24 mm
Set of sockets and attachments
Spark plug spanner with rubber insert – 10, 12, or 14 mm as appropriate
Adjustable spanner
C-spanner/pin spanner
Torque wrench (same size drive as sockets)
Set of screwdrivers (flat blade)
Set of screwdrivers (cross-head)
Set of Allen keys 4 – 10 mm
Impact screwdriver and bits
Ball pein hammer – 2 lb
Hacksaw (junior)
Self-locking pliers – Mole grips or vice grips
Pliers – combination
Pliers – needle nose
Wire brush (small)
Soft-bristled brush
Tyre pump
Tyre pressure gauge
Tyre tread depth gauge
Oil can
Fine emery cloth
Funnel (medium size)
Drip tray
Grease gun
Set of feeler gauges
Brake bleeding kit
Strobe timing light
Continuity tester (dry battery and bulb)
Soldering iron and solder
Wire stripper or craft knife
PVC insulating tape
Assortment of split pins, nuts, bolts, and washers

Repair and overhaul toolkit

The tools in this list are virtually essential for anyone undertaking major repairs to a motorcycle and are additional to the tools listed above. Concerning Torx driver bits, Torx screws are encountered on some of the more modern machines where their use is restricted to fastening certain components inside the engine/gearbox unit. It is therefore recommended that if Torx bits cannot be borrowed from a local dealer, they are purchased individually as the need arises. They are not in regular use in the motor trade and will therefore only be available in specialist tool shops.

Plastic or rubber soft-faced mallet
Torx driver bits
Pliers – electrician's side cutters
Circlip pliers – internal (straight or right-angled tips are available)
Circlip pliers – external
Cold chisel
Centre punch
Pin punch
Scriber
Scraper (made from soft metal such as aluminium or copper)
Soft metal drift
Steel rule/straight edge

Assortment of files
Electric drill and bits
Wire brush (large)
Soft wire brush (similar to those used for cleaning suede shoes)
Sheet of plate glass
Hacksaw (large)
Valve grinding tool
Valve grinding compound (coarse and fine)
Stud extractor set (E-Z out)

Specialized tools

This is not a list of the tools made by the machine's manufacturer to carry out a specific task on a limited range of models. Occasional references are made to such tools in the text of this manual and, in general, an alternative method of carrying out the task without the manufacturer's tool is given where possible. The tools mentioned in this list are those which are not used regularly and are expensive to buy in view of their infrequent use. Where this is the case it may be possible to hire or borrow the tools against a deposit from a local dealer or tool hire shop. An alternative is for a group of friends or a motorcycle club to join in the purchase.

Valve spring compressor
Piston ring compressor
Universal bearing puller
Cylinder bore honing attachment (for electric drill)
Micrometer set
Vernier calipers
Dial gauge set
Cylinder compression gauge
Vacuum gauge set
Multimeter
Dwell meter/tachometer

Care and maintenance of tools

Whatever the quality of the tools purchased, they will last much longer if cared for. This means in practice ensuring that a tool is used for its intended purpose; for example screwdrivers should not be used as a substitute for a centre punch, or as chisels. Always remove dirt or grease and any metal particles but remember that a light film of oil will prevent rusting if the tools are infrequently used. The common tools can be kept together in a large box or tray but the more delicate, and more expensive, items should be stored separately where they cannot be damaged. When a tool is damaged or worn out, be sure to renew it immediately. It is false economy to continue to use a worn spanner or screwdriver which may slip and cause expensive damage to the component being worked on.

Fastening systems

Fasteners, basically, are nuts, bolts and screws used to hold two or more parts together. There are a few things to keep in mind when working with fasteners. Almost all of them use a locking device of some type; either a lock washer, lock nut, locking tab or thread adhesive. All threaded fasteners should be clean, straight, have undamaged threads and undamaged corners on the hexagon head where the spanner fits. Develop the habit of replacing all damaged nuts and bolts with new ones.

Rusted nuts and bolts should be treated with a rust penetrating fluid to ease removal and prevent breakage. After applying the rust penetrant, let it 'work' for a few minutes before trying to loosen the nut or bolt. Badly rusted fasteners may have to be chiseled off or removed with a special nut breaker, available at tool shops.

Flat washers and lock washers, when removed from an assembly should always be replaced exactly as removed. Replace any damaged washers with new ones. Always use a flat washer between a lock washer and any soft metal surface (such as aluminium), thin sheet metal or plastic. Special lock nuts can only be used once or twice before they lose their locking ability and must be renewed.

If a bolt or stud breaks off in an assembly, it can be drilled out and removed with a special tool called an E-Z out. Most dealer service departments and motorcycle repair shops can perform this task, as well as others (such as the repair of threaded holes that have been stripped out).

Spanner size comparison

Jaw gap (in)	Spanner size	Jaw gap (in)	Spanner size
0.250	$\frac{1}{4}$ in AF	0.945	24 mm
0.276	7 mm	1.000	1 in AF
0.313	$\frac{5}{16}$ in AF	1.010	$\frac{9}{16}$ in Whitworth; $\frac{5}{8}$ in BSF
0.315	8 mm	1.024	26 mm
0.344	$\frac{11}{32}$ in AF; $\frac{1}{8}$ in Whitworth	1.063	$1\frac{1}{16}$ in AF; 27 mm
0.354	9 mm	1.100	$\frac{5}{16}$ in Whitworth; $\frac{11}{16}$ in BSF
0.375	$\frac{3}{8}$ in AF	1.125	$1\frac{1}{8}$ in AF
0.394	10 mm	1.181	30 mm
0.433	11 mm	1.200	$\frac{11}{16}$ in Whitworth; $\frac{3}{4}$ in BSF
0.438	$\frac{7}{16}$ in AF	1.250	$1\frac{1}{4}$ in AF
0.445	$\frac{3}{16}$ in Whitworth; $\frac{1}{4}$ in BSF	1.260	32 mm
0.472	12 mm	1.300	$\frac{3}{4}$ in Whitworth; $\frac{7}{8}$ in BSF
0.500	$\frac{1}{2}$ in AF	1.313	$1\frac{5}{16}$ in AF
0.512	13 mm	1.390	$\frac{13}{16}$ in Whitworth; $\frac{15}{16}$ in BSF
0.525	$\frac{1}{4}$ in Whitworth; $\frac{5}{16}$ in BSF	1.417	36 mm
0.551	14 mm	1.438	$1\frac{7}{16}$ in AF
0.563	$\frac{9}{16}$ in AF	1.480	$\frac{7}{8}$ in Whitworth; 1 in BSF
0.591	15 mm	1.500	$1\frac{1}{2}$ in AF
0.600	$\frac{5}{16}$ in Whitworth; $\frac{3}{8}$ in BSF	1.575	40 mm; $\frac{15}{16}$ in Whitworth
0.625	$\frac{5}{8}$ in AF	1.614	41 mm
0.630	16 mm	1.625	$1\frac{5}{8}$ in AF
0.669	17 mm	1.670	1 in Whitworth; $1\frac{1}{8}$ in BSF
0.686	$\frac{11}{16}$ in AF	1.688	$1\frac{11}{16}$ in AF
0.709	18 mm	1.811	46 mm
0.710	$\frac{3}{8}$ in Whitworth; $\frac{7}{16}$ in BSF	1.813	$1\frac{13}{16}$ in AF
0.748	19 mm	1.860	$1\frac{1}{8}$ in Whitworth; $1\frac{1}{4}$ in BSF
0.750	$\frac{3}{4}$ in AF	1.875	$1\frac{7}{8}$ in AF
0.813	$\frac{13}{16}$ in AF	1.969	50 mm
0.820	$\frac{7}{16}$ in Whitworth; $\frac{1}{2}$ in BSF	2.000	2 in AF
0.866	22 mm	2.050	$1\frac{1}{4}$ in Whitworth; $1\frac{3}{8}$ in BSF
0.875	$\frac{7}{8}$ in AF	2.165	55 mm
0.920	$\frac{1}{2}$ in Whitworth; $\frac{9}{16}$ in BSF	2.362	60 mm
0.938	$\frac{15}{16}$ in AF		

Standard torque settings

Specific torque settings will be found at the end of the specifications section of each chapter. Where no figure is given, bolts should be secured according to the table below.

Fastener type (thread diameter)	kgf m	lbf ft
5mm bolt or nut	0.45 – 0.6	3.5 – 4.5
6 mm bolt or nut	0.8 – 1.2	6 – 9
8 mm bolt or nut	1.8 – 2.5	13 – 18
10 mm bolt or nut	3.0 – 4.0	22 – 29
12 mm bolt or nut	5.0 – 6.0	36 – 43
5 mm screw	0.35 – 0.5	2.5 – 3.6
6 mm screw	0.7 – 1.1	5 – 8
6 mm flange bolt	1.0 – 1.4	7 – 10
8 mm flange bolt	2.4 – 3.0	17 – 22
10 mm flange bolt	3.0 – 4.0	22 – 29

Choosing and fitting accessories

The range of accessories available to the modern motorcyclist is almost as varied and bewildering as the range of motorcycles. This Section is intended to help the owner in choosing the correct equipment for his needs and to avoid some of the mistakes made by many riders when adding accessories to their machines. It will be evident that the Section can only cover the subject in the most general terms and so it is recommended that the owner, having decided that he wants to fit, for example, a luggage rack or carrier, seeks the advice of several local dealers and the owners of similar machines. This will give a good idea of what makes of carrier are easily available, and at what price. Talking to other owners will give some insight into the drawbacks or good points of any one make. A walk round the motorcycles in car parks or outside a dealer will often reveal the same sort of information.

The first priority when choosing accessories is to assess exactly what one needs. It is, for example, pointless to buy a large heavy-duty carrier which is designed to take the weight of fully laden panniers and topbox when all you need is a place to strap on a set of waterproofs and a lunchbox when going to work. Many accessory manufacturers have ranges of equipment to cater for the individual needs of different riders and this point should be borne in mind when looking through a dealer's catalogues. Having decided exactly what is required and the use to which the accessories are going to be put, the owner will need a few hints on what to look for when making the final choice. To this end the Section is now sub-divided to cover the more popular accessories fitted. Note that it is in no way a customizing guide, but merely seeks to outline the practical considerations to be taken into account when adding aftermarket equipment to a motorcycle.

Fairings and windscreens

A fairing is possibly the single, most expensive, aftermarket item to be fitted to any motorcycle and, therefore, requires the most thought before purchase. Fairings can be divided into two main groups: front fork mounted handlebar fairings and windscreens, and frame mounted fairings.

The first group, the front fork mounted fairings, are becoming far more popular than was once the case, as they offer several advantages over the second group. Front fork mounted fairings generally are much easier and quicker to fit, involve less modification to the motorcycle, do not as a rule restrict the steering lock, permit a wider selection of handlebar styles to be used, and offer adequate protection for much less money than the frame mounted type. They are also lighter, can be swapped easily between different motorcycles, and are available in a much greater variety of styles. Their main disadvantages are that they do not offer as much weather protection as the frame mounted types, rarely offer any storage space, and, if poorly fitted or naturally incompatible, can have an adverse effect on the stability of the motorcycle.

The second group, the frame mounted fairings, are secured so rigidly to the main frame of the motorcycle that they can offer a substantial amount of protection to motorcycle and rider in the event of a crash. They offer almost complete protection from the weather and, if double-skinned in construction, can provide a great deal of useful storage space. The feeling of peace, quiet and complete relaxation encountered when riding behind a good full fairing has to be experienced to be believed. For this reason full fairings are considered essential by most touring motorcyclists and by many people who ride all year round. The main disadvantages of this type are that fitting can take a long time, often involving removal or modification of standard motorcycle components, they restrict the steering lock and they can add up to about 40 lb to the weight of the machine. They do not usually affect the stability of the machine to any great extent once the front tyre pressure and suspension have been adjusted to compensate for the extra weight, but can be affected by sidewinds.

The first thing to look for when purchasing a fairing is the quality of the fittings. A good fairing will have strong, substantial brackets constructed from heavy-gauge tubing; the brackets must be shaped to fit the frame or forks evenly so that the minimum of stress is imposed on the assembly when it is bolted down. The brackets should be properly painted or finished – a nylon coating being the favourite of the better manufacturers – the nuts and bolts provided should be of the same thread and size standard as is used on the motorcycle and be properly plated. Look also for shakeproof locking nuts or locking washers to ensure that everything remains securely tightened down. The fairing shell is generally made from one of two materials: fibreglass or ABS plastic. Both have their advantages and disadvantages, but the main consideration for the owner is that fibreglass is much easier to repair in the event of damage occurring to the fairing. Whichever material is used, check that it is properly finished inside as well as out, that the edges are protected by beading and that the fairing shell is insulated from vibration by the use of rubber grommets at all mounting points. Also be careful to check that the windscreen is retained by plastic bolts which will snap on impact so that the windscreen will break away and not cause personal injury in the event of an accident.

Having purchased your fairing or windscreen, read the manufacturer's fitting instructions very carefully and check that you have all the necessary brackets and fittings. Ensure that the mounting brackets are located correctly and bolted down securely. Note that some manufacturers use hose clamps to retain the mounting brackets; these should be discarded as they are convenient to use but not strong enough for the task. Stronger clamps should be substituted; car exhaust pipe clamps of suitable size would be a good alternative. Ensure that the front forks can turn through the full steering lock available without fouling the fairing. With many types of frame-mounted fairing the handlebars will have to be altered or a different type fitted and the steering lock will be restricted by stops provided with the fittings. Also check that the fairing does not foul the front wheel or mudguard, in any steering position, under full fork compression. Re-route any cables, brake pipes or electrical wiring which may snag on the fairing and take great care to protect all electrical connections, using insulating tape. If the manufacturer's instructions are followed carefully at every stage no serious problems should be encountered. Remember that hydraulic pipes that have been disconnected must be carefully re-tightened and the hydraulic system purged of air bubbles by bleeding.

Two things will become immediately apparent when taking a motorcycle on the road for the first time with a fairing – the first is the tendency to underestimate the road speed because of the lack of wind pressure on the body. This must be very carefully watched until one has grown accustomed to riding behind the fairing. The second thing is the alarming increase in engine noise which is an unfortunate but inevitable by-product of fitting any type of fairing or windscreen, and is caused by normal engine noise being reflected, and in some cases amplified, by the flat surface of the fairing.

Luggage racks or carriers

Carriers are possibly the commonest item to be fitted to modern motorcycles. They vary enormously in size, carrying capacity, and durability. When selecting a carrier, always look for one which is made specifically for your machine and which is bolted on with as few separate brackets as possible. The universal-type carrier, with its mass of brackets and adaptor pieces, will generally prove too weak to be of any real use. A good carrier should bolt to the main frame, generally using the two suspension unit top mountings and a mudguard mounting bolt as attachment points, and have its luggage platform as low and as far forward as possible to minimise the effect of any load on the machine's stability. Look for good quality, heavy gauge tubing, good welding and good finish. Also ensure that the carrier does not prevent opening of the seat, sidepanels or tail compartment, as appropriate. When using a carrier, be very careful not to overload it. Excessive weight placed so high and so far to the rear of any motorcycle will have an adverse effect on the machine's steering and stability.

Luggage

Motorcycle luggage can be grouped under two headings: soft and hard. Both types are available in many sizes and styles and have advantages and disadvantages in use.

Soft luggage is now becoming very popular because of its lower cost and its versatility. Whether in the form of tankbags, panniers, or strap-on bags, soft luggage requires in general no brackets and no modification to the motorcycle. Equipment can be swapped easily from one motorcycle to another and can be fitted and removed in seconds. Awkwardly shaped loads can easily be carried. The disadvantages of soft luggage are that the contents cannot be secure against the casual thief, very little protection is afforded in the event of a crash, and waterproofing is generally poor. Also, in the case of panniers, carrying capacity is restricted to approximately 10 lb, although this amount will vary considerably depending on the manufacturer's recommendation. When purchasing soft luggage, look for good quality material, generally vinyl or nylon, with strong, well-stitched attachment points. It is always useful to have separate pockets, especially on tank bags, for items which will be needed on the journey. When purchasing a tank bag, look for one which has a separate, well-padded, base. This will protect the tank's paintwork and permit easy access to the filler cap at petrol stations.

Hard luggage is confined to two types: panniers, and top boxes or tail trunks. Most hard luggage manufacturers produce matching sets of these items, the basis of which is generally that manufacturer's own heavy-duty luggage rack. Variations on this theme occur in the form of separate frames for the better quality panniers, fixed or quickly-detachable luggage, and in size and carrying capacity. Hard luggage offers a reasonable degree of security against theft and good protection against weather and accident damage. Carrying capacity is greater than that of soft luggage, around 15 – 20 lb in the case of panniers, although top boxes should never be loaded as much as their apparent capacity might imply. A top box should only be used for lightweight items, because one that is heavily laden can have a serious effect on the stability of the machine. When purchasing hard luggage look for the same good points as mentioned under fairings and windscreens, ie good quality mounting brackets and fittings, and well-finished fibreglass or ABS plastic cases. Again as with fairings,

always purchase luggage made specifically for your motorcycle, using as few separate brackets as possible, to ensure that everything remains securely bolted in place. When fitting hard luggage, be careful to check that the rear suspension and brake operation will not be impaired in any way and remember that many pannier kits require re-siting of the indicators. Remember also that a non-standard exhaust system may make fitting extremely difficult.

Handlebars

The occupation of fitting alternative types of handlebar is extremely popular with modern motorcyclists, whose motives may vary from the purely practical, wishing to improve the comfort of their machines, to the purely aesthetic, where form is more important than function. Whatever the reason, there are several considerations to be borne in mind when changing the handlebars of your machine. If fitting lower bars, check carefully that the switches and cables do not foul the petrol tank on full lock and that the surplus length of cable, brake pipe, and electrical wiring are smoothly and tidily disposed of. Avoid tight kinks in cable or brake pipes which will produce stiff controls or the premature and disastrous failure of an overstressed component. If necessary, remove the petrol tank and re-route the cable from the engine/gearbox unit upwards, ensuring smooth gentle curves are produced. In extreme cases, it will be necessary to purchase a shorter brake pipe to overcome this problem. In the case of higher handlebars than standard it will almost certainly be necessary to purchase extended cables and brake pipes. Fortunately, many standard motorcycles have a custom version which will be equipped with higher handlebars and, therefore, factory-built extended components will be available from your local dealer. It is not usually necessary to extend electrical wiring, as switch clusters may be used on several different motorcycles, some being custom versions. This point should be borne in mind however when fitting extremely high or wide handlebars.

When fitting different types of handlebar, ensure that the mounting clamps are correctly tightened to the manufacturer's specifications and that cables and wiring, as previously mentioned, have smooth easy runs and do not snag on any part of the motorcycle throughout the full steering lock. Ensure that the fluid level in the front brake master cylinder remains level to avoid any chance of air entering the hydraulic system. Also check that the cables are adjusted correctly and that all handlebar controls operate correctly and can be easily reached when riding.

Crashbars

Crashbars, also known as engine protector bars, engine guards, or case savers, are extremely useful items of equipment which can contribute protection to the machine's structure if a crash occurs. They do not, as has been inferred in the US, prevent the rider from crashing, or necessarily prevent rider injury should a crash occur.

It is recommended that only the smaller, neater, engine protector type of crashbar is considered. This type will offer protection while restricting, as little as is possible, access to the engine and the machine's ground clearance. The crashbars should be designed for use specifically on your machine, and should be constructed of heavy-gauge tubing with strong, integral mounting brackets. Where possible, they should bolt to a strong lug on the frame, usually at the engine mounting bolts.

The alternative type of crashbar is the larger cage type. This type is not recommended in spite of their appearance which promises some protection to the rider as well as to the machine. The larger amount of leverage imposed by the size of this type of crashbar increases the risk of severe frame damage in the event of an accident. This type also decreases the machine's ground clearance and restricts access to the engine. The amount of protection afforded the rider is open to some doubt as the design is based on the premise that the rider will stay in the normally seated position during an accident, and the crash bar structure will not itself fail. Neither result can in any way be guaranteed.

As a general rule, always purchase the best, ie usually the most expensive, set of crashbars you an afford. The investment will be repaid by minimising the amount of damage incurred, should the machine be involved in an accident. Finally, avoid the universal type of crashbar. This should be regarded only as a last resort to be used if no alternative exists. With its usual multitude of separate brackets and spacers, the universal crashbar is far too weak in design and construction to be of any practical value.

Exhaust systems

The fitting of aftermarket exhaust systems is another extremely popular pastime amongst motorcyclists. The usual motive is to gain more performance from the engine but other considerations are to gain more ground clearance, to lose weight from the motorcycle, to obtain a more distinctive exhaust note or to find a cheaper alternative to the manufacturer's original equipment exhaust system. Original equipment exhaust systems often cost more and may well have a relatively short life. It should be noted that it is rare for an aftermarket exhaust system alone to give a noticeable increase in the engine's power output. Modern motorcycles are designed to give the highest power output possible allowing for factors such as quietness, fuel economy, spread of power, and long-term reliability. If there were a magic formula which allowed the exhaust system to produce more power without affecting these other considerations you can be sure that the manufacturers, with their large research and development facilities, would have found it and made use of it. Performance increases of a worthwhile and noticeable nature only come from well-tried and properly matched modifications to the entire engine, from the air filter, through the carburettors, port timing or camshaft and valve design, combustion chamber shape, compression ratio, and the exhaust system. Such modifications are well outside the scope of this manual but interested owners might refer to specialist books produced by the publisher of this manual which go into the whole subject in great detail.

Whatever your motive for wishing to fit an alternative exhaust system, be sure to seek expert advice before doing so. Changes to the carburettor jetting will almost certainly be required for which you must consult the exhaust system manufacturer. If he cannot supply adequately specific information it is reasonable to assume that insufficient development work has been carried out, and that particular make should be avoided. Other factors to be borne in mind are whether the exhaust system allows the use of both centre and side stands, whether it allows sufficient access to permit oil and filter changing and whether modifications are necessary to the standard exhaust system. Many two-stroke expansion chamber systems require the use of the standard exhaust pipe; this is all very well if the standard exhaust pipe and silencer are separate units but can cause problems if the two, as with so many modern two-strokes, are a one-piece unit. While the exhaust pipe can be removed easily by means of a hacksaw it is not so easy to refit the original silencer should you at any time wish to return the machine to standard trim. The same applies to several four-stroke systems.

On the subject of the finish of aftermarket exhausts, avoid black-painted systems unless you enjoy painting. As any trail-bike owner will tell you, rust has a great affinity for black exhausts and re-painting or rust removal becomes a task which must be carried out with monotonous regularity. A bright chrome finish is, as a general rule, a far better proposition as it is much easier to keep clean and to prevent rusting. Although the general finish of aftermarket exhaust systems is not always up to the standard of the original equipment the lower cost of such systems does at least reflect this fact.

When fitting an alternative system always purchase a full set of new exhaust gaskets, to prevent leaks. Fit the exhaust first to the cylinder head or barrel, as appropriate, tightening the retaining nuts or bolts by hand only and then line up the exhaust rear mountings. If the new system is a one-piece unit and the rear mountings do not line up exactly, spacers must be fabricated to take up the difference. Do not force the system into place as the stress thus imposed will rapidly cause cracks and splits to appear. Once all the mountings are loosely fixed, tighten the retaining nuts or bolts securely, being careful not to overtighten them. Where the motorcycle manufacturer's torque settings are available, these should be used. Do not forget to carry out any carburation changes recommended by the exhaust system's manufacturer.

Electrical equipment

The vast range of electrical equipment available to motorcyclists is so large and so diverse that only the most general outline can be given here. Electrical accessories vary from electric ignition kits fitted to replace contact breaker points, to additional lighting at the front and rear, more powerful horns, various instruments and gauges, clocks, anti-theft systems, heated clothing, CB radios, radio-cassette players, and intercom systems, to name but a few of the more popular items of equipment.

As will be evident, it would require a separate manual to cover this subject alone and this section is therefore restricted to outlining a few basic rules which must be borne in mind when fitting electrical equipment. The first consideration is whether your machine's electrical system has enough reserve capacity to cope with the added demand of the accessories you wish to fit. The motorcycle's manufacturer or importer should be able to furnish this sort of information and may also be able to offer advice on uprating the electrical system. Failing this, a good dealer or the accessory manufacturer may be able to help. In some cases, more powerful generator components may be available, perhaps from another motorcycle in the manufacturer's range. The second consideration is the legal requirements in force in your area. The local police may be prepared to help with this point. In the UK for example, there are strict regulations governing the position and use of auxiliary riding lamps and fog lamps.

When fitting electrical equipment always disconnect the battery first to prevent the risk of a short-circuit, and be careful to ensure that all connections are properly made and that they are waterproof. Remember that many electrical accessories are designed primarily for use in cars and that they cannot easily withstand the exposure to vibration and to the weather. Delicate components must be rubber-mounted to insulate them from vibration, and sealed carefully to prevent the entry of rainwater and dirt. Be careful to follow exactly the accessory manufacturer's instructions in conjunction with the wiring diagram at the back of this manual.

Accessories – general

Accessories fitted to your motorcycle will rapidly deteriorate if not cared for. Regular washing and polishing will maintain the finish and will provide an opportunity to check that all mounting bolts and nuts are securely fastened. Any signs of chafing or wear should be watched for, and the cause cured as soon as possible before serious damage occurs.

As a general rule, do not expect the re-sale value of your motorcycle to increase by an amount proportional to the amount of money and effort put into fitting accessories. It is usually the case that an absolutely standard motorcycle will sell more easily at a better price than one that has been modified. If you are in the habit of exchanging your machine for another at frequent intervals, this factor should be borne in mind to avoid loss of money.

Fault diagnosis

Contents

1 Introduction

This Section provides an easy reference-guide to the more common ailments that are likely to afflict your machine. Obviously, the opportunities are almost limitless for faults to occur as a result of obscure failures, and to try and cover all eventualities would require a book. Indeed, a number have been written on the subject.

Successful fault diagnosis is not a mysterious 'black art' but the application of a bit of knowledge combined with a systematic and logical approach to the problem. Approach any fault diagnosis by first accurately identifying the symptom and then checking through the list of possible causes, starting with the simplest or most obvious and progressing in stages to the most complex. Take nothing for granted, but above all apply liberal quantities of common sense.

The main symptom of a fault is given in the text as a major heading below which are listed, as Section headings, the various systems or areas which may contain the fault. Details of each possible cause for a fault and the remedial action to be taken are given, in brief, in the paragraphs below each Section heading. Further information should be sought in the relevant Chapter.

Starter motor problems

2 Starter motor not rotating

● Engine stop switch off.
● Fuse blown. Check the main fuse located behind the battery side cover.
● Battery voltage low. Switching on the headlamp and operating the horn will give a good indication of the charge level. If necessary recharge the battery from an external source.
● Neutral gear not selected.
● Faulty neutral indicator switch or clutch interlock switch. Check the switch wiring and switches for correct operation.
● Ignition switch defective. Check switch for continuity and connections for security.
● Engine stop switch defective. Check switch for continuity in 'Run' position. Fault will be caused by broken, wet or corroded switch contacts. Clean or renew as necessary.
● Starter button switch faulty. Check continuity of switch. Faults as for engine stop switch.
● Starter relay (solenoid) faulty. If the switch is functioning correctly a pronounced click should be heard when the starter button is depressed. This presupposes that current is flowing to the solenoid when the button is depressed.
● Wiring open or shorted. Check first that the battery terminal connections are tight and corrosion free. Follow this by checking that all wiring connections are dry, tight and corrosion free. Check also for frayed or broken wiring. Occasionally a wire may become trapped between two moving components, particularly in the vicinity of the steering head, leading to breakage of the internal core but leaving the softer but more resilient outer cover intact. This can cause mysterious intermittent or total power loss.
● Starter motor defective. A badly worn starter motor may cause high current drain from a battery without the motor rotating. If current is found to be reaching the motor, after checking the starter button and starter relay, suspect a damaged motor. The motor should be removed for inspection.

3 Starter motor rotates but engine does not turn over

● Starter motor clutch defective. Suspect jammed or worn engagement rollers, plungers and springs.
● Damaged starter motor drive train. Inspect and renew component where necessary. Failure in this area is unlikely.

4 Starter motor and clutch function but engine will not turn over

● Engine seized. Seizure of the engine is always a result of damage to internal components due to lubrication failure, or component breakage resulting from abuse, neglect or old age. A seizing or partially seized component may go un-noticed until the engine has cooled down and an attempt is made to restart the engine. Suspect first seizure of the valves, valve gear and the pistons. Instantaneous seizure whilst the engine is running indicates component breakage. In either case major dismantling and inspection will be required.

Engine does not start when turned over

5 No fuel flow to carburettor

● No fuel or insufficient fuel in tank.
● Fuel tap lever position incorrectly selected.
● Float chambers require priming after running dry (vacuum taps only).
● Tank filler cap air vent obstructed. Usually caused by dirt or water. Clean the vent orifice.
● Fuel tap or filter blocked. Blockage may be due to accumulation of rust or paint flakes from the tank's inner surface or of foreign matter from contaminated fuel. Remove the tap and clean it and the filter. Look also for water droplets in the fuel.
● Fuel line blocked. Blockage of the fuel line is more likely to result from a kink in the line rather than the accumulation of debris.

6 Fuel not reaching cylinder

● Float chamber not filling. Caused by float needle or floats sticking in up position. This may occur after the machine has been left standing for an extended length of time allowing the fuel to evaporate. When this occurs a gummy residue is often left which hardens to a varnish-like substance. This condition may be worsened by corrosion and crystaline deposits produced prior to the total evaporation of contaminated fuel. Sticking of the float needle may also be caused by wear. In any case removal of the float chamber will be necessary for inspection and cleaning.
● Blockage in starting circuit, slow running circuit or jets. Blockage of these items may be attributable to debris from the fuel tank by-passing the filter system or to gumming up as described in paragraph 1. Water droplets in the fuel will also block jets and passages. The carburettor should be dismantled for cleaning.
● Fuel level too low. The fuel level in the float chamber is controlled by float height. The float height may increase with wear or damage but will never reduce, thus a low float height is an inherent rather than developing condition. Check the float height and make any necessary adjustment.

7 Engine flooding

● Float valve needle worn or stuck open. A piece of rust or other debris can prevent correct seating of the needle against the valve seat thereby permitting an uncontrolled flow of fuel. Similarly, a worn needle or needle seat will prevent valve closure. Dismantle the carburettor float bowl for cleaning and, if necessary, renewal of the worn components.
● Fuel level too high. The fuel level is controlled by the float height which may increase due to wear of the float needle, pivot pin or operating tang. Check the float height, and make any necessary adjustment. A leaking float will cause an increase in fuel level, and thus should be renewed.
● Accelerator pump. On those models so equipped, repeated operation of the throttle prior to starting will cause flooding due to too much raw fuel being injected into the venturi.
● Cold starting mechanism. Check the choke (starter mechanism) for correct operation. If the mechanism jams in the 'On' position subsequent starting of a hot engine will be difficult.
● Blocked air filter. A badly restricted air filter will cause flooding. Check the filter and clean or renew as required. A collapsed inlet hose will have a similar effect.

8 No spark at plugs

● Ignition switch not on.
● Engine stop switch off.
● Fuse blown. Check fuse for ignition circuit. See wiring diagram.
● Battery voltage low. The current draw required by a starter motor is sufficiently high that an under-charged battery may not have enough spare capacity to provide power for the ignition circuit during starting.
● Starter motor inefficient. A starter motor with worn brushes and a worn or dirty commutator will draw excessive amounts of current causing power starvation in the ignition system. See the preceding paragraph. Starter motor overhaul will be required.
● Spark plug failure. Clean the spark plugs thoroughly and reset the electrode gap. Refer to the spark plug section
        in Routine Maintenance. If a spark plug shorts internally or has sustained visible damage to the electrodes, core or ceramic insulator it should be renewed. On rare occasions a plug that appears to spark vigorously will fail to do so when refitted to the engine and subjected to the compression pressure in the cylinder.
● Spark plug caps or high tension (HT) leads faulty. Check condition and security. Replace if deterioration is evident.
● Spark plug caps loose. Check that the spark plug caps fit securely over the plugs and, where fitted, the screwed terminals on the plug ends are secure.
● Shorting due to moisture. Certain parts of the ignition system are susceptible to shorting when the machine is ridden or parked in wet weather. Check particularly the area from the spark plug caps back to the ignition coil. A water dispersant spray may be used to dry out waterlogged components. Recurrence of the problem can be prevented by using an ignition sealant spray after drying out and cleaning.
● Ignition or stop switch shorted. May be caused by water, corrosion or wear. Water dispersant and contact cleaning sprays may be used. If this fails to overcome the problem dismantling and visual inspection of the switches will be required.
● Shorting or open circuit in wiring. Failure in any wire connecting any of the ignition components will cause ignition malfunction. Check also that all connections are clean, dry and tight.
● Ignition coil failure. Check the coil, referring to Chapter 4.
● Fault in ignition system. Refer to Chapter 4.

9 Weak spark at plugs

● Feeble sparking at the plugs may be caused by any of the faults mentioned in the preceding Section other than those items in paragraphs 1 and 2. Check first the spark plugs, these being the most likely culprits.

10 Compression low

● Spark plugs loose. This will be self-evident on inspection, and may be accompanied by a hissing noise when the engine is turned over. Remove the plugs and check that the threads in the cylinder head are not damaged. Check also that the plug sealing washers are in good condition.
● Cylinder head gasket leaking. This condition is often accompanied by a high pitched squeak from around the cylinder head and oil loss, and may be caused by insufficiently tightened cylinder head fasteners, a warped cylinder head or mechanical failure of the gasket material. Re-torqueing the fasteners to the correct specification may seal the leak in some instances but if damage has occurred this course of action will provide, at best, only a temporary cure.
● Valve not seating correctly. The failure of a valve to seat may be caused by insufficient valve clearance, pitting of the valve seat or face, carbon deposits on the valve seat or seizure of the valve stem or valve gear components. Valve spring breakage will also prevent correct valve closure. The valve clearances should be checked first and then, if these are found to be in order, further dismantling will be required to inspect the relevant components for failure.
● Cylinder, piston and ring wear. Compression pressure will be lost if any of these components are badly worn. Wear in one component is invariably accompanied by wear in another. A top end overhaul will be required.

● Piston rings sticking or broken. Sticking of the piston rings may be caused by seizure due to lack of lubrication or heating as a result of poor carburation or incorrect fuel type. Gumming of the rings may result from lack of use, or carbon deposits in the ring grooves. Broken rings result from over-revving, overheating or general wear. In either case a top-end overhaul will be required.

Engine stalls after starting

11 General causes

● Improper cold start mechanism operation. Check that the operating controls function smoothly and, where applicable, are correctly adjusted. A cold engine may not require application of an enriched mixture to start initially but may baulk without choke once firing. Likewise a hot engine may start with an enriched mixture but will stop almost immediately if the choke is inadvertently in operation.
● Ignition malfunction. See Section 9, 'Weak spark at plug'.
● Carburettors incorrectly adjusted. Maladjustment of the mixture strength or idle speed may cause the engine to stop immediately after starting. See Chapter 2.
● Fuel contamination. Check for filter blockage by debris or water which reduces, but does not completely stop, fuel flow or blockage of the slow speed circuit in the carburettor by the same agents. If water is present it can often be seen as droplets in the bottom of the float bowl. Clean the filter and, where water is in evidence, drain and flush the fuel tank and float bowl.
● Intake air leak. Check for security of the carburettor mounting and hose connections, and for cracks or splits in the hoses. Check also that the carburettor top is secure and that the vacuum gauge adaptor plug is tight.
● Air filter blocked or omitted. A blocked filter will cause an over-rich mixture; the omission of a filter will cause an excessively weak mixture. Both conditions will have a detrimental effect on carburation. Clean or renew the filter as necessary.
● Fuel filler cap air vent blocked. Usually caused by dirt or water. Clean the vent orifice.

Poor running at idle and low speed

12 Weak spark at plugs or erratic firing

● Battery voltage low. In certain conditions low battery charge, especially when coupled with a badly sulphated battery, may result in misfiring. If the battery is in good general condition it should be recharged; an old battery suffering from sulphated plates should be renewed.
● Spark plugs fouled, faulty or incorrectly adjusted. See Section 8 or refer to Routine Maintenance.
● Spark plug caps or high tension lead shorting. Check the condition of both these items ensuring that they are in good condition and dry and that the cap is fitted correctly.
● Spark plug types incorrect. Fit plugs of the correct type and heat range as given in Specifications. In certain conditions plugs of hotter or colder types may be required for normal running.
● Ignition timing incorrect. Check the ignition timing statically and dynamically, ensuring that the advance is functioning correctly.
● Faulty ignition coil. Partial failure of the coil internal insulation will diminish the performance of the coil. No repair is possible, a new component must be fitted.

13 Fuel/air mixture incorrect

● Intake air leak. See Section 11.
● Mixture strength incorrect. Adjust slow running mixture strength using pilot adjustment screw.
● Carburettor synchronisation.
● Pilot jet or slow running circuit blocked. The carburettors should be removed and dismantled for thorough cleaning. Blow through all jets and air passages with compressed air to clear obstructions.

● Air cleaner clogged or omitted. Clean or fit air cleaner element as necessary. Check also that the element and air filter cover are correctly seated.
● Cold start mechanism in operation. Check that the choke has not been left on inadvertently and the operation is correct. Where applicable check the operating cable free play.
● Fuel level too high or too low. Check the float height and adjust as necessary. See Section 7.
● Fuel tank air vent obstructed. Obstruction usually caused by dirt or water. Clean vent orifice.
● Valve clearance incorrect. Check, and if necessary, adjust, the clearances.

14 Compression low

● See Section 10.

Acceleration poor

15 General causes

● All items as for previous Section.
● Accelerator pump defective. Where so equipped, check that the accelerator pump injects raw fuel into the carburettor venturi, when the throttle is open fully. If this does not occur check the condition of the pump components and that the feed passage to the pump is not obstructed.
● Timing not advancing. On models with transistorised ignition, this is caused by a sticking or damaged automatic timing unit (ATU). Cleaning and lubrication of the ATU will usually overcome sticking, failing this, and in any event if damage is evident, renewal of the ATU will be required.
● Sticking throttle vacuum piston.
● Brakes binding. Usually caused by maladjustment or partial seizure of the operating mechanism due to poor maintenance. Check brake adjustment (where applicable). A bent wheel spindle or warped brake disc can produce similar symptoms.

Poor running or lack of power at high speeds

16 Weak spark at plugs or erratic firing

● All items as for Section 12.
● HT lead insulation failure. Insulation failure of the HT leads and spark plug caps due to old age or damage can cause shorting when the engine is driven hard. This condition may be less noticeable, or not noticeable at all at lower engine speeds.

17 Fuel/air mixture incorrect

● All items as for Section 13, with the exception of items 2 and 4.
● Main jet blocked. Debris from contaminated fuel, or from the fuel tank, and water in the fuel can block the main jet. Clean the fuel filter, the float bowl area, and if water is present, flush and refill the fuel tank.
● Main jet is the wrong size. The standard carburettor jetting is for sea level atmospheric pressure. For high altitudes, usually above 5000 ft, a smaller main jet will be required.
● Jet needle and needle jet worn. These can be renewed individually but should be renewed as a pair. Renewal of both items requires partial dismantling of the carburettor.
● Air bleed holes blocked. Dismantle carburettor and use compressed air to blow out all air passages.
● Reduced fuel flow. A reduction in the maximum fuel flow from the fuel tank to the carburettor will cause fuel starvation, proportionate to the engine speed. Check for blockages through debris or a kinked fuel line.
● Vacuum diaphragm split. Renew.

18 Compression low

● See Section 10.

Knocking or pinking

19 General causes

● Carbon build-up in combustion chamber. After high mileages have been covered large accumulation of carbon may occur. This may glow red hot and cause premature ignition of the fuel/air mixture, in advance of normal firing by the spark plug. Cylinder head removal will be required to allow inspection and cleaning.
● Fuel incorrect. A low grade fuel, or one of poor quality may result in compression induced detonation of the fuel resulting in knocking and pinking noises. Old fuel can cause similar problems. A too highly leaded fuel will reduce detonation but will accelerate deposit formation in the combustion chamber and may lead to early pre-ignition as described in item 1.
● Spark plug heat range incorrect. Uncontrolled pre-ignition can result from the use of a spark plug the heat range of which is too hot.
● Weak mixture. Overheating of the engine due to a weak mixture can result in pre-ignition occurring where it would not occur when engine temperature was within normal limits. Maladjustment, blocked jets or passages and air leaks can cause this condition.

Overheating

20 Firing incorrect

● Spark plug fouled, defective or maladjusted. See Section 6.
● Spark plug type incorrect. Refer to the Specifications and ensure that the correct plug type is fitted.
● Incorrect ignition timing. Timing that is far too much advanced or far too much retarded will cause overheating. Check the ignition timing is correct and that the advance mechanism is functioning.

21 Fuel/air mixture incorrect

● Slow speed mixture strength incorrect. Adjust pilot air screw.
● Main jet wrong size. The carburettor is jetted for sea level atmospheric conditions. For high altitudes, usually above 5000 ft, a smaller main jet will be required.
● Air filter badly fitted or omitted. Check that the filter element is in place and that it and the air filter box cover are sealing correctly. Any leaks will cause a weak mixture.
● Induction air leaks. Check the security of the carburettor mountings and hose connections, and for cracks and splits in the hoses. Check also that each carburettor top is secure and that the vacuum gauge adaptor plug is tight.
● Fuel level too low. See Section 6.
● Fuel tank filler cap air vent obstructed. Clear blockage.

22 Lubrication inadequate

● Engine oil too low. Not only does the oil serve as a lubricant by preventing friction between moving components, but it also acts as a coolant. Check the oil level and replenish.
● Engine oil overworked. The lubricating properties of oil are lost slowly during use as a result of changes resulting from heat and also contamination. Always change the oil at the recommended interval.
● Engine oil of incorrect viscosity or poor quality. Always use the recommended viscosity and type of oil.
● Oil filter and filter by-pass valve blocked. Renew filter and clean the by-pass valve.

23 Miscellaneous causes

● Faulty thermostat. If the thermostat fails in the closed position the engine will rapidly overheat. Renew the thermostat: See Chapter 2.
● Radiator fins clogged. Accumulated debris in the radiator core will gradually reduce its ability to dissipate the heat generated by the engine. It is worth noting that during the summer months dead insects can cause as many problems in this respect as road dirt and mud during the winter. Cleaning is best carried out by dislodging the debris with a high pressure hose from the back of the radiator. Once cleaned it is worth painting the matrix with a heat-dispersant matt black paint both to assist cooling and to prevent corrosion.

Clutch operating problems

24 Clutch slip

● No clutch lever play. Adjust clutch lever end play according to the procedure in Routine Maintenance.
● Friction plates worn or warped. Overhaul clutch assembly, replacing plates out of specification.
● Steel plates worn or warped. Overhaul clutch assembly, replacing plates out of specification.
● Clutch springs broken or wear. Old or heat-damaged (from slipping clutch) springs should be replaced with new ones.
● Clutch inner cable snagging. Caused by a frayed cable or kinked outer cable. Replace the cable with a new one. Repair of a frayed cable is not advised.
● Clutch release mechanism defective. Worn or damaged parts in the clutch release mechanism could include the shaft, thrust piece, or ball bearing. Replace parts as necessary.
  Clutch centre and outer drum worn. Severe indentation by the clutch plate tangs of the channels in the centre and drum will cause snagging of the plates preventing correct engagement. If this damage occurs, renewal of the worn components is required.
● Lubricant incorrect. Use of a transmission lubricant other than that specified may allow the plates to slip.

25 Clutch drag

● Clutch lever play excessive. Adjust lever at bars or at cable end if necessary. See Routine maintenance.
● Clutch plates warped or damaged. This will cause a drag on the clutch, causing the machine to creep. Overhaul clutch assembly.
● Clutch spring tension uneven. Usually caused by a sagged or broken spring. Check and replace springs.
● Engine oil deteriorated. Badly contaminated engine oil and a heavy deposit of oil sludge and carbon on the plates will cause plate sticking. The oil recommended for this machine is of the detergent type, therefore it is unlikely that this problem will arise unless regular oil changes are neglected.
● Engine oil viscosity too high. Drag in the plates will result from the use of an oil with too high a viscosity. In very cold weather clutch drag may occur until the engine has reached operating temperature.
● Clutch centre and outer drum worn. Indentation by the clutch plate tangs of the channels in the centre and drum will prevent easy plate disengagement. If the damage is light the affected areas may be dressed with a fine file. More pronounced damage will necessitate renewal of the components.
● Clutch outer drum seized to shaft. Lack of lubrication, severe wear or damage can cause the outer drum to seize to the shaft. Overhaul of the clutch, and perhaps the transmission, may be necessary to repair damage.
● Clutch release mechanism defective. Worn or damaged release mechanism parts can stick and fail to provide leverage. Overhaul clutch cover components.
● Loose clutch centre nut. Causes drum and centre misalignment, putting a drag on the engine. Engagement adjustment continually varies. Overhaul clutch assembly.

Gear selection problems

26 Gear lever does not return

● Weak or broken return spring. Renew the spring.
● Gearchange shaft bent or seized. Distortion of the gearchange shaft often occurs if the machine is dropped heavily on the gear lever. Provided that damage is not severe straightening of the shaft is permissible.

27 Gear selection difficult or impossible

● Clutch not disengaging fully. See Section 25.
● Gearchange shaft bent. This often occurs if the machine is dropped heavily on the gear lever. Straightening of the shaft is permissible if the damage is not too great.
● Gearchange arms, or pins worn or damaged. Wear or breakage of any of these items may cause difficulty in selecting one or more gears. Overhaul the selector mechanism.
● Gearchange arm spring broken. Renew spring.
● Selector drum stopper cam or detent roller arm damage. Failure, rather than wear, of these items may jam the drum thereby preventing gearchanging. The damaged items must be renewed.
● Selector forks bent or seized. This can be caused by dropping the machine heavily on the gearchange lever or as a result of lack of lubrication. Though rare, bending of a shaft can result from a missed gearchange or false selection at high speed.
● Selector fork claw end and guide pin wear. Pronounced wear of these items and the channels in the selector drum can lead to imprecise selection and, eventually, no selection. Renewal of the worn components will be required.
● Structural failure. Failure of any one component of the selector mechanism will result in improper or fouled gear selection.

28 Jumping out of gear

● Detent roller arm assembly worn or damaged. Wear of the arm roller and the cam with which it locates and breakage of the detent spring can cause imprecise gear selection resulting in jumping out of gear. Renew the damaged components.
● Gear pinion dogs worn or damaged. Rounding off the dog edges and the mating recesses in adjacent pinion can lead to jumping out of gear when under load. The gears should be inspected and renewed. Attempting to reprofile the dogs is not recommended.
● Selector forks, selector drum and pinion channels worn. Extreme wear of these interconnected items can occur after high mileages especially when lubrication has been neglected. The worn components must be renewed.
● Gear pinions, bushes and shafts worn. Renew the worn components.
● Bent gearchange shaft. Often caused by dropping the machine on the gear lever.
● Gear pinion tooth broken. Chipped teeth are unlikely to cause jumping out of gear once the gear has been selected fully; a tooth which is completely broken off, however, may cause problems in this respect and in any event will cause transmission noise.

29 Overselection

● Claw arm spring weak or broken. Renew the spring.
● Detent roller arm worn or broken. Renew the damaged items.
● Stopper arm spring worn or broken. Renew the spring.
● Selector claw arm stop pads worn. Repairs can be made by welding and reprofiling with a file.

Abnormal engine noise.

30 Knocking or pinking

● See Section 19.

31 Piston slap or rattling from cylinder

● Cylinder bore/piston clearance excessive. Resulting from wear, partial seizure or improper boring during overhaul. This condition can often be heard as a high, rapid tapping noise when the engine is under little or no load, particularly when power is just beginning to be applied. Reboring to the next correct oversize should be carried out and a new oversize piston fitted.
● Connecting rod bent. This can be caused by over-revving, trying to start a very badly flooded engine (resulting in a hydraulic lock in the cylinder) or by earlier mechanical failure such as a dropped valve. Attempts at straightening a bent connecting rod from a high performance engine are not recommended. Careful inspection of the crankshaft should be made before renewing the damaged connecting rod.
● Gudgeon pin, piston boss bore or small-end bearing wear or seizure. Excess clearance or partial seizure between normal moving parts of these items can cause continuous or intermittent tapping noises. Rapid wear or seizure is caused by lubrication starvation resulting from an insufficient engine oil level or oilway blockage.
● Piston rings worn, broken or sticking. Renew the rings after careful inspection of the piston and bore.

32 Valve noise or tapping from the cylinder head

● Valve clearance incorrect. Adjust the clearances with the engine cold.
● Valve spring broken or weak. Renew the spring set.
● Rocker arm or spindle wear. Rapid wear of a rocker arm, and the resulting need for frequent valve clearance adjustment, indicates breakthrough or failure of the surface hardening on the rocker arm tips. Similar wear in the cam lobes can be expected. Renew the worn components after checking for lubrication failure.
● Worn camshaft drive components. A light rattle or tapping which is not improved by correct re-adjustment of the cam chain tension can be emitted by defective tensioner components, a worn cam chain or worn sprockets and chain. If uncorrected, subsequent cam chain breakage may cause extensive damage. The worn components must be renewed before wear becomes too far advanced.

33 Other noises

● Big-end bearing wear. A pronounced knock from within the crankcase which worsens rapidly is indicative of big-end bearing failure as a result of extreme normal wear or lubrication failure. Remedial action in the form of a bottom end overhaul should be taken; continuing to run the engine will lead to further damage including the possibility of connecting rod breakage.
● Main bearing failure. Extreme normal wear or failure of the main bearings is characteristically accompanied by a rumble from the crankcase and vibration felt through the frame and footrests. Renew the worn bearings and carry out a very careful examination of the crankshaft.
● Crankshaft excessively out of true. A bent crank may result from over-revving or damage from an upper cylinder component or gearbox failure. A replacement item should be fitted.
● Engine mounting loose. Tighten all the engine mounting nuts and bolts.
● Cylinder head gasket leaking. The noise most often associated with a leaking head gasket is a high pitched squeaking, although any other noise consistent with gas being forced out under pressure from a small orifice can also be emitted. Gasket leakage is often accompanied by oil seepage from around the mating joint. Leakage into the pushrod

tunnel or oil return passages will increase crankcase pressure and may cause oil leakage at joints and oil seals. Also, oil contamination will be accelerated. Leakage results from insufficient or uneven tightening of the cylinder head fasteners, or from random mechanical failure. Retightening to the correct torque figure will, at best, only provide a temporary cure. The gasket should be renewed at the earliest opportunity.
● Exhaust system leakage. Popping or crackling in the exhaust system, particularly when it occurs with the engine on the overrun, indicates a poor joint either at the cylinder port or at the exhaust pipe/silencer connection. Failure of the gasket or looseness of the clamp should be looked for.

Abnormal transmission noise

34 Clutch noise

● Clutch outer drum/friction plate tang clearance excessive.
● Clutch outer drum/centre sleeve clearance excessive.
● Clutch outer drum/thrust washer clearance excessive.
● Primary drive gear teeth worn or damaged.
● Clutch shock absorber assembly worn or damaged.

35 Transmission noise

● Bearing or bushes worn or damaged. Renew the affected components.
● Gear pinions worn or chipped. Renew the gear pinions.
● Metal chips jam in gear teeth. This can occur when pieces of metal from any failed component are picked up by a meshing pinion. The condition will lead to rapid bearing wear or early gear failure.
● Engine/transmission oil level too low. Top up immediately to prevent damage to gearbox and engine.
● Gearchange mechanism worn or damaged. Wear or failure of certain items in the selection and change components can induce mis-selection of gears (see Section 27) where incipient engagement of more than one gear set is promoted. Remedial action, by the overhaul of the gearbox, should be taken without delay.
● Worn or damaged final gear case teeth. If the teeth or any component of the gear case are thought to be worn or damaged, take the machine to a Honda Service Agent for attention.
● Teeth that are worn or wrongly preloaded will whine loudly; if any other noises are heard stop the machine immediately and investigate.

Exhaust smokes excessively

36 White/blue smoke (caused by oil burning)

● Piston rings worn or broken. Breakage or wear of any ring, but particularly the oil control ring, will allow engine oil past the piston into the combustion chamber. Overhaul the cylinder bores and pistons.
● Cylinder liner cracked, worn or scored. These conditions may be caused by overheating, lack of lubrication, component failure or advanced normal wear. The cylinders should be rebored and the next oversize pistons fitted.
● Valve oil seal damaged or worn. This can occur as a result of valve guide failure or old age. The emission of smoke is likely to occur when the throttle is closed rapidly after acceleration, for instance, when changing gear. Renew the valve oil seals and, if necessary, the valve guides.
● Valve guides worn. See the preceding paragraph.
● Engine oil level too high. This increases the crankcase pressure and allows oil to be forced past the piston rings. Often accompanied by seepage of oil at joints and oil seals.
● Cylinder head gasket blown between pushrod tunnel or oil return passage. Renew the cylinder head gasket.
● Abnormal crankcase pressure. This may be caused by blocked breather passages or hoses causing back-pressure at high engine revolutions.

37 Black smoke (caused by over-rich mixture)

● Air filter element clogged. Clean or renew the element.
● Main jet loose or too large. Remove the float chamber to check for tightness of the jet. If the machine is used at high altitudes rejetting will be required to compensate for the lower atmospheric pressure.
● Cold start mechanism jammed on. Check that the mechanism works smoothly and correctly and that, where fitted, the operating cable is lubricated and not snagged.
● Fuel level too high. The fuel level is controlled by the float height which can increase as a result of wear or damage. Remove the float bowl and check the float height. Check also that floats have not punctured; a punctured float will loose buoyancy and allow an increased fuel level.
● Float valve needle stuck open. Caused by dirt or a worn valve. Clean the float chamber or renew the needle and, if necessary, the valve seat.

Oil pressure indicator lamp goes on

38 Engine lubrication system failure

● Engine oil defective. Oil pump shaft or locating pin sheared off from ingesting debris or seizing from lack of lubrication (low oil level).
● Engine oil screen clogged. Change oil and filter and service pickup screen.
● Engine oil level too low. Inspect for leak or other problem causing low oil level and add recommended lubricant.
● Engine oil viscosity too low. Very old, thin oil, or an improper weight of oil used in engine. Change to correct lubricant.
● Crankshaft and/or bearings worn. Overhaul lower end.
● Relief valve stuck open. This causes the oil to be dumped back into the sump. Repair or replace (Chapter 3).

39 Electrical system failure

● Oil pressure switch defective. Check switch according to the procedures in Chapter 7. Replace if defective.
● Oil pressure indicator lamp wiring system defective. Check for pinched, shorted, disconnected or damaged wiring.

Poor handling or roadholding

40 Directional instability

● Steering head bearing adjustment too tight. This will cause rolling or weaving at low speeds. Re-adjust the bearings.
● Steering head bearing worn or damaged. Correct adjustment of the bearing will prove impossible to achieve if wear or damage has occurred. Inconsistent handling will occur including rolling or weaving at low speed and poor directional control at indeterminate higher speeds. The steering head bearing should be dismantled for inspection and renewed if required. Lubrication should also be carried out.
● Bearing races pitted or dented. Impact damage caused, perhaps, by an accident or riding over a pot-hole can cause indentation of the bearing, usually in one position. This should be noted as notchiness when the handlebars are turned. Renew and lubricate the bearings.
● Steering stem bent. This will occur only if the machine is subjected to a high impact such as hitting a curb or a pot-hole. The bottom yoke/stem should be renewed; do not attempt to straighten the stem.
● Front or rear tyre pressures too low.
● Front or rear tyre worn. General instability, high speed wobbles and skipping over white lines indicates that tyre renewal may be required. Tyre induced problems, in some machine/tyre combinations, can occur even when the tyre in question is by no means fully worn.
● Swinging arm bearings worn. Difficulties in holding line, particularly when cornering or when changing power settings indicates wear in the swinging arm bearings. The swinging arm should be checked and adjusted if necessary. Renew the bearings if worn.
● Swinging arm flexing. The symptoms given in the preceding paragraph will also occur if the swinging arm fork flexes badly. This can be caused by structural weakness as a result of corrosion, fatigue or impact damage, or because the rear wheel spindle is slack.
● Wheel bearings worn. Renew the worn bearings.
● Tyres unsuitable for machine. Not all available tyres will suit the characteristics of the frame and suspension, indeed, some tyres or tyre combinations may cause a transformation in the handling characteristics. If handling problems occur immediately after changing to a new tyre type or make, revert to the original tyres to see whether an improvement can be noted. In some instances a change to what are, in fact, suitable tyres may give rise to handling deficiences. In this case a thorough check should be made of all frame and suspension items which affect stability.

41 Steering bias to left or right

● Wheels out of alignment. This can be caused by impact damage to the frame, swinging arm, wheel spindles or front forks. Although occasionally a result of material failure or corrosion it is usually as a result of a crash.
● Front forks twisted in the steering yokes. A light impact, for instance with a pot-hole or low curb, can twist the fork legs in the steering yokes without causing structural damage to the fork legs or the yokes themselves. Re-alignment can be made by loosening the yoke pinch bolts, wheel spindle and mudguard bolts. Re-align the wheel with the handlebars and tighten the bolts working upwards from the wheel spindle. This action should be carried out only when there is no chance that structural damage has occurred.

42 Handlebar vibrates or oscillates

● Tyres worn or out of balance. Either condition, particularly in the front tyre, will promote shaking of the fork assembly and thus the handlebars. A sudden onset of shaking can result if a balance weight is displaced during use.
● Tyres badly positioned on the wheel rims. A moulded line on each wall of a tyre is provided to allow visual verification that the tyre is correctly positioned on the rim. A check can be made by rotating the tyre; any misalignment will be immediately obvious.
● Wheels rims warped or damaged. Inspect the wheels for runout as described in Routine Maintenance.
● Swinging arm bearings worn. Adjust and/or renew the bearings.
● Wheel bearings worn. Renew the bearings.
● Steering head bearings incorrectly adjusted. Vibration is more likely to result from bearings which are too loose rather than too tight. Re-adjust the bearings.
● Loosen fork component fasteners. Loose nuts and bolts holding the fork legs, wheel spindle, mudguards or steering stem can promote shaking at the handlebars. Fasteners on running gear such as the forks and suspension should be check tightened occasionally to prevent dangerous looseness of components occurring.
● Engine mounting bolts loose. Tighten all fasteners.

43 Poor front fork performance

● Damping fluid level incorrect. If the fluid level is too low poor suspension control will occur resulting in a general impairment of roadholding and early loss of tyre adhesion when cornering and braking. Too much oil is unlikely to change the fork characteristics unless severe overfilling occurs when the fork action will become stiffer and oil seal failure may occur.
● Damping oil viscosity incorrect. The damping action of the fork is directly related to the viscosity of the damping oil. The lighter the oil used, the less will be the damping action imparted. For general use, use the recommended viscosity of oil, changing to a slightly higher or heavier oil only when a change in damping characteristic is required. Overworked oil, or oil contaminated with water which has found its way past the seals, should be renewed to restore the correct damping performance and to prevent bottoming of the forks.
● Damping components worn or corroded. Advanced normal wear of the fork internals is unlikely to ocur until a very high mileage has

been covered. Continual use of the machine with damaged oil seals which allows the ingress of water, or neglect, will lead to rapid corrosion and wear. Dismantle the forks for inspection and overhaul. See Chapter 5.

● Weak fork springs. Progressive fatigue of the fork springs, resulting in a reduced spring free length, will occur after extensive use. This condition will promote excessive fork dive under braking, and in its advanced form will reduce the at-rest extended length of the forks and thus the fork geometry. Renewal of the springs as a pair is the only satisfactory course of action.

● Bent stanchions or corroded stanchions. Both conditions will prevent correct telescoping of the fork legs, and in an advanced state can cause sticking of the fork in one position. In a mild form corrosion will cause stiction of the fork thereby increasing the time the suspension takes to react to an uneven road surface. Bent fork stanchions should be attended to immediately because they indicate that impact damage has occurred, and there is a danger that the forks will fail with disastrous consequences.

44 Front fork judder when braking (see also Section 56)

● Wear between the fork stanchions and the fork legs. Renewal of the affected components is required.
● Slack steering head bearings. Re-adjust the bearings.
● Warped brake disc or drum. If irregular braking action occurs fork judder can be induced in what are normally serviceable forks. Renew the damaged brake components.

45 Poor rear suspension performances

● Rear suspension unit damper worn out or leaking. The damping performance of most rear suspension units falls off with age. This is a gradual process, and thus may not be immediately obvious. Indications of poor damping include hopping of the rear end when cornering or braking, and a general loss of positive stability. See Chapter 4.
● Weak rear springs. If the suspension unit springs fatigue they will promote excessive pitching of the machine and reduce the ground clearance when cornering. Although replacement springs are available separately from the rear suspension damper unit it is probable that if spring fatigue has occurred the damper units will also require renewal.
● Swinging arm flexing or bearings worn. See Sections 40 and 41.
● Bent suspension unit damper rod. This is likely to occur only if the machine is dropped or if seizure of the piston occurs. If either happens the suspension units should be renewed as a pair.
● Worn suspension linkage pivot bearings. Overhaul the rear suspension, renewing worn components. See Chapter 5.

Abnormal frame and suspension noise

46 Front end noise

● Oil level low or too thin. This can cause a 'spurting' sound and is usually accompanied by irregular fork action.
● Spring weak or broken. Makes a clicking or scraping sound. Fork oil will have a lot of metal particles in it.
● Steering head bearings loose or damaged. Clicks when braking. Check, adjust or replace.
● Fork clamps loose. Make sure all fork clamp pinch bolts are tight.
● Fork stanchion bent. Good possibility if machine has been dropped. Repair or replace tube.

47 Rear suspension noise

● Fluid level too low. Leakage of a suspension unit, usually evident by oil on the outer surfaces, can cause a spurting noise. The suspension units should be renewed as a pair.
● Defective rear suspension unit with internal damage. Renew the suspension units as a pair.
● Worn suspension linkage pivot bearings. See Section 45.

Brake problems

48 Brakes are spongy or ineffective – disc brakes

● Air in brake circuit. This is only likely to happen in service due to neglect in checking the fluid level or because a leak has developed. The problem should be identified and the brake system bled of air.
● Pad worn. Check the pad wear against the wear lines provided and renew the pads if necessary.
● Contaminated pads. Cleaning pads which have been contaminated with oil, grease or brake fluid is unlikely to prove successful; the pads should be renewed.
● Pads glazed. This is usually caused by overheating. The surface of the pads may be roughened using glass-paper or a fine file.
● Brake fluid deterioration. A brake which on initial operation is firm but rapidly becomes spongy in use may be failing due to water contamination of the fluid. The fluid should be drained and then the system refilled and bled.
● Master cylinder seal failure. Wear or damage of master cylinder internal parts will prevent pressurisation of the brake fluid. Overhaul the master cylinder unit.
● Caliper seal failure. This will almost certainly be obvious by loss of fluid, a lowering of fluid in the master cylinder reservoir and contamination of the brake pads and caliper. Overhaul the caliper assembly.
● Caliper stuck with corrosion. Overhaul the caliper assemblies and thoroughly clean all components. Lubricate on reassembly.

49 Brakes drag – disc brakes

● Disc warped. The disc must be renewed.
● Caliper piston, caliper or pads corroded. The brake caliper assembly is vulnerable to corrosion due to water and dirt, and unless cleaned at regular intervals and lubricated in the recommended manner, will become sticky in operation.
  Piston seal deteriorated. The seal is designed to return the piston in the caliper to the retracted position when the brake is released. Wear or old age can affect this function. The caliper should be overhauled if this occurs.
● Brake pad damaged. Pad material separating from the backing plate due to wear or faulty manufacture. Renew the pads. Faulty installation of a pad also will cause dragging.
● Wheel spindle bent. The spindle may be straightened if no structural damage has occurred.
● Brake lever or pedal not returning. Check that the lever or pedal works smoothly throughout its operating range and does not snag on any adjacent cycle parts. Lubricate the pivot if necessary.
● Twisted caliper support bracket. This is likely to occur only after impact in an accident. No attempt should be made to re-align the caliper; the bracket should be renewed.

50 Brake lever or pedal pulsates in operation – disc brakes

● Disc warped or irregularly worn. The disc must be renewed.
● Wheel spindle bent. The spindle may be straightened provided no structural damage has occurred.

51 Disc brake noise

● Brake squeal. This can be caused by the omission or incorrect installation of the anti-squeal shim fitted to the rear of one pad. The arrow on the shim should face the direction of wheel normal rotation. Squealing can also be caused by dust on the pads, usually in combination with glazed pads, or other contamination from oil, grease, brake fluid or corrosion. Persistent squealing which cannot be traced to any of the normal causes can often be cured by applying a thin layer of high temperature silicone grease to the rear of the pads. Make absolutely certain that no grease is allowed to contaminate the braking surface of the pads.

● Glazed pads. This is usually caused by high temperatures or contamination. The pad surfaces may be roughened using glass-paper or a fine file. If this approach does not effect a cure the pads should be renewed.
● Disc warped. This can cause a chattering, clicking or intermittent squeal and is usually accompanied by a pulsating brake lever or pedal or uneven braking. The disc must be renewed.
● Brake pads fitted incorrectly or undersize. Longitudinal play in the pads due to omission of the locating springs (where fitted) or because pads of the wrong size have been fitted will cause a single tapping noise every time the brake is operated. Inspect the pads for correct installation and security.

52 Brakes are spongy or ineffective – drum brakes

● Worn brake linings. Determine lining wear using the external brake wear indicator on the brake backplate, or by removing the wheel and withdrawing the brake backplate. Renew the shoes as a pair if the linings are worn below the recommended limit.
● Worn brake camshaft. Wear between the camshaft and the bearing surface will reduce brake feel and reduce operating efficiency. Renewal of one or both items will be required to rectify the fault.
● Worn brake cam and shoe ends. Renew the worn components.
● Linings contaminated with dust or grease. Any accumulations of dust should be cleaned from the brake assembly and drum using a petrol dampened cloth. Do not blow or brush off the dust because it is asbestos based and thus harmful if inhaled. Light contamination from grease can be removed from the surface of the brake linings using a solvent; attempts at removing heavier contamination are less likely to be successful because some of the lubricant will have been absorbed by the lining material which will severely reduce the braking performance.

53 Brake drag – drum brakes

● Incorrect adjustment. Re-adjust the brake operating mechanism.
● Drum warped or oval. This can result from overheating or impact. The condition is difficult to correct, although if slight ovality only occurs, skimming the surface of the brake drum can provide a cure. This is work for a specialist engineer. Renewal of the complete wheel is normally the only satisfactory solution.
● Weak brake shoe return springs. This will prevent the brake lining/shoe units from pulling away from the drum surface once the brake is released. The springs should be renewed.
● Brake camshaft or lever pivot poorly lubricated. Failure to attend to regular lubrication of these areas will increase operating resistance which, when compounded, may cause tardy operation and poor release movement.

54 Brake pedal pulsates in operation – drum brakes

● Drums warped or oval. This can result from overheating or impact. This condition is difficult to correct, although if slight ovality only occurs skimming the surface of the drum can provide a cure. This is work for a specialist engineer. Renewal of the wheel is normally the only satisfactory solution.

55 Drum brake noise

● Drum warped or oval. This can cause intermittent rubbing of the brake linings against the drum. See the preceding Section.
● Brake linings glazed. This condition, usually accompanied by heavy lining dust contamination, often induces brake squeal. The surface of the linings may be roughened using glass-paper or a fine file.

56 Brake induced fork judder

● Worn front fork stanchions and legs, or worn or badly adjusted steering head bearings. These conditions, combined with uneven or pulsating braking as described in Section 50 will induce more or less judder when the brakes are applied, dependent on the degree of wear and poor brake operation. Attention should be given to both areas of malfunction. See the relevant Sections.

Electrical problems

57 Battery dead or weak

● Battery faulty. Battery life should not be expected to exceed 3 to 4 years. Gradual sulphation of the plates and sediment deposits will reduce the battery performance. Plate and insulator damage can often occur as a result of vibration. Complete power failure, or intermittent failure, may be due to a broken battery terminal. Lack of electrolyte will prevent the battery maintaining charge.
● Battery leads making poor contact. Remove the battery leads and clean them and the terminals, removing all traces of corrosion and tarnish. Reconnect the leads and apply a coating of petroleum jelly to the terminals.
● Load excessive. If additional items such as spot lamps, are fitted, which increase the total electrical load above the maximum alternator output, the battery will fail to maintain full charge. Reduce the electrical load to suit the electrical capacity.
● Regulator/rectifier failure.
● Alternator generating coils open-circuit or shorted.
● Charging circuit shorting or open circuit. This may be caused by frayed or broken wiring, dirty connectors or a faulty ignition switch. The system should be tested in a logical manner. See Section 60.

58 Battery overcharged

● Rectifier/regulator faulty. Overcharging is indicated if the battery becomes hot or it is noticed that the electrolyte level falls repeatedly between checks. In extreme cases the battery will boil causing corrosive gases and electrolyte to be emitted through the vent pipes.
● Battery wrongly matched to the electrical circuit. Ensure that the specified battery is fitted to the machine.

59 Total electrical failure

● Fuse blown. Check the main fuse. If a fault has occurred, it must be rectified before a new fuse is fitted.
● Battery faulty. See Section 57.
● Earth failure. Check that the frame main earth strap from the battery is securely affixed and is making a good contact.
● Ignition switch or power circuit failure. Check for current flow through the battery positive lead (red) to the ignition switch. Check the ignition switch for continuity.

60 Circuit failure

● Cable failure. Refer to the machine's wiring diagram and check the circuit for continuity. Open circuits are a result of loose or corroded connections, either at terminals or in-line connectors, or because of broken wires. Occasionally, the core of a wire will break without there being any apparent damage to the outer plastic cover.
● Switch failure. All switches may be checked for continuity in each switch position, after referring to the switch position boxes incorporated in the wiring diagram for the machine. Switch failure may be a result of mechanical breakage, corrosion or water.
● Fuse blown. Refer to the wiring diagram to check whether or not a circuit fuse is fitted. Replace the fuse, if blown, only after the fault has been identified and rectified.

61 Bulbs blowing repeatedly

● Vibration failure. This is often an inherent fault related to the natural vibration characteristics of the engine and frame and is, thus, difficult to resolve. Modifications of the lamp mounting, to change the damping characteristics may help.

● Intermittent earth. Repeated failure of one bulb, particularly where the bulb is fed directly from the generator, indicates that a poor earth exists somewhere in the circuit. Check that a good contact is available at each earthing point in the circuit.

● Reduced voltage. Where a quartz-halogen bulb is fitted the voltage to the bulb should be maintained or early failure of the bulb will occur. Do not overload the system with additional electrical equipment in excess of the system's power capacity and ensure that all circuit connections are maintained clean and tight.

Routine maintenance

Specifications

Engine

	NGK	ND
Spark plugs:		
All UK 500 models, CX500 C 1982, GL500 and GL500 I 1982	DR8ES-L	X24ESR-U
All other US 500 models	D8EA	X24ES-U
650 models	DPR8EA-9	X24EPR-U9

Spark plug gap:		
500 models	0.6 – 0.7 mm (0.024 – 0.028 in)	
650 models	0.8 – 0.9 mm (0.032 – 0.035 in)	

	Inlet	Exhaust
Valve clearances – engine cold:		
CX500-A, CX500-B, CX500 C-B, all 650 models	0.10 mm (0.004 in)	0.12 mm (0.005 in)
All other 500 models	0.08 mm (0.003 in)	0.10 mm (0.004 in)
Throttle cable free play – approx	2 – 6 mm (0.08 – 0.24 in)	
Idle speed	1100 ± 100 rpm	
Clutch cable free play	10 – 20 mm (0.4 – 0.8 in)	

Cycle parts

Drum brake pedal free play	20 – 30 mm (0.8 – 1.2 in)	

	Front	Rear
Tyre pressures – tyres cold:		
CX500 (UK), CX500 1978, 1979:		
Solo	25 psi (1.75 kg/cm²)	28 psi (2.00 kg/cm²)
Pillion	25 psi (1.75 kg/cm²)	36 psi (2.50 kg/cm²)
CX500-A, CX500-B, CX500 E-C, all GL500 models:		
UK models solo, US models up to 90 kg (198 lb) load*	28 psi (2.00 kg/cm²)	28 psi (2.00 kg/cm²)
UK models pillion, US models over 90 kg (198 lb) load*	28 psi (2.00 kg/cm²)	36 psi (2.50 kg/cm²)
All CX500 C and CX500 D models:		
UK model solo, US models up to 90 kg (198 lb) load*	28 psi (2.00 kg/cm²)	28 psi (2.00 kg/cm²)
UK model pillion, US models over 90 kg (198 lb) load*	28 psi (2.00 kg/cm²)	32 psi (2.25 kg/cm²)
All 650 models:		
UK models solo, US models up to 90 kg (198 lb) load*	32 psi (2.25 kg/cm²)	32 psi (2.25 kg/cm²)
UK models pillion, US models over 90 kg (198 lb) load*	32 psi (2.25 kg/cm²)	40 psi (2.80 kg/cm²)

*Loads given are the total weight of rider, passenger and any accessories or luggage

Front fork air pressure:	
CX500 C 1981, 1982, CX500 D 1981	10 – 16 psi (0.7 – 1.1 kg/cm²)
CX500 E-C, all GL500 models	11 – 17 psi (0.8 – 1.2 kg/cm²)
CX650 C, CX650 E-D	0 – 6 psi (0 – 0.4 kg/cm²)
All GL650 models	6 – 17 psi (0.4 – 1.2 kg/cm²)
All other models	N/App.
Rear suspension air pressure:	
CX500 E-C, GL500 1981, 1982, CX650 E-D, GL650	0 – 71 psi (0 – 5.0 kg/cm²)
GL500 D-C, GL500 I 1981, 1982, GL650 D2-E, GL650 I	14 – 71 psi (1.0 – 5.0 kg/cm²)
All other models	N/App

Recommended lubricants

Engine oil:
- Recommended type .... Good quality SAE10W/40 engine oil, API class SE or SF

  Capacity – at oil and filter change:
  - 500 models .... 2.6 lit (2.75 US qt/4.58 Imp pint)
  - 650 models .... 3.1 lit (3.28 US qt/5.46 Imp pint)

Final drive case oil:
- Recommended type .... Good quality hypoid gear oil, API class GL-5

  Viscosity:
  - Above 5°C (41°F) .... SAE 90
  - Below 5°C (41°F) .... SAE 80
- Capacity .... 170 ± 10 cc (5.75 ± 0.34 US fl oz/5.98 ± 0.35 Imp fl oz)

Final drive shaft joint and splines:
- Recommended grease .... Lithium based multipurpose NLG1 No 2 grease with molybdenum disulphide additive (e.g. Dow Corning Molykote BR2-S or Mitsubishi Oil Multipurpose M2)
  - Amount – approx .... 20 cc (0.67 US fl oz/0.70 Imp fl oz)

Front fork oil:
- Recommended type .... Automatic Transmission Fluid (ATF) or similar

  Capacity – per leg:
  - CX500 (UK), CX500-A, CX500-B, CX500 1978, 1979, CX500 C and CX500 D 1979, 1980, CX500 C-B .... 135 cc (4.6 US fl oz/4.8 Imp fl oz)
  - CX500 D 1981 .... 185 cc (6.3 US fl oz/6.5 Imp fl oz)
  - CX500 C 1981, 1982 .... 220 cc (7.4 US fl oz/7.7 Imp fl oz)
  - All GL500 models .... 210 cc (7.1 US fl oz/7.4 Imp fl oz)
  - CX650 C .... 480 cc (16.2 US fl oz/16.9 Imp fl oz)
  - All GL650 models .... 275 cc (9.3 US fl oz/9.7 Imp fl oz)

	Left-hand leg	Right-hand leg
CX500 E-C	265 cc (9.0 US fl oz/ 9.3 Imp fl oz)	250 cc (8.5 US fl oz/ 8.8 Imp fl oz)
CX650 E-D	290 cc (9.8 US fl oz/ 10.2 Imp fl oz)	275 cc (9.3 US fl oz/ 9.7 Imp fl oz)

Swinging arm bearings and suspension linkage pivots:
- Recommended grease .... Molybdenum paste grease containing at least 45% molybdenum disulphide (eg Dow Corning Molykote G or G-n Paste, Sumico Lubricant Rocol Paste)
- Brake fluid .... DOT3 (US) or SAEJ1703 (UK) hydraulic fluid

Introduction

Periodic routine maintenance is a continuous process which should commence immediately the machine is used. The object is to maintain all adjustments and to diagnose and rectify minor defects before they develop into more extensive, and often more expensive, problems.

It follows that if the machine is maintained properly, it will both run and perform with optimum efficiency, and be less prone to unexpected breakdowns. Regular inspection of the machine will show up any parts which are wearing, and with a little experience, it is possible to obtain the maximum life from any one component, renewing it when it becomes so worn that it is liable to fail.

Regular cleaning can be considered as important as mechanical maintenance. This will ensure that all the cycle parts are inspected regularly and are kept free from accumulations of road dirt and grime.

Cleaning is especially important during the winter months, despite its appearance of being a thankless task which very soon seems pointless. On the contrary, it is during these months that the paintwork, chromium plating, and the alloy casings suffer the ravages of abrasive grit, rain and road salt. A couple of hours spent weekly on cleaning the machine will maintain its appearance and value, and highlight small points, like chipped paint, before they become a serious problem.

The various maintenance tasks are described under their respective mileage and calender headings, and are accompanied by diagrams and photographs where pertinent.

It should be noted that the intervals between each maintenance task serve only as a guide. As the machine gets older, or if it is used under particularly arduous conditions, it is advisable to reduce the period between each check.

For ease of reference, most service operations are described in detail under the relevant heading. However, if further general information is required, this can be found under the pertinent Section heading and Chapter in the main text.

Although no special tools are required for routine maintenance, a good selection of general workshop tools is essential. Included in the tools must be a range of metric ring or combination spanners, a selection of crosshead screwdrivers, and two pairs of circlip pliers, one external opening and the other internal opening.

Note: This section on routine maintenance is divided into five basic service intervals which are the pre-ride check, the monthly check, the minor service, the major service, and one or two additional items. The first two and the last must be carried out at the specified time interval as they concern items which deteriorate irrespective of whether the machine is used a lot or hardly at all. The two services should be carried out at the specified time or mileage interval, **whichever applies first.** Note that different mileages are specified for some models.

Commence all service operations by carrying out the tasks listed under the previous mileage/time headings.

Daily (pre-ride check)

It is recommended that the following items are checked whenever the machine is about to be used. This is important to prevent the risk of unexpected failure of any component while riding the machine and, with experience, can be reduced to a simple checklist which will only take a few moments to complete. For those owners who are not inclined to check all items with such frequency, it is suggested that the best course is to carry out the checks in the form of a service which can be undertaken each week or before any long journey. It is essential that all items are checked and serviced with reasonable frequency.

1 Check the engine oil level

With the machine standing upright on its centre stand on level ground, start the engine and allow it to idle for a few seconds so that the oil can circulate, then stop the engine. Wait one or two minutes for the level to settle and unscrew the dipstick/filler plug from the crankcase left-hand side. Wipe it clean and insert it into the filler orifice; **do not screw** it in, but allow it to rest. Withdraw the dipstick; the oil level should be between the maximum and minimum level lines, ie in the cross-hatched area.

If topping up is necessary use only good quality SAE10W/40 engine oil of the specified type. Do not allow the level to rise above the top of the cross-hatched area on the dipstick, and never use the machine if the level is found to be in the plain area below the cross-hatching; top up immediately.

Tighten the dipstick securely and wash off any spilt oil.

2 Check the coolant level

Although the cooling system is semi-sealed and should not require frequent topping up, it is still necessary to check the level at regular intervals. A separate expansion tank is fitted to allow for expansion of the coolant when the engine is hot, the displaced liquid being drawn back into the system when it cools. It is therefore the level of coolant in the expansion tank which is to be checked; the tank is constructed of translucent plastic so that the coolant level can be seen easily in relation to the upper and lower level lines marked on the tank side. Check only when the engine is at normal operating temperature.

The tank is on the left-hand side immediately in front of the swinging arm pivot.

The coolant level must be between the higher ('Full' or 'F') and lower ('Low' or 'L') level marks at all times. Although the level will vary with engine temperature, Honda state that if the level is ever found to be below the lower mark it must be topped up to the higher level mark regardless of engine temperature. If the level is significantly above the

Use dipstick as described to check engine oil level ...

... and use only good quality engine oil if topping-up is necessary

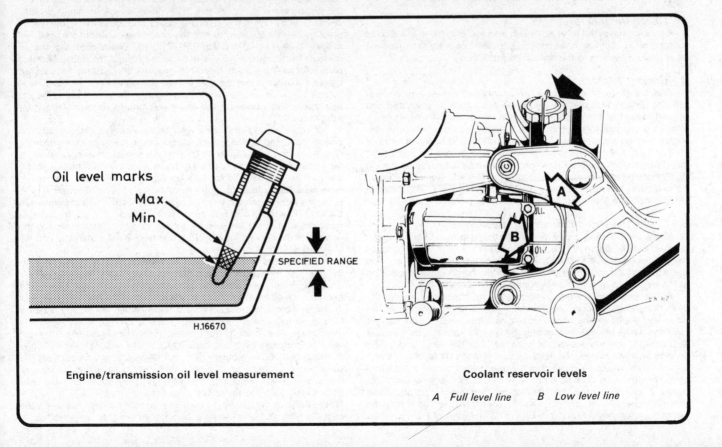

Oil level marks

Max

Min

SPECIFIED RANGE

H.16670

Engine/transmission oil level measurement

Coolant reservoir levels

A Full level line   B Low level line

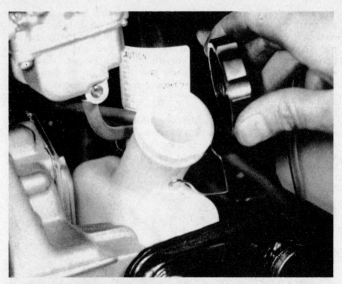

Coolant level is checked and topped-up at expansion tank

higher level mark at any time the surplus should be siphoned off to prevent coolant being blown over the rear of the machine via the tank breather.

Use only the specified ingredients to make coolant of the required strength, as described in Chapter 2, and always have a supply prepared for topping up. In cases of real emergency distilled water or **clean** rainwater may be used, but remember that this will dilute the coolant and reduced the degree of protection against freezing.

If the level falls steadily, check the system very carefully for leaks, as described in Chapter 2. Also, do not forget to check that the radiator matrix is clean, unblocked and free from damage of any sort.

3 Check the fuel level

Checking the petrol level may seem obvious, but it is all too easy to forget. Ensure that you have enough petrol to complete your journey, or at least to get you to the nearest petrol station.

4 Check the brakes

Check that the front and rear brakes work effectively and without binding. Ensure that the rod linkage as applicable, is lubricated and properly adjusted. Check the fluid level in the master cylinder reservoir, where appropriate, and ensure that there are no fluid leaks. Should topping-up be required, use only the recommended hydraulic fluid.

The height of the rear brake pedal can be adjusted by means of a stop bolt and locknut fitted at the pedal rear end; the final setting being up to the individual owner's requirement. When adjustment is correct, tighten the locknut securely; if the pedal height was altered significantly check the rear brake and stop lamp rear switch settings.

5 Check the tyre pressures and tread wear

Check the tyre pressures with a gauge that is known to be accurate. It is worthwhile purchasing a pocket gauge for this purpose because the gauges on garage forecourt airlines are notoriously inaccurate. The pressures, which should be checked with the tyres cold, are specified at the beginning of Routine Maintenance and Chapter 6.

At the same time as the tyre pressures are checked, examine the tyres themselves. Check them for damage, especially splitting of the sidewalls. Remove any small stones or other road debris caught between the treads. When checking the tyres for damage, they should be examined for tread depth in view of both the legal and safety aspects. It is vital to keep the tread depth within the UK legal limits of 1 mm of depth over three-quarters of the tread breadth around the entire circumference with no bald patches. Many riders, however, consider nearer 2 mm to be the limit for secure roadholding, traction, and braking, especially in adverse weather conditions, and it should be noted that Honda recommend minimum tread depths of 1.5 mm (0.06

in) for the front tyre and 2.0 mm (0.08 in) for the rear; these measurements to be taken at the centre of the tread. Renew any tyre that is found to be damaged or excessively worn.

6 Check the suspension settings

On some later models (see Specifications), the front forks are provided with an air valve in the top of each leg to enable the fork's effective spring rate to be altered to suit the individual owner's needs. The pressure must be checked frequently to ensure that it is at the desired setting and that it is exactly the same for both legs. Where the fork legs are linked this is unnecessary as there is only one valve.

Two tools are essential for setting the air pressure; a gauge capable of reading the low pressures involved, and a low-pressure pump. The gauge must be finely calibrated to ensure that both legs can be set to the same pressure, and must cause only a minimal drop in pressure whenever a reading is taken; as the total air volume is so small, an ordinary gauge, such as a tyre pressure gauge, will cause a large drop in pressure because of the amount of air required to operate it. Gauges for use on suspension components are now supplied by several companies and should be available through any good motorcycle dealer. The pump must be of the hand- or foot-operated type, a bicycle pump being ideal; aftermarket pumps are available for use on suspension systems and are very useful, but expensive. **Never** use a compressor-powered air line; it is all too easy to exceed the maximum recommended pressure which may cause damage to the fork oil seals and may even result in personal injury. Add air very carefully, by means of a hand-operated pump only, and in small amounts at a time.

The air pressure must be set when the forks are cold, and with the machine securely supported so that its front wheel is clear of the ground, thus ensuring that the air pressure is not artificially increased.

Set the pressure to the required amount within the specified range; on models with unlinked fork legs, be careful to ensure that each leg is at exactly the same pressure within the tolerance of 1.5 psi (0.1 kg/cm²). This is essential as any imbalance in pressures will impair fork performance and may render the machine unsafe to ride. Note that good quality aftermarket kits are now available to link separate air caps; the use of one of these, when correctly installed, will ensure that the pressures in the legs are equal at all times and will aid the task of setting the pressure in the future.

Models with 'Pro-Link' rear suspension are fitted with an air/oil suspension unit. If the oil viscosity and quantity are varied as well as the air pressure, units of this type have almost infinitely variable characteristics. With the standard grade and quantity of oil the air pressure can be adjusted to suit the rider's needs within the specified range. Never exceed 5.0 kg/cm² (71 psi). When checking the air pressure note that the valve is to be found behind the right-hand side panel. As with the front forks, the pressure should only be checked before a journey when the unit is cold and only when the machine is supported on its centre stand. The procedure is carried out using the equipment and observing the same precautions as given above for front forks.

All other models are fitted with coil-spring suspension units on which the spring preload is adjustable through five positions using a C-spanner or a screwdriver. In general the softest (lowest) position is for a solo rider using the machine at low speed on good roads, while the hardest (highest) is for two-up riding or when the machine is carrying a heavy load of luggage, or being used at high speed. The positions in between are to suit needs between the two extremes. On CX500 E-C and CX650 E-D models the anti-dive is set by using a screwdriver to rotate the selector so that its punch mark is next to the number of the desired setting; No 1 is softest, No 2 is standard and No 4 is hardest.

It is **essential** that any suspension settings are exactly the same on the left- and right-hand sides of the machine (where applicable); any discrepancy in air pressure or spring preload may upset the performance of the suspension to the point that the machine becomes unstable. This worsens markedly if the machine is carrying a heavy load or is being used on a rough road or at high speed. Similarly settings must be matched front to rear, if the front is stiffened, the rear must be stiffened by a proportionate amount and vice-versa. If any handling or stability problems are encountered at any time, check the suspension settings first. Similarly if any suspension faults are discovered while adjusting the settings, they must be put right immediately. Refer to the relevant Sections of Chapter 5.

Note: Refer to the machine's owners's manual for information on the carrying of luggage. Maximum permissible weights are specified in each case, and the suspension and tyre pressures must be adjusted to compensate for the addition or removal of any load. Note carefully the information on balancing the load from side to side. While this applies principally to GL models with Interstate fairings and panniers, it should be of concern to owners of all models; if the machine is unstable when laden, either reduce the load to within specified limits, re-distribute the load, or unload the machine and test-ride it unladen before looking for any faults.

7 Check the final drive

Make a quick examination of the driveshaft gaiter, swinging arm and final drive case. If any oil leaks are seen or if any strange noises become evident, the machine should be taken to a Honda Service Agent for prompt attention.

8 Check the battery

The battery is located on the left-hand side of the machine, behind the side panel. Remove the side panel to check the electrolyte level; if the battery is to be removed disconnect the terminals (negative terminal first, always) and remove the nut securing the retaining strap.

On all models, whenever the battery is disconnected, remember to disconnect the negative (–) terminal first, to prevent the possibility of short circuits. The electrolyte level, visible through the translucent casing, must be between the two level marks. If necessary remove the cell caps and top up to the upper level using only distilled water. Check that the terminals are clean and apply a thin smear of petroleum jelly (not grease) to each to prevent corrosion. On refitting, check that the vent hose is not blocked and that it is correctly routed with no kinks, also that it hangs well below any other component, particularly the chain or exhaust system. Remember always to connect the negative (–) terminal last when refitting the battery.

Always check that the terminals are tight and that the rubber covers are correctly refitted, also that the fuse connections are clean and tight, that the fuse is of the correct rating and in good condition, and that a spare is available on the machine should the need arise.

At regular intervals remove the battery and check that there is no pale grey sediment deposited at the bottom of the casing. This is caused by sulphation of the plates as a result of re-charging at too high a rate or as a result of the battery being left discharged for long periods. A good battery should have little or no sediment visible and its plates should be straight and pale grey or brown in colour. If sediment deposits are deep enough to reach the bottom of the plates, or if the plates are buckled and have whitish deposits on them, the battery is faulty and must be renewed. Remember that a poor battery will give rise to a large number of minor electrical faults.

If the machine is not in regular use, disconnect the battery and give it a refresher charge ever month to six weeks, as described in Chapter 7.

9 Check the controls

Check the throttle and clutch cables and levers, the gear lever and the footrests to ensure that they are adjusted correctly, functioning correctly, and that they are securely fastened. If a bolt is going to work loose, or a cable snap, it is better that it is discovered at this stage with the machine at a standstill, rather than when it is being ridden.

10 Legal check

Check that all lights, turn signals, horn and speedometer are working correctly to make sure that the machine complies with all legal requirements in this respect. Check also that the headlamp is correctly aimed to comply with local legislation. Horizontal aim is adjusted by means of a spring-loaded screw. On models with headlamp fairings or nacelles a second spring-loaded screw is provided to adjust vertical aim; this is situated under the headlamp bottom edge. Models with Interstate fairings are provided with an adjusting knob on the inside of the fairing. On models with conventional headlamps slacken the mounting bolts and tilt the headlamp to the correct angle. Note that reference marks are stamped on the headlamp and mounting bracket to provide an initial setting.

Maintain electrolyte level between level marks on battery case

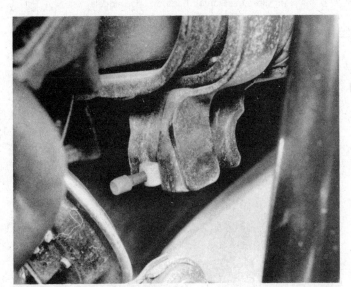

Headlamp vertical aim is adjusted at screw underneath headlamp on some models ...

... and by tilting headlamp shell on others. Note alignment marks (arrowed) for initial setting

Monthly check

1 Check the battery

Honda recommend that the battery be checked carefully at least once a month. Proceed as described under the pre-ride check.

2 Check the brake fluid level

Again Honda recommend that the fluid levels be checked at this interval to prevent any risk of the level dropping unnoticed to a dangerous degree. Top up if necessary (refer to the brake inspection described under the major service).

Minor service

This service to be conducted every six months or at the following mileage, whichever is the sooner.

Every 3600 miles (6000 km) – all UK CX500 and CX650 models, all US models up to 1981
Every 4000 miles (6400 km) – UK GL500 D-C, GL650 D2-E, all US models from 1982 on

1 Additional engine/transmission oil change

Since the engine relies so heavily on the quantity and quality of its oil, and since the oil in any motorcycle engine is worked far harder than in other vehicles, it is recommended that the engine oil is changed at more frequent intervals than those specified by the manufacturer. This is particularly important if the machine is used at very high speeds for long periods of time, and even more important if the machine is used only at very slow speed or for very short journeys.

Follow the instructions given under the major service interval, but note that there will be no need to change the oil filter, and that therefore, a smaller amount (See Chapter 3) of oil will be necessary to refill the crankcase.

2 Clean the air filter

On all CX500, CX500 C and CX500 D models remove the seat, either by releasing the locking catches or by unscrewing the two mounting bolts. On all GL models and the CX500 E-C and CX650 E-D models carefully remove the right-hand side panel. To remove the element, twist the air filter box lid a few degrees in an anti-clockwise direction until it is released and then lift it away. The element can be lifted out.

On CX650 C models carefully remove the right-hand side panel, unscrew the four retaining screws and withdraw the filter cover. Pull out the element.

Tap the element gently to remove any loose dust and then use an air hose to remove the remainder of the dust. Apply the air current from the inside of the element only. If an air hose is not available, a tyre pump can be utilised instead. If the corrugated paper element is damp, oily or beginning to disintegrate, it must be renewed. Never run the engine with the element removed as the weak mixture caused may result in engine overheating and damage to the cylinders and pistons. A weak mixture can also result if the rubber sealing rings on the element are perished or omitted. When replacing the filter assembly, note that the filter box cover should be fitted with the arrow pointing forwards.

3 Clean the crankcase breather

Of the various breather tubes which terminate below and to the rear of the swinging arm cross-member, two are fitted with plastic caps. These are the crankcase breather tubes. Remove the cap from each and allow any accumulated sludge or deposits to drain away. Refit the caps after draining has been completed, ensuring that they are pressed firmly into position. If the machine is ridden in the rain at full throttle service these components more frequently. Similarly, if the deposit level can ever be seen in either tube's transparent section, more frequent cleaning is necessary.

Brake fluid level is checked against level lines on reservoir body (early models) ...

... or via sight glass in reservoir body on later models

Air filter element can be cleaned using compressed air only – renew if badly clogged

4 Check or renew the spark plugs

Note: On all US models from 1982 on, Honda recommend that the spark plugs are renewed at every service interval. Proceed as described under the major service.

The spark plugs supplied as original equipment will prove satisfactory in most operating conditions; alternatives are available to allow for varying altitudes, climatic conditions and the use to which the machine is put. If a spark plug is suspected of being faulty it can be tested only by the substitution of a brand new (not second-hand) plug of the correct make, type, and heat range; always carry a spare on the machine.

Note that the advice of a competent Honda Service Agent or similar expert should be sought before the plug heat range is altered from standard. The use of too cold, or hard, a grade of plug will result in fouling and the use of too hot, or soft, a grade of plug will result in engine damage due to excess heat being generated. If the correct grade of plug is fitted, however, it will be possible to use the condition of the spark plug electrodes to diagnose a fault in the engine or to decide whether the engine is operating efficiently or not. The accompanying series of colour photographs will show this clearly. Also, always ensure that the plug is of the resistor type (indicated by the letter 'R'), where applicable, so that its resistance value is correct for the ignition system. The same applies to the suppressor cap; if a cap or plug of the wrong type is fitted, thus producing a much greater or lesser resistance value than that for which the ignition system was designed, one or more components of the system may break down.

It is advisable to carry a new spare spark plug on the machine, having first set the electrodes to the correct gap. Whilst spark plugs do not often fail, a new replacement is well worth having if a breakdown does occur. Ensure that the spare is of the correct heat range and type.

The electrode gap can be assessed using feeler gauges. If necessary, alter the gap by bending the outer electrode, preferably using a proper electrode tool. **Never** bend the centre electrode, otherwise the ceramic insulator will crack, and may cause damage to the engine if particles break away whilst the engine is running. If the outer electrode is seriously eroded as shown in the photographs, or if the spark plug is heavily fouled, it should be renewed. Clean the electrodes using a wire brush or a sharp-pointed knife, followed by rubbing a strip of fine emery across the electrodes. If a sand-blaster is used, check carefully that there are no particles of sand trapped inside the plug body to fall into the engine at a later date. For this reason such cleaning methods are no longer recommended; if the plug is so heavily fouled it should be renewed.

Whenever the sparking plugs are removed, take the opportunity to clear out the drain channels which pass from the plug wells to the underside of the cylinder heads. If these become blocked and the machine is used in heavy rain the wells will fill up causing shorting in the suppressor caps and complete ignition failure.

Before refitting a spark plug into the cylinder head; coat the threads sparingly with a graphited grease to aid future removal. Use the correct size spanner when tightening the plug, otherwise the spanner may slip and damage the ceramic insulator. The plug should be tightened by hand only at first and then secured with a quarter turn of the spanner so that it seats firmly on its sealing ring.

Never overtighten a spark plug otherwise there is risk of stripping the thread from the cylinder head, especially as it is cast in light alloy. A stripped thread can be repaired without having to scrap the cylinder head by using a 'Helicoil' thread insert. This is a low-cost service, operated by a number of dealers.

5 Check the valve clearances

This task must be carried out at the machine's first 600 mile (1000 km) service, then at its first minor service, and again at its first major service. Thereafter the valve clearances should only be checked (as part of Routine Maintenance) at the major service. If appropriate, carry out the task as described under the major service, at this interval.

6 Adjust the cam chain tension

Obviously, this task applies only to those models fitted with a manual cam chain tensioner (ie all 500 models except the CX500 E-C and GL500 D-C).

Rotate the engine until the left-hand piston is at TDC on the compression stroke, ie with both valves closed and the TL mark on the alternator rotor in alignment with the index pointer. It will be necessary

to remove the left-hand cylinder head cover to check, but if valve clearance adjustment has just been carried out the engine will already be close to the correct position.

Loosen the cam chain tensioner adjustment bolt, the head of which protrudes from the engine rear casing, just above the timing mark inspection aperture. When the bolt is loosened – approximately two full turns – the tensioner automatically adjusts the tensioner blade to the correct value. Tighten the bolt to a torque setting of 0.8 – 1.2 kgf m (6 – 9 lbf ft). On some machines, due to the casing thickness, the bolt will tighten up as it butts against the casing, before being loosened a full two turns. Do not force the bolt to unscrew because it may shear.

If any unusual noises ever lead you to suspect that all is not well with the cam chain tensioner, either have the machine listened to by a Honda expert, or remove the engine from the frame and withdraw the engine rear cover to check carefully the tensioner assembly. Since the potential for serious damage is so great, no odd engine noises should be allowed to go un-checked on these machines.

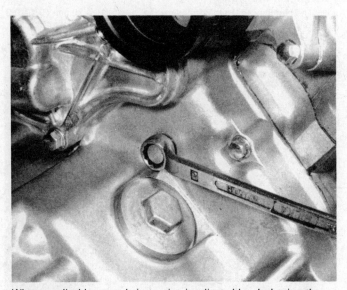

Where applicable, cam chain tension is adjusted by slackening the bolt shown

7 Clean the fuel tap filter

This applies only to CX500 E-C, GL500 D-C, CX500 C 1982, GL500 and GL500 I 1982 and all 650 models, which are fitted with the vacuum fuel tap that incorporates a separate filter bowl.

Switch the tap to the 'Off' position and unscrew the filter bowl from the tap base, then remove the O-ring and filter gauze. Check the condition of the sealing O-ring and renew, if it is seriously compressed, distorted or damaged. Clean the filter gauze using a fine-bristled toothbrush or similar; remove all traces of dirt or debris and renew the gauze if it is split or damaged. Thoroughly clean the filter bowl; if excessive signs of dirt or water are found in the petrol, remove the tank as described in Chapter 3, empty the petrol into a clean container and remove the fuel tap by unscrewing its retaining gland nut.

Remove the tubular filter gauze from the tap stack pipe, noting the presence of a small spacer and of the sealing O-ring, and clean it using a fine-bristled brush; if the gauze is split, twisted or damaged it should be renewed. Flush the tank until all traces of dirt or water are removed. The tap cannot be dismantled further and must be renewed as a complete assembly if the lever is leaking or defective in any way. If its passages are blocked, use compressed air to blow them clear.

On reassembly, renew the sealing O-ring if damaged or worn. Fit the filter gauze, spacer and O-ring into the tap and refit the assembly to the tank. Check that the tap is correctly aligned before tightening the gland nut; do not overtighten the gland nut or its threads may be stripped, necessitating the renewal of the tap.

Fit the filter gauze to the tap, ensuring that it is located correctly, then press the O-ring into place to retain it. Use only a close-fitting ring spanner to tighten the filter bowl, which should be secured by just enough to nip up the O-ring; do not overtighten it as this will only damage the filter bowl, distort the O-ring and promote fuel leaks. The recommended torque setting is only 0.3 – 0.5 kgf m (2 – 3.5 lbf ft). If any leaks are found in the tap they can be cured only by the renewal of the tap assembly or the defective seal, where separate.

8 Check the carburettor idle speed

Start the engine and warm it up to normal operating temperature. Check that the engine idles evenly and steadily at between 1000 – 1200 rpm. If adjustment is required, rotate the (usually black plastic) heavily-knurled adjusting screw set between the carburettors, on their underside.

If any difficulty is experienced in achieving a smooth idle, check the air filter, spark plugs, ignition timing, valve clearances and exhaust system as described elsewhere in this Manual, then check the carburettor adjustment and synchronisation as described in Chapter 3.

9 Adjust the clutch

The clutch is adjusted correctly when there is 10 – 20 mm (0.4 – 0.8 in) of free play in the cable, measured at the handlebar lever ball end, and the clutch operates smoothly with no sign of slip or drag.

There is no provision for adjustment of the clutch itself or of the release mechanism; adjustment can only be made at the cable. Normal adjustment is made only at the cable lower end, reserving the handlebar adjuster for quick roadside adjustments. Slacken the adjuster locknuts and screw in fully the handlebar adjuster (if necessary), then turn the lower adjusting nut until the correct setting is achieved. Tighten both adjuster locknuts and apply a few drops of oil to all exposed lengths of inner cable and all lever pivots and cable nipples.

If adjustment is used up on the lower adjuster, both adjusters may be used together; if both are used up the cable must be renewed. Do not allow the upper adjuster to project by more than 8 mm (0.3 in) from the lever clamp or it may break. If cable adjustment does not eliminate slip or drag, or if any other problems are encountered, the clutch must be dismantled for examination as described in Chapter 1.

10 Check the brake shoes and pads

At this interval the brake fluid level must be checked and the brake shoes or pads must be checked for wear using the external wear indicators. Proceed as described under the brake inspection operation given at the major service interval.

Adjust clutch at cable lower adjuster

Pad wear is checked through aperture in caliper rear face on single-piston calipers ...

... or through slot indicated by cast arrow on twin-piston calipers

Both pads must be renewed if either is worn to red wear limit line or beyond

Electrode gap check - use a wire type gauge for best results

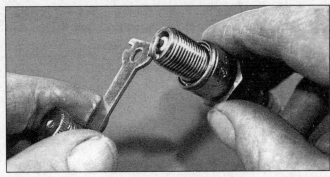

Electrode gap adjustment - bend the side electrode using the correct tool

Normal condition - A brown, tan or grey firing end indicates that the engine is in good condition and that the plug type is correct

Ash deposits - Light brown deposits encrusted on the electrodes and insulator, leading to misfire and hesitation. Caused by excessive amounts of oil in the combustion chamber or poor quality fuel/oil

Carbon fouling - Dry, black sooty deposits leading to misfire and weak spark. Caused by an over-rich fuel/air mixture, faulty choke operation or blocked air filter

Oil fouling - Wet oily deposits leading to misfire and weak spark. Caused by oil leakage past piston rings or valve guides (4-stroke engine), or excess lubricant (2-stroke engine)

Overheating - A blistered white insulator and glazed electrodes. Caused by ignition system fault, incorrect fuel, or cooling system fault

Worn plug - Worn electrodes will cause poor starting in damp or cold weather and will also waste fuel

Major service

This service to be conducted annually, or at the following mileage, whichever is the sooner.

Every 7200 miles (12 000 km) – UK CX500 and CX650 models, all US models up to 1981
Every 8000 miles (12 800 km) – UK GL500 D-C, GL650 D2-E, all US models from 1982 on

Start by carrying out all operations listed under the previous service headings.

1   Change the engine oil and filter element

Place the machine on its centre stand on level ground. Drain the old oil after the engine has been running, or after bringing the engine up to normal running temperature. This will thin the oil and improve the draining rate. Place a container of suitable capacity below the front of the engine. Remove the oil filler plug/dipstick followed by the oil drain plug, which is located below the filter housing on 500 models, and in the sump left-hand side on 650 models. Allow the oil to drain completely and then remove the oil filter housing, complete with element by unscrewing the filter centre bolt.

Pull the old element off the filter bolt, noting the washer and spring behind it. Check the O-rings sealing the filter cover and centre bolt, and the drain plug sealing washer. While in practice these can be re-used a few times, they must be renewed if worn, compressed, distorted or otherwise damaged. Wipe away any surplus oil, clean the drain plug threads and refit its sealing washer, then refit the drain plug tightening it to a torque setting of 2.5 – 3.5 kgf m (18 – 25 lbf ft).

Thoroughly clean the filter cover, centre bolt and the crankcase area around the filter location. Check that the bypass valve components in the filter centre bolt are clean and free from obstructions; swill the bolt in a high flash-point solvent to clean it, then apply a blast of compressed air to dry it.

Fit a new O-ring to the centre bolt, grease the O-ring and insert the bolt into the filter cover. Fit the light coil spring, then the washer to the bolt. Smear oil on the filter centre grommets and push it carefully on to the centre bolt; do not displace the grommets, and ensure that the coil spring and washer are fitted as described to prevent damage to the filter and to ensure that the bypass valve can operate, if necessary. Fit a new O-ring to the filter cover, then refit the filter assembly. Tighten the centre bolt to a torque setting of 2.0 – 2.5 kgf m (14.5 – 18 lbf ft). Wash off any spilt oil.

Refill the crankcase with the specified amount and type of oil, then fit the dipstick, start the engine and allow it to idle for a few minutes until the new oil is fully distributed around the engine. Stop the engine and recheck the level (see pre-ride check).

Note: While this is not specified as a regular maintenance item, the oil pump pick-up filter gauze must be cleaned with reasonable frequency. On 500 models, remove the engine front cover and the oil pump (see Chapter 1), then unscrew the two bolts securing the pick-up to the rear of the pump. Wash the unit in a high flash-point solvent and use a fine-bristled toothbrush to clean any large particles off the gauze. Wipe all old oil from the inside of the crankcase using a clean, lint-free rag. On reassembly, renew the sealing O-ring and tighten securely the pick up/oil pump mounting bolts. Refit the oil pump and engine front cover, then fill the crankcase with oil as described above. Note that a larger amount of oil than normal will be required (see Chapter 3 Specifications).

On 650 models, drain the engine oil as described above, then remove the detachable oil pan (sump). Working in a diagonal sequence from the outside inwards, unscrew the eight sump retaining bolts, tap the sump with a soft-faced mallet to break the seal and withdraw the sump. Remove and discard the gasket, then thoroughly clean the gasket surfaces when the sump and crankcase are cleaned. Remove the pick-up from the pump inlet and clean it as described above. On reassembly, renew the sealing O-ring around the pick-up feed pipe and install it so that its projecting tabs align with the grooves on the raised lug in the sump centre. Fit a new sump gasket then refit the sump. Tighten its bolts securely and evenly, working progressively in a diagonal sequence from the centre outwards. Refit the drain plug and oil filter as desribed above, then refill the crankcase with oil. Note that a larger amount of oil than normal will be required (see Chapter 3 Specifications).

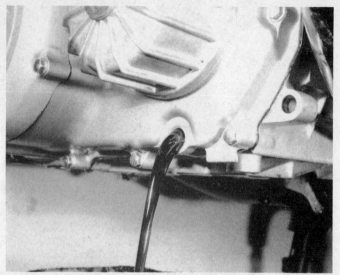

Remove drain plug and allow engine oil to drain ...

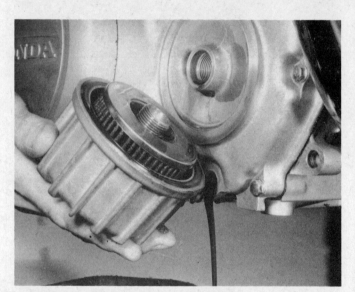

... and unscrew filter assembly

Withdraw element, noting washer and spring underneath it

Thoroughly clean all components before refitting

2 Renew the air filter element

The manufacturers recommend that the air filter element be renewed at this interval. If the machine is used continually in dusty areas the useful life of the filter may be reduced and therefore filter renewal should be made at correspondingly shorter intervals.

3 Renew the spark plugs

The spark plugs should be renewed at this interval, regardless of their apparent condition, as they will have passed peak efficiency. Check that the new plugs are of the correct type and heat range and that they are gapped correctly before they are fitted.

4 Check and adjust the valve clearances

The engine must be cold before the valve clearances can be checked accurately. Disconnect their caps and remove both spark plugs, then remove the cylinder head covers; each is retained by two bolts. To enable the covers to be displaced the fuel tank may have to be raised about one inch. This may be accomplished by detaching the dualseat and removing the single retaining bolt from the rear of the tank. Remove the inspection plug from the top right-hand side of the engine rear cover so that the timing marks on the alternator rotor can be seen.

Select top gear and turn the rear wheel in its normal direction of rotation to turn the crankshaft. Watch the right-hand cylinder inlet valves; when they have opened fully and just closed (sunk down and then raised up again) watch the alternator rotor. Turn the engine carefully until the TR mark on the alternator periphery aligns with the index pointer in the rear cover. With the engine in the specified position the piston in the right-hand cylinder will be at TDC on the compression stroke. Check that all the valves are fully closed, (free play at both rockers). Check the clearance between the valve stem head and the rocker of each valve using a feeler gauge of the correct thickness. (See Specifications.)

If the gap on any valve is incorrect, loosen the locknut on the adjuster and screw the adjuster in or out as necessary. When the gap is correct, prevent the screw rotating by using a small spanner and tighten the locknut to the specified torque setting. When the gap is correct, the feeler gauge will be a light sliding fit. Some care should be taken when resetting the clearance because even slightly loose rockers will increase tappet noise dramatically. Do not overtighten the adjuster locknuts; this will merely distort the threads and make future adjustment very awkward.

When the clearances on the right-hand cylinder are correct rotate the crankshaft forwards so that the TL mark is aligned with the index pointer and the left-hand piston is on the compression stroke. Repeat the valve clearance check and adjustment on the left-hand cylinder.

Reassemble the engine components by reversing the dismantling procedure. It is recommended that oil be smeared on the rocker cover sealing ring and the two conical bolt seals before replacement; this will help seat the seals and prevent oil leakage. Do not overtighten the retaining bolts; they are easily sheared. The specified torque setting is 0.8 – 1.2 kgf m (6 – 9 lbf ft). Where applicable, do not refit the timing mark inspection plug or the sparking plugs until completion of cam chain tension adjustment.

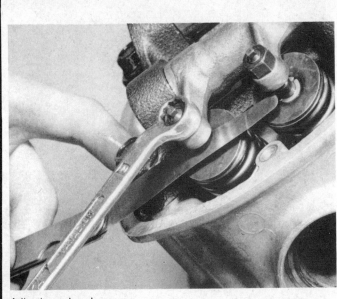

Adjusting valve clearances

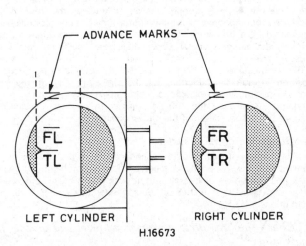

H.16673

Alternator rotor timing marks – CDI ignition (static marks similar for transistorised ignition)

5 Check the fuel feed pipes and clean the filter gauze

Give the pipe(s) which connects the fuel tap and carburettor a
close visual examination and check for cracks or any signs of leakage.
In time, the synthetic rubber pipe will tend to deteriorate, and will
eventually leak. Apart from the obvious fire risk, the leaking fuel will
affect fuel economy. The pipe will usually split only at the ends; if there
is sufficient spare length the damaged portion can be cut off and the
pipe refitted. The seal is effected by the interference fit of the pipe on
the spigot; although the wire clips are only an additional security
measure they should always be refitted correctly and should be
renewed if damaged, twisted or no longer effective. If the pipe is to be
renewed, always use the correct replacement type and size of neoprene
tubing to ensure a good leak-proof fit. Never use natural rubber tubing,
as this breaks up when in contact with petrol and will obstruct the
carburettor jets, or clear plastic tubing which stiffens to the point of
being brittle when in contact with petrol and will produce leaks that
are difficult to cure.

Earlier models with no separate fuel tap filter bowls are only fitted
with a filter gauze on the tap stack pipe inside the tank. At regular
intervals the tank must be drained and the tap must be removed so that
the gauze can be cleaned. Proceed as described for the later models
under the minor service heading.

6 Check the carburettor settings

If rough running of the engine has developed, some adjustment of
the carburettor pilot setting and tick over speed may be required. If this
is the case refer to Chapter 3 for details. Do not make these
adjustments unless they are obviously required; there is little to be
gained by unwarranted attention to the carburettor. Complete
carburettor maintenance by removing the drain plug on the float
chamber, turning the petrol on, and allowing a small amount of fuel to
drain through, thus flushing any water or dirt from the carburettor. Refit
the drain plug securely and switch the petrol off.

Once the carburettor has been checked and reset if necessary, the
throttle cable free play can be checked. Open and close the throttle
several times, allowing it to snap shut under its own pressure. Ensure
that it is able to shut off quickly and fully at all handlebar positions,
then check that there is 2 – 6 mm (0.08 – 0.24 in) free play measured in
terms of twistgrip rotation (ie 10 – 15°). Use the opening cable adjuster
at the twistgrip to achieve the correct setting then open and close the
throttle again to settle the cables and check that the adjustment has not
altered. If a major alteration is required, screw in fully the twistgrip
adjuster nut and set the required free play at the cable lower end
adjuster, on the carburettors; this will require the removal of the fuel
tank.

To adjust the choke cable, check that the choke knob or lever is
pressed fully into its open (off) position, and slacken the single screw
securing the choke cable clamp to the side of the carburettor body. Pull
the cable outer slowly through the clamp towards the front of the
machine, stopping when all but the slightest trace of free play has been
eliminated from the exposed length of the cable inner, then tighten
securely the retaining screw. Check the choke operation. Finally on
most models, note that a simple friction device is fitted at the choke
cable handlebar end to permit the cable to be held in any desired
position. To adjust the friction, pull the choke knob out to the fully
closed position and peel off the rubber cover fitted over the cable
retaining nut. Rotate the knurled adjuster to produce the desired
amount of friction, then refit the rubber cover.

7 Check the cooling system

The cooling system should be checked for signs of leakage or
damage and any suspect hoses renewed. The used coolant can be put
back into the system if it is clean, but note that it must be renewed
every two years as a matter of course. Refer to Chapter 2 for full
instructions.

8 Check the final drive gear case oil level and grease the shaft joint

Place the machine on the centre stand and remove the oil filler plug
from the final drive gear case. The oil level should be up to the bottom
edge of the filler orifice. Top up, if necessary, using oil of the specified
type.

The driveshaft joint in the final drive casing is lubricated separately,
on 500 models only, by means of the grease nipple provided on the
casing. Three or four energetic pumps on a grease gun filled with a

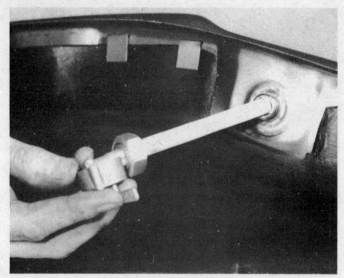

On earlier models, tank must be drained and tap removed to clean
fuel filter

Throttle opening cable only can be adjusted to set free play

Final drive oil level should be up to bottom edge of filler opening
with machine on centre stand

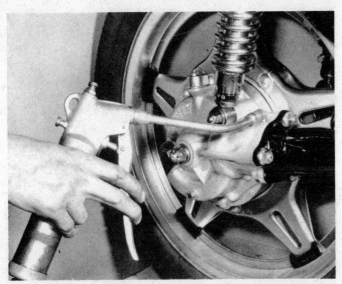

On 500 models, driveshaft splines can be lubricated via grease nipple provided

multi-purpose lithium-based grease (see Specifications) should prove adequate. On 650 models no grease nipple is fitted and so great care should be taken to clean the driveshaft components, to check them for wear and to pack them with grease on reassembly whenever they are disturbed.

9 Check the brakes
Fluid level check
The hydraulic front brake requires no regular adjustments; pad wear is compensated for by the automatic entry of more fluid into the system from the handlebar reservoir. All that is necessary is to maintain a regular check on the fluid level and the degree of pad wear.

To check the fluid level, turn the handlebars until the reservoir is horizontal and check that the fluid level, as seen through the translucent reservoir body, or, where applicable, the sight glass in the front or rear face of the reservoir body, is not below the lower level mark on the body. Remember that while the fluid level will fall steadily as the pad friction material is used up, if the level falls below the lower level mark there is a risk of air entering the system; it is therefore sufficient to maintain the fluid level above the lower level mark, by topping-up if necessary. Do not top up the higher level mark (formed by a cast line on the inside of the reservoir where it is not made of translucent plastic) unless this is necessary after new pads have been fitted. If topping up is necessary, wipe any dirt off the reservoir, remove the retaining screws, where fitted, and lift away the reservoir cover and diaphragm. Use only good quality brake fluid of the recommended type and ensure that it comes from a freshly opened sealed container; brake fluid is hygroscopic, which means that it absorbs moisture from the air, therefore old fluid may have become contaminated to such an extent that its boiling point has been lowered to an unsafe level. Remember also that brake fluid is an excellent paint stripper and will attack plastic components; wash away any spilled fluid immediately with copious quantities of water. When the level is correct, clean and dry the diaphragm, fold it into its compressed state and fit it to the reservoir. Refit the reservoir cover (and gasket, where fitted) and tighten securely, but do not overtighten, the retaining screws (where fitted).

On CX500 E-C and CX650 E-D models remove carefully the right-hand side panel to expose the rear brake master cylinder, then check the level as described above.

Pad wear check
On models with single-piston calipers remove the inspection window cap from the caliper rear face; the pads can be seen through the aperture, with the aid of a torch.

On models with twin-piston calipers, the pads can be seen with the aid of a torch through the slot indicated by the cast arrow on the caliper body. On CX500 E-C and CX650 E-D models the same applies to the rear brake, but it may be necessary to remove the dust cover to gain an adequate view of the pads.

Wear limit marks are provided in the form of deep notches cut in the top and bottom edges of the friction material or as red painted lines cut around the outside of the material. If either pad is worn at any point so that the inside end of the mark (next to the metal backing) is in contact with the disc, or if the wear limit marks have been removed completely, both pads must be renewed as a set. If the pads are so fouled with dirt that the marks cannot be seen, or if oil or grease is seen on them, they must be removed for cleaning and examination.

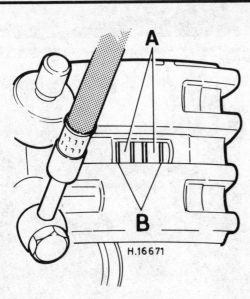

Checking pad wear – single piston caliper

A   Brake pad   *B   Red wear limit line*

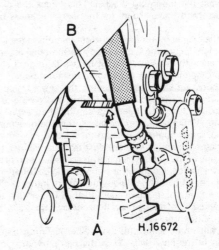

Checking pad wear – front twin piston caliper

A   Inspection window   *B   Wear limit lines*

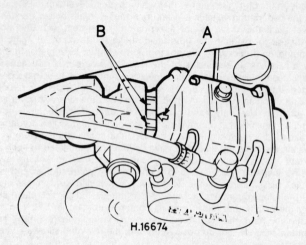

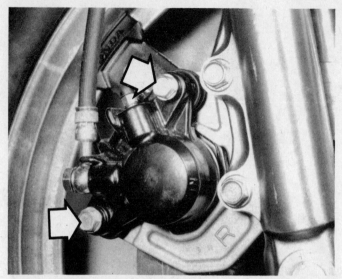

Pad removal, single-piston caliper – remove caliper mounting bolts (arrowed) ...

Checking pad wear – rear twin piston caliper

A   Inspection window   *B   Wear limit lines*

Pad removal and refitting – single-piston caliper

Unscrew the two bolts that pass into the caliper body and secure the body to the mounting bracket. Lift the caliper body off the mounting bracket, still interconnected with the hydraulic hose.

Lift the old pads out and clean the mounting bracket thoroughly. Install the new pads and also the shim which fits against the outer face of the outer pad. The shim must be fitted so that the arrow is pointing in the direction of wheel rotation. Refit the caliper and replace the bolts, tightening them to a torque setting of 1.5 – 2.0 kgf m (11 – 14.5 lbf ft). It may be necessary to push the caliper cylinder piston inwards to give the necessary clearance.

Pad removal and refitting – twin-piston caliper

Remove the pad pin retainer by unscrewing its retaining bolt. This is located inboard of the brake hose union. The retainer has keyhole slots which locate over the ends of the pad pins and can be disengaged and removed once the bolt has been released. Push the caliper body inwards against the mounting bracket. This forces the pistons back into their bores and makes room for the extra thickness of the new pads.

Remove the caliper mounting bolt. Note that this is the lower of the two bolts which secure the caliper body to the bracket, and that on machines fitted with anti-dive units it passes through a short torque link. The upper bolt is, in fact, a pivot and need not be removed. Pivot the caliper body upwards and clear of the disc. Using a pair of pointed-nosed pliers, grasp the ends of the pad pins and withdraw them. The pads will now be freed and can be lifted away, as can the anti-rattle shim. When removing the latter, note the direction in which it was fitted to avoid confusion during installation.

Clean off any accumulated brake dust from the caliper, taking care not to inhale any of the dust, which has an asbestos content and is thus toxic. Check the caliper carefully for signs of leakage round the pistons; if there are traces of hydraulic fluid which might indicate a leak, trace and rectify the fault before proceeding further. Check that the pistons are fully retracted into their bores. If necessary, they can be pushed inwards using thumb pressure.

Refit the anti-rattle shim into the caliper ensuring that it locates correctly then fit the new pads. Slide the pad pins into position and refit the pin retainer and its mounting bolt. Swing the caliper down over the disc and fit the mounting bolt, tightening it to its recommended torque setting.

On CX500 E-C and CX650 E-D models, the rear pads can be dealt with in a similar manner to that described above. Note that the dust cover at the rear of the caliper should be removed before the mounting bolt is released and the caliper body swung away from the disc.

... and lift caliper away to expose pads – note correct position of anti-squeal shim

Thoroughly clean all components and lubricate caliper axles before refitting

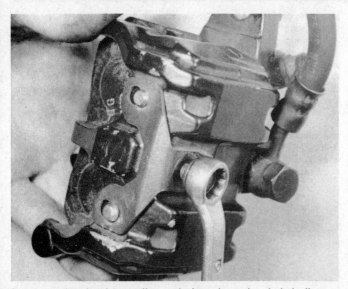

Pad removal, twin-piston caliper – slacken pin retainer bolt (caliper shown removed for clarity) ...

... and unscrew both mounting bolts (arrowed) to remove caliper fully ...

... noting slightly different arrangement (bolts arrowed) for anti-dive assembly

Refit anti-rattle shim as shown and position moving pad over retaining pins ...

... then refit fixed pad

Pad examination and refitting – all models

Thoroughly clean all components, removing all traces of dirt, grease and old friction material then use fine abrasive paper to carefully polish clean any corroded items. Check carefully that the caliper body slides easily on its two axle bolts and that there is no damage to any of the caliper components, especially the rubber seals. If it is stiff, clean the axle bolts and check them for wear or damage (which can be cured only by the renewal of the bolts or mounting bracket, as applicable) then smear a good quantity of silicone or PBC (Poly Butyl Cuprysil) based caliper grease over the bolts and caliper or mounting bracket bores before refitting. It is essential that the caliper body can move smoothly and easily on the mounting bracket for the brake to be effective. **Warning:** do not use ordinary high-melting point grease; this will melt and foul the pads, rendering the brake ineffective.

If the pads are worn to the limit marks, fouled with oil or grease, or heavily scored or damaged by dirt and debris, they must be renewed as a set; there is no satisfactory way of degreasing friction marerial. Note that all four pads of a twin disc system should be renewed together. If the pads can be used again, clean them carefully using a fine wire brush that is completely free of oil or grease. Remove all traces of road dirt and corrosion, then use a pointed instrument to clean out the groove(s) in the friction material and to dig out any embedded particles of foreign matter. Any areas of glazing may be removed using emery cloth.

On reassembly, if new pads are to be fitted, the caliper pistons must now be pushed back as far as possible into the caliper bores to provide the clearance necessary to accommodate the unworn pads. It should be possible to do this with hand pressure alone. If any undue stiffness is encountered the caliper assembly should be dismantled for examination as described in Chapter 6. While pushing the pistons back, maintain a careful watch on the fluid level in the handlebar reservoir. If the reservoir has been overfilled, the surplus fluid will prevent the pistons returning fully and must be removed by soaking it up with a clean cloth. Take care to prevent fluid spillage. Apply a thin smear of caliper grease to the outer edge and rear surface of the moving pad (next to the piston) and to the pad retaining pins (where fitted). Take care to apply caliper grease to the metal backing of the pad only and not to allow any grease to contaminate the friction material.

When the caliper has ben refitted, apply the brake lever or pedal gently and repeatedly to bring the pads firmly into contact with the disc until full brake pressure is restored. Be careful to watch the fluid level in the reservoir; if the pads have been re-used it will suffice to keep the level above the lower level mark by topping-up if necessary, but if new pads have been fitted the level must be restored to the upper level line described above by topping up or removing surplus fluid as necessary. Refit the reservoir cover, gasket (where fitted) and diaphragm as described above.

Before taking the machine out on the road, be careful to check for fluid leaks from the system, and that the front brake is working correctly. Remember also that new pads, and to a lesser extent, cleaned pads will require a bedding-in period before they will function at peak efficiency. Where new pads are fitted use the brake gently but firmly for the first 50 – 100 miles to enable the pads to bed in fully.

Rear drum brake – adjusting and checking for wear

Drum brakes will require regular adjustment to compensate for shoe wear; adjustment is made by rotating the nut at the rear end of the operating rod.

Place the machine on its centre stand so that the wheel is clear of the ground, spin the wheel and tighten the adjusting nut until a rubbing sound is heard as the shoes begin to contact the drum, then slacken the nut by one or two turns until the sound ceases. Spin the wheel and apply the brake hard to settle the components. Check that the adjustment has not altered and tighten securely all disturbed fasteners. This should approximate the specified setting, which is that there should be 20 – 30 mm (0.8 – 1.2 in) of free play before the brake begins to engage the drum, measured at the pedal tip. Switch on the ignition and check that the stop lamp lights just as all free play has been taken up and the brake is beginning to engage. This is adjusted by holding steady the stop lamp rear switch body and rotating the plastic sleeve nut to raise or lower the switch as necessary; do not

allow the body to rotate or its terminal will be damaged. Check the switch setting whenever the rear brake is adjusted.

Complete brake maintenance by checking that the wheels are free to rotate easily, and then lubricate all lever or linkage pivots and the stoplamp switch. To prevent the risk of oil finding its way on to the tyres or the brake friction material do not oil excessively the brake components; a few drops of oil at each point will suffice. Dismantle the rear brake pedal pivot and grease it at regular intervals. Note that the operating mechanism is at its most efficient when, with the brake correctly adjusted and applied fully, the angle between the rod and the operating arm on the brake backplate does not exceed 90°. This can be adjusted by removing the operating arm from the brake camshaft and rotating it by one or two splines until the angle is correct. Ensure that all components are correctly secured on reassembly.

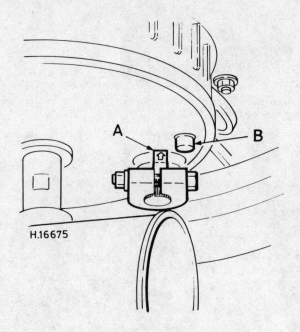

H.16675

Rear drum brake wear indicator marks – typical

A   Wear indicator arrow   B   Brake panel mark

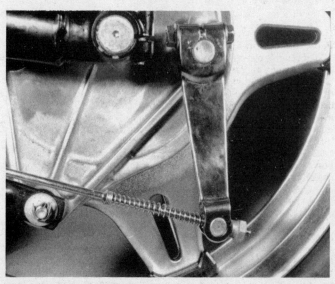

Drum rear brakes are adjusted at nut at end of operating rod

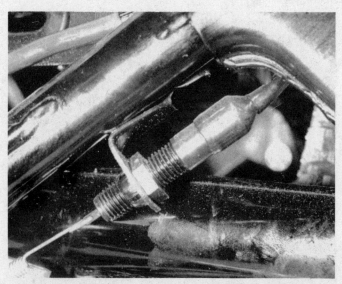

Rotate plastic sleeve nut to alter stop lamp rear switch setting

Drum brake shoe wear is checked by applying the brake firmly and looking at the wear indicator marks on the backplate. If the indicator pointer on the camshaft is aligned with, or has moved beyond, the fixed index mark cast on the backplate, the shoes are worn out and must be renewed as a pair. This involves the removal of the wheel from the machine as described in Chapter 6 so that the brake components can then be dismantled, cleaned, checked for wear, and reassembled following the instructions given in the same Chapter. It is important that moving parts such as the brake camshaft are lubricated with a smear of high melting-point grease on reassembly.

10 Check the stands

Examine the main (centre) stand and side stand for cracks or bending and lubricate the pivots with a multi-purpose or graphited grease. The centre stand pivots on a hollow shaft, retained at one end by a split pin. The side stand pivots about a single shouldered bolt. Check the return springs and renew them, if weak or strained.

Inspect the rubber pad on the side stand for wear. If it is worn down to or past the wear mark, it should be renewed. Renew with a pad marked '260 lbs'.

Bent or cracked stands can usually be repaired by heating or welding. It is important that the stands on a machine of this weight are in good order. This is particularly so as the cylinder heads are considerably outboard of the frame and consequently vulnerable, if the machine is dropped.

11 Check the suspension and steering

Support the machine so that it is secure with the front wheel clear of the ground, then grasp the front fork legs near the wheel spindle and push and pull firmly in a fore and aft direction. If play is evident between the top and bottom fork yokes and the steering head, the steering head bearings are in need of adjustment. Imprecise handling or a tendency for the front forks to judder may be caused by this fault.

Bearing adjustment is correct when the adjuster ring is tightened, until resistance to movement is felt and then loosened $1/8$ to $1/4$ a turn. The adjuster ring should be rotated by means of a C-spanner after slackening the steering stem nut and clamp bolt (where fitted).

Take great care not to overtighten the adjuster ring. It is possible to place a pressure of several tons on the head bearings by over-tightening even though the handlebars may seem to turn quite freely. Overtight bearings will cause the machine to roll at low speeds and give imprecise steering. Adjustment is correct if there is no play in the bearings and the handlebars swing to full lock either side when the machine is supported with the front wheel clear of the ground. Only a light tap on each end should cause the handlebars to swing. Secure the adjuster ring by tightening the steering stem top nut to the specified torque setting, then check that the setting has not altered. Do not forget to tighten the clamp bolt, if fitted.

At the same time as the steering head bearings are checked, take the opportunity to examine closely the front and rear suspension. Ensure that the front forks work smoothly and progressively by pumping them up and down whilst the front brake is held on. Any faults revealed by this check should be investigated further, as any deterioration in the stability of the machine can have serious consequences. Check carefully for signs of leaks around the front fork oil seals. If any damage is found, it must be repaired immediately as described in the relevant Sections of Chapter 5. Inspect the stanchions, looking for signs of chips or other damage, then lift the dust excluder at the top of each fork lower leg and wipe away any dirt from its sealing lips or above the fork oil seal. Pack grease above the seal and refit the dust excluder. Note that none of this would be necessary, and fork stiction would be reduced, if gaiters are fitted; they are available from any good motorcycle dealer.

The fork oil **must** be changed at regular intervals to prevent the inevitable reduction in fork performance which results as the oil deteriorates in service. Refer to the relevant part of Chapter 5 for full information.

To check the swinging arm support the machine so that the rear wheel is clear of the ground then pull and push horizontally at the rear end of the swinging arm; there should be no discernible play at the pivot. If play is found the bearings can be adjusted but it is preferable to remove the swinging arm so that they can be cleaned, checked for wear and renewed if necessary, and packed with new grease on reassembly. Note that Honda specify the use of a special grease for this application.

On models with 'Pro-Link' rear suspension, the linkage consists of several highly-stressed bearing surfaces which are not fitted with grease nipples and which are very exposed to all the water, dirt and salt thrown up by the rear wheel. Regular cleaning and greasing is essential; owners should note that this is greatly simplified if the linkage were dismantled for all bearings to be fitted with grease nipples. A local Honda Service Agent should be able to tell you of someone competent to undertake such work. Note that not only are the linkage components expensive to renew if allowed to wear out through lack of attention, but the machine will almost certainly fail its DOT certificate test if such wear is found.

To check the linkage components, two people are required; one to sit on the machine and bounce the rear suspension while the other watches closely the action of the various components. If any squeaks or other noises are heard, if the linkage appears stiff, or if any signs of dry bearings or other wear or damage are detected, the rear suspension should be removed as a complete assembly and dismantled for thorough cleaning, checking and greasing. At regular intervals the linkage must be dismantled so that the bearings can be cleaned and packed with new grease. Refer to Chapter 5.

12 Lubricate the cables, stands and controls

At regular intervals, all control and instrument drive cables and all control pivots should be checked for wear or damage and lubricated, dismantling them where necessary and removing all traces of dirt or corrosion. This operation must be carried out to prevent excessive wear and to ensure that the various components can be operated smoothly and easily, in the interests of safety. The opportunity should be taken to examine closely each component, renewing any that show signs of excessive wear or of any damage.

The twistgrip is removed by unscrewing the screws which fasten both halves of the handlebar right-hand switch assembly. Each throttle cable upper end nipple can be detached from the twistgrip with a suitable pair of pliers and the twistgrip slid off the handlebar end. Carefully clean and examine the handlebar end, the internal surface of the twistgrip, and the two halves of the switch cluster. Remove any rough burrs with a fine file, and apply a coating of grease to all the bearing surfaces. Slide the twistgrip back over the handlebar end, insert the throttle cable end nipples into the twistgrip flange, and reassemble the switch cluster. Check that the twistgrip rotates easily and that the throttle snaps shut as soon as it is released. Tighten the switch retaining screws securely, but do not overtighten them.

Although the regular daily checks will ensure that the control cables are lubricated and maintained in good order, it is recommended that a positive check is made on each cable at this mileage/time interval to ensure that any faults will not develop unnoticed to the point where smooth and safe control operation is impaired. If any doubt exists about the condition of any of the cables, the component in question should be removed from the machine for close examination. Check the outer cables for signs of damage, then examine the exposed portions of the inner cables. Any signs of kinking or fraying will indicate that renewal is required. To obtain maximum life and reliability from the cables they should be thoroughly lubricated using light machine oil. To do the job properly and quickly use one of the hydraulic cable oilers available from most motorcycle shops. Free one end of the cable and assemble the cable oiler as described by the manufacturer's instructions. Operate the oiler until oil emerges from the lower end, indicating that the cable is lubricated throughout its length. This process will expel any dirt or moisture and will prevent its subsequent ingress.

If a cable oiler is not available, an alternative is to remove the cable from the machine. Hang the cable upright and make up a small funnel arrangement using plasticine or by taping a plastic bag around the upper end. Fill the funnel with oil and leave it overnight to drain through. Note that where nylon-lined cables are fitted, they should be used dry or lubricated with a silicone-based lubricant suitable for this application. On no account use ordinary engine oil because this will cause the liner to swell, pinching the cable.

Each instrument drive cable is secured at its upper mounting by a knurled, threading ring and at its bottom mounting by a single retaining screw in the speedometer drive gearbox or crankcase, as appropriate. Remove the cable by using a pair of pliers to unscrew the knurled ring, and a screwdriver, to slacken and remove the retaining screw. It should then be possible to withdraw the cable carefully from its mountings and remove it from the machine. Remove the inner cable by pulling it out from the bottom of the outer. Carefully examine the inner cable for signs of fraying, kinking, or for any shiny areas which will indicate tight spots, and the outer cable for signs of cracking, kinking or any other damage. Renew either cable if necessary. To lubricate the cable, smear a small quantity of grease on to the lower length only of the inner. Do not allow any grease on the top six inches of the cable as the grease will work its way rapidly up the length of the cable as it rotates and get into the instrument itself. This will rapidly ruin the instrument which will then have to be renewed. Insert the inner cable in the outer and refit the cable.

Check all pivots and control levers, cleaning and lubricating them to prevent wear or corrosion. Where necessary, dismantle and clean any moving part which may have become stiff in operation.

When refitting the cables onto the machine, ensure that they are routed in easy curves and that full use is made of any guide or clamps that have been provided to secure the cable out of harm's way. Adjustment of the individual cables is described under other Routine Maintenance tasks.

Be very careful to ensure that all controls are correctly adjusted and are functioning correctly before taking the machine out on the road.

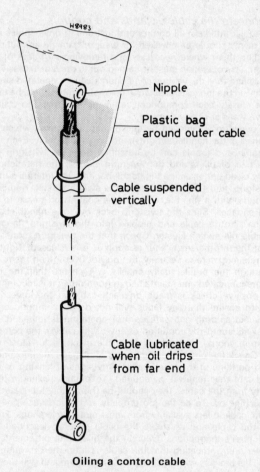

H8983

Nipple

Plastic bag around outer cable

Cable suspended vertically

Cable lubricated when oil drips from far end

Oiling a control cable

13 Check all fittings, fasteners, lights and signals

Check around the machine, looking for loose nuts, bolts or screws, retightening them as necessary.

It is advisable to lubricate the handlebar switches and stop lamp switches with WD40 or a similar water dispersant lubricant. This will keep the switches working properly and prolong their life, especially if the machine is used in adverse weather conditions. Check that all lights, turn signals, the horn and speedometer are working properly and that their mountings and connections are securely fastened.

14 Check the wheels, bearings and tyres

To check the wheels, support the machine securely so that the wheel to be checked is clear of the ground and can spin freely. If necessary slacken the brake adjustment. Proceed as follows:

Comstar wheels

Spin the wheel and check for rim alignment by placing a pointer close to the rim edge. If the total radial or axle alignment variation is greater than 2.0 mm (0.08 in) the manufacturer recommends that the wheel is renewed. This policy is, however, a counsel of perfection and in practice a larger runout may not affect the handling properites excessively.

Although Honda do not offer any form of wheel rebuilding facility, a number of private engineering firms offer this service. It should be noted however, that Honda do not approve of this course of action.

Check the rim for localised damage in the form of dents or cracks. The existence of even a small crack renders the wheel unfit for further use unless it is found that a permanent repair is possible using arc-welding. This method of repair is highly specialised and therefore the advice of a wheel repair specialist should be sought.

Because tubeless tyres are used, dents may prevent complete sealing between the rim and tyre bead. This may not be immediately obvious until the tyre strikes a severely irregular surface, when the unsupported tyre wall may be deflected away from the rim, causing rapid deflation of the tyre. Honda recommend that the wheel be renewed if the bead seating surface of the rim is scratched to a depth of 0.5 mm (0.02 in) or more. Again, if in doubt, seek specialist advice whether continued use of the wheel is advisable.

Inspect the spoke blades for cracking and security. Check carefully the area immediately around the rivets which pass through the spokes and into the rim. In certain circumstances corrosion may occur between the spokes, rivets and rim due to the use of different metals.

Cast wheels

Carefully check the complete wheel for cracks and chipping, particularly at the spoke roots and the edge of the rim. As a general rule a damaged wheel must be renewed as cracks will cause stress points which may lead to sudden failure under heavy load. Small nicks may be radiused carefully with a fine file and emery paper (No 600 – No 1000) to relieve the stress. If there is any doubt as to the condition of a wheel, advice should be sought from a reputable dealer or specialist repairer.

Each wheel is covered with a coating of lacquer, to prevent corrosion. If damage occurs to the wheel and the lacquer finish is penetrated, the bared aluminium alloy will soon start to corrode. A whitish grey oxide will form over the damaged area, which in itself is a protective coating. This deposit, however, should be removed carefully as soon as possible and a new protective coating of lacquer applied.

Check the lateral run out at the rim by spinning the wheel and placing a fixed pointer close to the rim edge. If the maximum run out is greater than 2.0 mm (0.080 in) the manufacturer recommends that the wheel be renewed. This is, however, a counsel of perfection; a run out somewhat greater than this can probably be accommodated without noticeable effect on steering. No means is available for straightening a warped wheel without resorting to the expense of having the wheel skimmed on all faces. If warpage was caused by impact during an accident, the safest measure is to renew the wheel complete. Worn wheel bearings may cause rim run out. These should be renewed.

Note that impact damage or serious corrosion on models fitted with tubeless tyres has wider implications in that it could lead to a loss of pressure from the tubeless tyres. If in any doubt as to the wheel's condition, seek professional advice.

All models

To check the wheel bearings, grasp each wheel firmly at the top and bottom and attempt to rock it from side to side; any free play indicates worn bearings which must be renewed as described in Chapter 6. Make a careful check of the tyres, looking for signs of damage to the tread or sidewalls and removing any embedded stones etc. Renew the tyre if the tread is excessively worn or if it is damaged in any way.

Additional routine maintenance

1 Renew the brake fluid

Since brake fluid deteriotates with age, as well as through use, it must be renewed every two years, or at the following mileage, whichever is the sooner:

Every 21 600 miles (36 000 km) – All UK CX500 and CX650 models, all US models up to 1981.

Every 24 000 miles (38 400 km) – GL500 D-C, GL650 D2-E, all US models from 1982 on.

Before starting work, obtain a new, full can of SAE J1703 or DOT3 hydraulic fluid and read carefully the Section on brake bleeding in Chapter 6. Prepare a clear plastic tube and glass jar in the same way as for bleeding the hydraulic system, open the bleed nipple by unscrewing it $1/4 - 1/2$ a turn with a spanner and apply the front brake lever gently and repeatedly. This will pump out the old fluid. **Keep the master cylinder reservoir topped up at all times**, otherwise air may enter the system and greatly lengthen the operation. The old brake fluid is invariably much darker in colour than the new, making it easier to see when it is pumped out and the new fluid has completely replaced it.

When the new fluid appears in the clear plastic tubing completely uncontaminated by traces of old fluid, close the bleed nipple, remove the plastic tubing and replace the rubber cap on the nipple. Repeat on the remaining caliper (twin-disc system). Top the master cylinder reservoir up to above the lower level mark, unless the brake pads have been renewed in which case the reservoir should be topped up to its higher level. Clean and dry the rubber diaphragm, fold it into its compressed state and refit the diaphragm and reservoir cover, tightening securely the retaining screws.

Wash off any surplus fluid and check that the brake is operating correctly before taking the machine out on the road.

2 Renew the coolant

To minimise the build-up of deposits in the cooling system and to ensure maximum protection against its freezing, the coolant should be drained completely, the system should be flushed out and checked for leaks or damage and new coolant mixed for refilling. This should be done at the mileage interval given for the previous item, and is described in Chapter 2.

3 Change the final gear case oil

This operation should also be carried out at the mileage interval specified for item 1 in this Section.

When ready, take the machine for a journey of sufficient length to warm up fully the oil in the final drive unit. The oil is thick and will not drain quickly or remove any impurities until it is fully warmed up.

Place the machine on its centre stand and place a container under the gear case with a sheet of paper or cardboard to keep oil off the wheel and tyre. Remove the oil level/filler and drain plugs and allow the oil to drain fully. Renew the plug sealing washer (or O-ring, as appropriate) if it is damaged or worn. When draining is complete, refit the drain plug, tightening it securely. Add sufficient oil of the recommended grade and viscosity to bring the level up to the bottom of the filler/level plug orifice; the amount required is given in the Specifications Section. Refit the filler/level plug, wash off any surplus oil and take the machine for a short journey to warm up the oil and distribute it, then stop the engine and allow a few minutes for the level to settle before rechecking it; top up as necessary. Tighten the filler/level plug securely and wash off all traces of oil from the outside of the swinging arm and casing.

4 Cleaning the machine

Keeping the motorcycle clean should be considered as an important part of the routine maintenance, to be carried out whenever the need arises. A machine cleaned regularly will not only succumb less speedily to the inevitable corrosion of external surfaces, and hence maintain its market value, but will be far more approachable when the time comes for maintenance or service work. Furthermore, loose or failing components are more readily spotted when not partially obscured by a mantle of road grime and oil.

Surface dirt should be removed using a sponge and warm, soapy water; the latter being applied copiously to remove the particles of grit which might otherwise cause damage to the paintwork and polished surfaces.

Oil and grease is removed most easily by the application of a cleaning solvent such as 'Gunk' or 'Jizer'. The solvent should be applied when the parts are still dry and worked in with a stiff brush. Large quantities of water should be used when rinsing off, taking care that water does not enter the carburettors, air cleaners or electrics.

If desired a polish such as Solvol Autosol can be applied to the aluminium alloy parts to restore the original lustre. This does not apply in instances, much favoured by Japanese manufacturers, where the components are lacquered. Application of a wax polish to the cycle parts and a good chrome cleaner to the chrome parts will also give a good finish. Always wipe the machine down if used in the wet, and make sure the chain is well oiled. There is less chance of water getting into control cables if they are regularly lubricated, which will prevent stiffness of action.

Chapter 1 Engine, clutch and gearbox

Contents

Specifications

Note: *Unless otherwise stated, information applies to all models*

Engine

	500 models	650 models
Type	Water-cooled, four stroke 80° V-twin cylinder	
Bore	78.0 mm (3.07 in)	82.5 mm (3.25 in)
Stroke	52.0 mm (2.05 in)	63.0 mm (2.48 in)
Capacity	497 cc (30.3 cu in)	673 cc (41.1 cu in)
Compression ratio	10.0:1	9.8:1
Compression pressure – at cranking speed	12 ± 2 kg/cm² (171 ± 28 psi)	

Valve clearances – engine cold

	Inlet	Exhaust
CX500-A, CX500-B, CX500 C-B, all 650 models	0.10 mm (0.0039 in)	0.12 mm (0.0047 in)
All other 500 models	0.08 mm (0.0032 in)	0.10 mm (0.0039 in)

Valve timing – at 1 mm (0.0394 in) lift

	CX500 E-C	All other 500 models	All 650 models
Inlet opens	5° BTDC	6° BTDC	7° BTDC
Inlet closes	30° ABDC	46° ABDC	53° ABDC
Exhaust opens	30° BBDC	46° BBDC	40° BBDC
Exhaust closes	5° ATDC	6° ATDC	15° ATDC

Rocker gear
Spindle OD:
   All CX500 models except CX500 E-C .... 13.982 – 14.000 mm (0.5505 – 0.5512 in)
   Service limit .... 13.960 mm (0.5496 in)
   All other models .... 14.966 – 14.984 mm (0.5892 – 0.5899 in)
   Service limit .... 14.950 mm (0.5886 in)
Arm ID:
   All CX500 models except CX500 E-C .... 14.016 – 14.027 mm (0.5518 – 0.5522 in)
   Service limit .... 14.050 mm (0.5532 in)
   All other models .... 15.000 – 15.018 mm (0.5906 – 0.5913 in)
   Service limit .... 15.040 mm (0.5921 in)
Carrier bracket ID:
   All CX500 models except CX500 E-C .... 14.000 – 14.027 mm (0.5512 – 0.5522 in)
   Service limit .... 14.050 mm (0.5532 in)
   All other models .... 14.988 – 15.006 mm (0.5901 – 0.5908 in)
   Service limit .... 15.030 mm (0.5917 in)

Valves, springs and guides
Valve stem OD:
   Inlet .... 6.580 – 6.590 mm (0.2591 – 0.2595 in)
   Exhaust .... 6.550 – 6.560 mm (0.2579 – 0.2583 in)
   Service limit – inlet and exhaust .... 6.540 mm (0.2575 in)
Valve guide ID – inlet and exhaust:
   Standard .... 6.600 – 6.620 mm (0.2598 – 0.2606 in)
   Service limit .... 6.700 mm (0.2638 in)
Stem/guide clearance:
   Inlet .... 0.010 – 0.040 mm (0.0004 – 0.0016 in)
   Exhaust .... 0.040 – 0.070 mm (0.0016 – 0.0028 in)
   Service limit – inlet and exhaust .... 0.100 mm (0.0039 in)
Valve spring free length – inlet and exhaust:
   Outer – all models .... 50.4 mm (1.9843 in)
   Service limit .... 48.5 mm (1.9095 in)
   Inner – CX650 C, CX650 E-D .... 49.5 mm (1.9488 in)
   Service limit .... 47.6 mm (1.8740 in)
   Inner – all other models .... 50.3 mm (1.9803 in)
   Service limit .... 48.4 mm (1.9055 in)

Cylinder head
Gasket face maximum warpage .... 0.10 mm (0.0039 in)
Valve seat width:
   Standard .... 1.1 – 1.3 mm (0.0433 – 0.0512 in)
   Service limit .... 2.0 mm (0.0787 in)
Valve seat cutting angles:

	CX500 models except CX500 E-C	All other models
At combustion chamber	37.5°	32°
At inlet or exhaust port	63.5°	60°
At valve seat face	45°	45°

Piston rings
End-gap installed:
   Compression rings – all CX500 models except CX500 E-C .... 0.10 – 0.30 mm (0.0039 – 0.0118 in)
   Compression rings – all GL models, CX500 E-C .... 0.10 – 0.25 mm (0.0039 – 0.0098 in)
   Compression rings – CX650 C, CX650 E-D .... 0.20 – 0.35 mm (0.0079 – 0.0138 in)
   Service limit – all models .... 0.60 mm (0.0236 in)
   Oil scraper ring side rail – all GL500 models, CX500 E-C .... 0.20 – 0.40 mm (0.0079 – 0.0158 in)
   Service limit .... 1.00 mm (0.0394 in)
   Oil scraper ring side rail – all other models .... 0.30 – 0.90 mm (0.0118 – 0.0354 in)
   Service limit .... 1.10 mm (0.0433 in)
Compression ring/piston groove clearance:
   Standard .... 0.015 – 0.050 mm (0.0006 – 0.0020 in)
   Service limit .... 0.100 mm (0.0039 in)

Pistons

	500 models	650 models
Piston OD	77.940 – 77.960 mm (3.0685 – 3.0693 in)	82.460 – 82.485 mm (3.2465 – 3.2474 in)
Service limit	77.860 mm (3.0654 in)	82.365 mm (3.2427 in)
Oversizes available	0.25 mm (0.010 in), 0.50 mm (0.020 in)	
Gudgeon pin bore ID	21.002 – 21.008 mm (0.8269 – 0.8271 in)	
Service limit	21.040 mm (0.8283 in)	
Gudgeon pin OD	20.994 – 21.000 mm (0.8265 – 0.8268 in)	
Service limit	20.984 mm (0.8261 in)	

Cylinder bores

	500 models	650 models
Bore ID	78.000 – 78.015 mm (3.0709 – 3.0715 in)	82.500 – 82.515 mm (3.2480 – 3.2486 in)
Service limit	78.100 mm (3.0748 in)	82.600 mm (3.2520 in)
Maximum ovality	0.10 mm (0.0039 in)	
Piston/cylinder maximum clearance	0.10 mm (0.0039 in)	

Connecting rods

Small-end ID .... 21.020 – 21.041 mm (0.8276 – 0.8284 in)
Service limit .... 21.068 mm (0.8295 in)
Big-end ID:
  500 models .... 43.000 – 43.024 mm (1.6929 – 1.6939 in)
  650 models .... 46.000 – 46.024 mm (1.8110 – 1.8120 in)

Crankshaft

Connecting rod big-end side clearance:
  500 models .... 0.150 – 0.170 mm (0.0059 – 0.0067 in)
  Service limit .... 0.350 mm (0.0138 in)
  650 models .... 0.150 – 0.350 mm (0.0059 – 0.0138 in)
  Service limit .... 0.500 mm (0.0197 in)
Crankpin OD:
  500 models – 1st version .... 39.966 – 39.990 mm (1.5735 – 1.5744 in)
  500 models – 2nd version .... 39.976 – 40.000 mm (1.5739 – 1.5748 in)
  650 models .... 42.966 – 42.990 mm (1.6916 – 1.6925 in)
Main bearing journal OD:
  500 models .... 42.980 – 43.000 mm (1.6921 – 1.6929 in)
  650 models .... 45.980 – 46.000 mm (1.8102 – 1.8110 in)

Big-end bearings – 500 models

Connecting rod size groups – big-end ID:
  Connecting rod marked 1 .... 43.000 – 43.008 mm (1.6929 – 1.6932 in)
  Connecting rod marked 2 .... 43.008 – 43.016 mm (1.6932 – 1.6935 in)
  Connecting rod marked 3 .... 43.016 – 43.024 mm (1.6935 – 1.6939 in)

Big-end bearing shell:

	Thickness	Part number
Code B (black)	1.503 – 1.507 mm (0.0592 – 0.0593 in)	13216-415-013
Code C (brown)	1.499 – 1.503 mm (0.0590 – 0.0592 in)	13217-415-013
Code D (green)	1.495 – 1.499 mm (0.0589 – 0.0590 in)	13218-415-013
Code E (yellow)	1.491 – 1.495 mm (0.0587 – 0.0589 in)	13219-415-013
Code F (pink)	1.487 – 1.491 mm (0.0585 – 0.0587 in)	13220-415-003

	1st version (see text)	2nd version (see text)
Bearing shell/crankpin clearance	0.028 – 0.052 mm (0.0011 – 0.0021 in)	0.020 – 0.044 mm (0.0008 – 0.0017 in)
Service limit	0.080 mm (0.0032 in)	0.080 mm (0.0032 in)
Crankpin size groups – OD:		
Crankshaft inner letter marked A	39.982 – 39.990 mm (1.5741 – 1.5744 in)	39.992 – 40.000 mm (1.5745 – 1.5748 in)
Crankshaft inner letter marked B	39.974 – 39.982 mm (1.5738 – 1.5741 in)	39.984 – 39.992 mm (1.5742 – 1.5745 in)
Crankshaft inner letter marked C	39.966 – 39.974 mm (1.5735 – 1.5738 in)	39.976 – 39.984 mm (1.5739 – 1.5742 in)

Big-end bearings – 650 models

Connecting rod size groups – big-end ID:
  Connecting rod marked 1 .... 46.000 – 46.008 mm (1.8110 – 1.8113 in)
  Connecting rod marked 2 .... 46.008 – 46.016 mm (1.8113 – 1.8116 in)
  Connecting rod marked 3 .... 46.016 – 46.024 mm (1.8116 – 1.8120 in)

Big-end bearing shell:

	Thickness	Part number
Code B (black)	1.503 – 1.507 mm (0.0592 – 0.0593 in)	13216-ME2-003
Code C (brown)	1.499 – 1.503 mm (0.0590 – 0.0592 in)	13217-ME2-003
Code D (green)	1.495 – 1.499 mm (0.0589 – 0.0590 in)	13218-ME2-003
Code E (yellow)	1.491 – 1.495 mm (0.0587 – 0.0589 in)	13219-ME2-003
Code F (pink)	1.487 – 1.491 mm (0.0585 – 0.0587 in)	13220-ME2-003

Bearing shell/crankpin clearance .... 0.028 – 0.052 mm (0.0011 – 0.0021 in)
Service limit .... 0.085 mm (0.0034 in)

Crankpin size groups – OD:
Crankshaft inner letter marked A .... 42.982 – 42.990 mm (1.6922 – 1.6925 in)
Crankshaft inner letter marked B .... 42.974 – 42.982 mm (1.6919 – 1.6922 in)
Crankshaft inner letter marked C .... 42.966 – 42.974 mm (1.6916 – 1.6919 in)

Crankshaft main bearings

	500 models	650 models
Crankcase/rear main bearing cap size groups – ID:		
Smaller group	47.000 – 47.010 mm (1.8504 – 1.8508 in)	50.000 – 50.010 mm (1.9685 – 1.9689 in)
Larger group	47.010 – 47.020 mm (1.8508 – 1.8512 in)	50.010 – 50.020 mm (1.9689 – 1.9693 in)
Crankshaft journal size groups – OD:		
Crankshaft outer letter marked A	42.990 – 43.000 mm (1.6925 – 1.6929 in)	45.990 – 46.000 mm (1.8106 – 1.8110 in)
Crankshaft outer letter marked B	42.980 – 42.990 mm (1.6921 – 1.6925 in)	45.980 – 45.990 mm (1.8102 – 1.8106 in)

Main bearing bush:	Thickness	Part number 500 models	Part number 650 models
Code A (blue)	1.999 – 2.009mm (0.0787 – 0.0791 in)	13325-415-305	13323-MC7-305
Code B (black)	1.994 – 2.004 mm (0.0785 – 0.0789 in)	13326-415-305	13324-MC7-305
Code C (brown)	1.989 – 1.999 mm (0.0783 – 0.0787 in)	13327-415-305	13325-MC7-305

Main bearing bush/crankshaft clearance – all models .... 0.020 – 0.060 mm (0.0008 – 0.0024 in)
Service limit .... 0.085 mm (0.0034 in)

Camshaft and followers
Intake cam lobe height:
  500 models .... 37.046 mm (1.4585 in)
  Service limit .... 36.058 mm (1.4196 in)
  650 models .... 37.988 mm (1.4956 in)
  Service limit .... 37.866 mm (1.4908 in)
Exhaust cam lobe height:
  500 models .... 37.015 mm (1.4573 in)
  Service limit .... 36.027 mm (1.4184 in)
  650 models .... 38.143 mm (1.5017 in)
  Service limit .... 38.021 mm (1.4969 in)
Camshaft front journal OD .... 21.959 – 21.980 mm (0.8645 – 0.8654 in)
Service limit .... 21.910 mm (0.8626 in)
Camshaft front bearing cap ID .... 22.000 – 22.021 mm (0.8661 – 0.8670 in)
Service limit .... 22.050 mm (0.8681 in)
Camshaft rear journal OD .... 25.959 – 26.980 mm (1.0220 – 1.0622 in)
Service limit .... 25.910 mm (1.0201 in)
Camshaft rear bearing surface ID .... 26.000 – 26.021 mm (1.0236 – 1.0245 in)
Service limit:
  CX500 (UK) .... 26.050 mm (1.0256 in)
  All other models .... 26.170 mm (1.0303 in)
Cam follower spindle OD .... 13.982 – 14.000 mm (0.5505 – 0.5512 in)
Service limit .... 13.966 mm (0.5498 in)
Cam follower bore ID .... 14.016 – 14.027 mm (0.5518 – 0.5522 in)
Service limit .... 14.046 mm (0.5530 in)

Primary drive
Type .... Gear, with anti-backlash assembly
Reduction ratio:
  500 models 2.242:1 (74/33T)
  650 models 2.114:1 (74/35T)

Clutch
Type .... Wet, multi-plate
Spring free length:
  500 models .... 33.9 mm (1.3346 in)
  Service limit .... 32.5 mm (1.2795 in)
  650 models .... 39.4 mm (1.5512 in)
  Service limit .... 37.9 mm (1.4921 in)
Friction plate thickness:
  Type A .... 2.62 – 2.78 mm (0.1032 – 0.1094 in)
  Service limit .... 2.30 mm (0.0906 in)
  Type B .... 3.50 mm (0.1378 in)
  Service limit .... 3.10 mm (0.1221 in)
Plate maximum warpage .... 0.20 mm (0.0079 in)
Clutch outer drum ID – at centre sleeve:
  CX500 1982, CX500 E-C, all GL500 models .... 32.000 – 32.025 mm (1.2598 – 1.2608 in)
  Service limit .... 32.070 mm (1.2626 in)
  All other CX500 models .... 33.000 – 33.016 mm (1.2992 – 1.2998 in)

Service limit	33.070 mm (1.3020 in)
650 models	N/Av
Clutch centre sleeve OD:	
CX500 C 1982, CX500 E-C, all GL500 models	31.959 – 31.975 mm (1.2582 – 1.2589 in)
Service limit	31.900 mm (1.2559 in)
All other CX500 models	32.950 – 32.975 mm (1.2972 – 1.2982 in)
Service limit	32.900 mm (1.2953 in)
650 models	31.987 – 32.000 mm (1.2593 – 1.2598 in)
Service limit	31.928 mm (1.2570 in)
Clutch centre sleeve ID:	
650 models	25.000 – 25.025 mm (0.9843 – 0.9852 in)
Service limit	25.070 mm (0.9870 in)
500 models	N/Av

Gearbox

Type	Five-speed, constant mesh

	500 models	650 models
Reduction ratios:		
1st	2.733:1 (41/15T)	2.500:1 (40/16T)
2nd	1.850:1 (37/20T)	1.714:1 (36/21T)
3rd	1.416:1 (34/24T)	1.280:1 (32/25T)
4th	1.148:1 (31/27T)	1.036:1 (29/28T)
5th	0.931:1 (27/29T)	0.839:1 (26/31T)

Input shaft dimensions – 500 models:	
Shaft OD at clutch, 4th and 5th gear pinions	24.940 – 24.959 mm (0.9819 – 0.9826 in)
Service limit	24.910 mm (0.9807 in)
4th and 5th gear pinion ID	25.020 – 25.041 mm (0.9850 – 0.9859 in)
Service limit	25.100 mm (0.9882 in)
Output shaft dimensions – 500 models:	
Shaft OD at 1st gear pinion	19.987 – 20.000 mm (0.7869 – 0.7874 in)
Service limit	19.960 mm (0.7858 in)
1st gear pinion bush ID	20.020 – 20.041 mm (0.7882 – 0.7890 in)
Service limit	20.060 mm (0.7898 in)
1st gear pinion bush OD	23.984 – 24.005 mm (0.9443 – 0.9451 in)
Service limit	23.950 mm (0.9429 in)
1st gear pinion ID	24.020 – 24.041 mm (0.9457 – 0.9465 in)
Service limit	24.100 mm (0.9488 in)
Shaft OD at 3rd gear pinion	24.959 – 24.980 mm (0.9826 – 0.9835 in)
Service limit	24.930 mm (0.9815 in)
3rd gear pinion ID	25.020 – 25.041 mm (0.9850 – 0.9859 in)
Service limit	25.100 mm (0.9882 in)
Shaft OD at 2nd gear pinion	27.459 – 27.480 mm (1.0811 – 1.0819 in)
Service limit	27.430 mm (1.0799 in)
2nd gear pinion ID	27.520 – 27.541 mm (1.0835 – 1.0843 in)
Service limit	27.600 mm (1.0866 in)
Input shaft dimensions – 650 models:	
Shaft OD at clutch	24.991 – 25.009 mm (0.9839 – 0.9846 in)
Service limit	24.960 mm (0.9827 in)
Shaft OD at 4th and 5th gear pinion	24.959 – 24.980 mm (0.9826 – 0.9835 in)
Service limit	24.930 mm (0.9815 in)
4th and 5th gear pinion bush OD	28.979 – 29.000 mm (1.1409 – 1.1417 in)
Service limit	28.950 mm (1.1398 in)
4th and 5th gear pinion ID	29.020 – 29.041 mm (1.1425 – 1.1433 in)
Service limit	29.100 mm (1.1457 in)
Output shaft dimensions – 650 models:	
Shaft OD at front bearing and 1st gear pinion	19.987 – 20.000 mm (0.7869 – 0.7874 in)
Service limit	19.960 mm (0.7858 in)
1st gear pinion bush ID	20.020 – 20.041 mm (0.7882 – 0.7890 in)
Service limit	20.060 mm (0.7898 in)
1st gear pinion bush OD	23.984 – 24.005 mm (0.9443 – 0.9451 in)
Service limit	23.950 mm (0.9429 in)
1st gear pinion ID	24.020 – 24.041 mm (0.9457 – 0.9465 in)
Service limit	24.100 mm (0.9488 in)
3rd gear pinion bush OD	28.979 – 29.000 mm (1.1409 – 1.1417 in)
Service limit	28.950 mm (1.1398 in)
3rd gear pinion ID	29.020 – 29.041 mm (1.1425 – 1.1433 in)
Service limit	29.100 mm (1.1457 in)
Shaft OD at 2nd gear pinion	27.459 – 27.480 mm (1.0811 – 1.0819 in)
Service limit	27.430 mm (1.0799 in)
2nd gear pinion bush ID	27.500 – 27.521 mm (1.0827 – 1.0835 in)
Service limit	27.540 mm (1.0843 in)
2nd gear pinion bush OD	30.985 – 31.010 mm (1.2199 – 1.2209 in)
Service limit	30.950 mm (1.2185 in)
2nd gear pinion ID	31.025 – 31.050 mm (1.2215 – 1.2224 in)
Service limit	31.100 mm (1.2244 in)

Pinion gear/bush maximum clearance	0.150 mm (0.0059 in)
Bush/shaft maximum clearance	0.100 mm (0.0039 in)
Selector drum OD – at front bearing	34.950 – 34.975 mm (1.3760 – 1.3770 in)
Service limit	34.900 mm (1.3740 in)
Selector drum front bearing ID – in gearbox end cover	35.000 – 35.025 mm (1.3780 – 1.3789 in)
Service limit	35.060 mm (1.3803 in)
Selector drum front bearing clearance	0.025 – 0.075 mm (0.0010 – 0.0030 in)
Service limit	0.160 mm (0.0063 in)
Selector fork claw end thickness .,..	5.930 – 6.000 mm (0.2335 – 0.2362 in)
Service limit	5.500 mm (0.2165 in)
Selector fork shaft OD	12.966 – 12.984 mm (0.5105 – 0.5112 in)
Service limit	12.950 mm (0.5098 in)
Selector fork bore ID	13.000 – 13.018 mm (0.5118 – 0.5125 in)
Service limit	13.050 mm (0.5138 in)
Final output shaft spring free length	73.0 mm (2.8740 in)
Service limit:	
All CX500 models except CX500 E-C	68.0 mm (2.6772 in)
CX500 E-C, all GL500 models, all 650 models	72.0 mm (2.8346 in)

Final drive

Type	Shaft
Reduction ratio	3.091:1 (34/11T)

Torque wrench settings

Component	kgf m	lbf ft
Cylinder head cover bolts	0.8 – 1.2	6.0 – 9.0
Cylinder head bolts:		
All CX500 models	5.0 – 5.5	36.0 – 40.0
All GL500 models, all 650 models	5.0 – 6.0	36.0 – 43.0
Valve clearance adjuster locknuts:		
500 models	1.5 – 1.8	11.0 – 13.0
650 models	2.0 – 2.5	14.5 – 18.0
Cooling fan bolt – 500 models only	2.0 – 2.5	14.5 – 18.0
Engine rear cover bolts:		
6 mm	0.8 – 1.2	6.0 – 9.0
8 mm	1.8 – 2.5	13.0 – 18.0
ATU retaining bolt – transistorised ignition	0.8 – 1.2	6.0 – 9.0
Alternator rotor bolt:		
All CX500 models except CX500 E-C	8.0 – 10.0	58.0 – 72.0
CX500 E-C, all GL500 models, all 650 models	9.0 – 10.5	65.0 – 76.0
Starter clutch body Torx bolts	1.8 – 2.5	13.0 – 18.0
Cam chain guide plate bolts	0.8 – 1.2	6.0 – 9.0
Manual cam chain tensioner adjuster bolt	0.8 – 1.2	6.0 – 9.0
Automatic cam chain tensioner mounting bolt	1.8 – 2.5	13.0 – 18.0
Camshaft driven sprocket/mounting flange bolts	1.6 – 2.0	11.5 – 14.5
Camshaft sprocket flange mounting nut	8.0 – 10.0	58.0 – 72.0
Rear main bearing cap bolts	2.0 – 2.4	14.5 – 17.0
Clutch centre nut	8.0 – 10.0	58.0 – 72.0
Primary drive gear bolt	8.0 – 9.5	58.0 – 69.0
Oil pump mounting bolts	0.8 – 1.2	6.0 – 9.0
Gearbox front cover bolts:		
6 mm x 20 mm (0.24 in x 0.79 in)	1.5 – 2.0	11.0 – 14.5
6 mm x 32 mm (0.24 in x 1.26 in)	1.0 – 1.4	7.0 – 10.0
Connecting rod cap bolts:		
500 models	2.8 – 3.2	20.0 – 24.0
650 models	4.1 – 4.5	29.5 – 32.5
Engine mounting bolts and nuts – all CX500 models except CX500 E-C:		
8 mm	1.8 – 2.5	13.0 – 18.0
10 mm	3.5 – 4.5	25.0 – 32.5
12 mm	6.0 – 7.0	43.0 – 50.5
Front mounting bracket nuts	3.5 – 4.5	25.0 – 32.5
Engine mounting bolts and nuts – CX500 E-C, all GL500 models, all 650 models:		
10 mm	4.5 – 7.0	32.5 – 50.5
12 mm	6.0 – 8.0	43.0 – 58.0
Front mounting bracket nuts	3.0 – 4.0	22.0 – 29.0
Final drive shaft pinch bolt:		
All CX500 models except CX500 E-C	1.8 – 2.5	13.0 – 18.0
CX500 E-C, all GL500 models, all 650 models	1.8 – 2.8	13.0 – 20.0
Gearchange pedal pinch bolt	1.0 – 1.4	7.0 – 10.0

1 General description

Although the engine/gearbox unit is extremely unconventional by Japanese standards, being a water-cooled V-twin mounted transversely in the frame and having a shaft final drive, the construction of the components which form the main sub-assemblies follows usual Japanese manufacturing practice and as such gives rise to little difficulty in servicing or overhaul.

The basis of the engine is a one-piece aluminium casting which incorporates separate and isolated chambers for the crankshaft and gearbox and contains the integral cylinder barrels. The steel cylinder liners, at 80° to each other, are cast into the unit during manufacture and are not removable.

The crankshaft is housed in the upper central chamber in the main engine casing, where it runs on two bush-type plain metal main bearings. The front bearing is pressed into the front wall of the chamber and the rear bearing is located similarly in a large diameter detachable housing which bolts to the rear of the case. The two big-end bearings are of the split shell type, lying side-by-side on the same crankpin. This arrangement gives a slight off-set in the cylinder line so that the left-hand cylinder is further forward than the right-hand cylinder. Each cylinder head is an unfinned aluminum casting with an integral water jacket interconnecting with the water jacket surrounding the cylinder sleeve. Two inlet and two exhaust valves are used in each cylinder head, each set of valves being operated via a pushrod and a forked rocker arm from a single four-lobe camshaft mounted between the cylinders, in line with the crankshaft and driven by a (Hy-vo) inverted-tooth chain from the crankshaft. The two-shaft gearbox is housed within a chamber below and to the right of the crankshaft, with the multi-plate clutch mounted on the input shaft at the front of the engine. A spur gear mounted on the crankshaft meshes directly with a gear on the rear of the clutch body to provide a primary drive. The large mass of the clutch and driven gear, rotating in the opposite direction to the crankshaft, is used to reduce the out of balance forces generated by the crankshaft and pistons. Note that the engine rotates anti-clockwise when viewed from the rear (alternator end), or clockwise when seen from the clutch end.

2 Operations with the engine/gearbox in the frame

Access to most components is not possible when the engine is in the frame. The following components can be removed with the engine still in position:

1 Cylinder heads
2 Carburettors
3 Radiator
4 Fan and water pump
5 Clutch
6 Starter motor
7 Oil pump and primary drive gear

3 Operations with the engine/gearbox removed

The engine must be removed for access to and removal of all remaining components, including the following:

1 Pistons and connecting rods
2 Crankshaft assembly
3 Gearbox components
4 Oil pump
5 Camshaft and followers
6 Alternator and ignition components. (See Section 11)
7 Final output shaft
8 Water pump mechanical seal

4 Carrying out a compression test

1 A good indication of the degree of wear in the engine top end components can be gained by carrying out a compression test. This will require the use of an accurate compression tester (gauge) which has an adaptor suitable for use with 12 mm spark plug threads and is extended sufficiently to reach the deeply-recessed spark plug locations.
2 The engine must be fully warmed up to normal operating temperature after the valve clearances have been checked, and adjusted if necessary, (see Routine Maintenance) and the battery must be in good condition and fully charged for the test results to be accurate.
3 Remove the spark plugs, fit them to the plug caps and lay the plugs on the cylinder heads so that the plug metal bodies are earthed to the engine. This is to prevent damage to the ignition system; as a further precaution switch off the engine kill switch.
4 Attach the compression tester following its manufacturer's instructions. Check that the choke is in the fully-open position (for normal running) and switch on the ignition.
5 Open the throttle twistgrip fully and turn the engine over on the starter motor. The gauge reading should rapidly increase to a maximum level and settle at that (after about 4 - 7 seconds); note the reading and repeat the operation on the remaining cylinder.
6 If the pressures obtained are outside the specified tolerances the cylinder heads must be removed for examination. Similarly if there is a marked discrepancy between the two cylinders careful examination will be necessary; the maximum permissible difference is 4 kg/cm² (57 psi). In the unlikely event of the recorded pressures being above those specified, it is probable that excessive amounts of carbon have built up in the combustion chambers; these must be removed as soon as possible.
7 If the pressures are too low, pour a small quantity of oil into the combustion chamber(s) and repeat the test. If the pressures recorded increase significantly, then the piston, piston rings and cylinder bore are excessively worn (the oil having provided a temporary seal around the piston rings). This will require a complete strip-down of the engine/gearbox unit to find the cause and rectify it.
8 If the pressures do not alter after the oil has been added, then the cylinder head gasket or valves are leaking. These can be examined after the cylinder heads have been removed, but the cylinder bores should be examined carefully (as far as possible) while the opportunity arises.

5 Removing the engine/gearbox unit from the frame

1 Place the machine firmly on its centre stand so that it is standing securely and there is no likelihood that it may fall over. This is extremely important as owing to the weight of the complete machine and the engine, any instability during dismantling will probably be uncontrollable. If possible, place the machine on a raised platform. This will improve accessibility and ease engine removal. Again, owing to the weight of the machine, ensure that the platform is sufficiently strong and well supported.
2 Drain the engine oil and remove the oil filter, as described in Routine Maintenance.
3 On GL500 D, GL500 I and GL650 I models only, remove the fairing as described in Chapter 5, then remove the horns and the fairing mounting bracket.
4 On all models, remove the seat and both side panels, then remove the fuel tank as described in Chapter 3.
5 Note that whenever any component is moved, all mounting nuts, bolts, or screws should be refitted in their original locations with their respective washers and mounting rubbers and/or spacers.
6 To prevent the possibility of short circuits, disconnect the battery at its negative (–) terminal. If the machine is to be out of service for some time, remove the battery and give it regular refresher charges as described in Chapter 7.
7 On CX650 C models only, remove its two retaining screws and withdraw the frame front cover.
8 On all GL models remove the left-hand footrest. On CX650 C models remove both footrests and the rear brake pedal.

9 On all models, remove the complete exhaust system as described in Chapter 3. Note than on CX500, CX500 C and CX500 D models it will probably be necessary to disconnect the rear brake operating rod and to disconnect the pedal return spring and stoplamp switch spring from the pedal so that the pedal can be swung down far enough to permit the balance chamber to be removed.

10 Disconnect the tachometer drive cable. On earlier models remove the single retaining bolt at the cable lower end, pull the cable out of its housing and lay it on the frame top tubes. On all later models unscrew the knurled retaining ring which secures the cable upper end to the instrument; the cable can then be pulled clear of all clamps or guides securing it to the frame.

11 Similarly the clutch cable can either be disconnected at the handlebar and removed with the engine, or it can be disconnected at its lower end and can be pulled up through the radiator shroud and laid on the frame top tubes..

12 Disconnect all electrical leads tracing them from the engine/gearbox unit up to the connector or connector blocks joining them to the main wiring loom and releasing any clamps or ties securing them to the frame. These include the alternator and ignition system leads, the water temperature gauge and oil pressure warning lamp sender unit wires, the neutral switch wire, the spark plug caps and, on 650 models only, the electric fan motor lead. Disconnect the starter motor lead at the motor terminal. On all GL500 models, the CX500 E-C and all 650 models disconnect the battery earth lead from the rear of the engine/gearbox unit.

13 On all CX500, CX500 C and CX500 D models remove the ignition HT coil rear mounting bolt; note that it may be necessary to disconnect their wires and to remove the coils completely. Unscrew the retaining nuts and remove the two engine top mounting plates and their bolts (CX500 models only, except the CX500 E-C and CX500 C 1982 models).

14 Disconnect the radiator/expansion tank pipe at the radiator filler neck, then disconnect the engine breather pipe(s).

15 The carburettors can be removed completely if required, as described in Chapter 3. In practice it is only necessary to disconnect them from the engine so that they remain with the frame; this is much quicker and easier but does carry a higher risk of damage as the engine is removed and refitted. To disconnect the carburettors either remove the inlet stub mounting bolts to release them from the cylinder heads or slacken the clamp screws to release the carburettors from the stub hoses. Check carefully that the carburettors are released before removing the engine.

16 If required, the radiator and cooling system components can

be dismantled at this stage to reduce the bulk of the unit. Refer to Chapter 2.

17 On all models peel back the rubber gaiter so that the driveshaft is exposed. Except on CX650 E, CX650 C and GL650 D2-E models, turn the rear wheel so that the pinch bolt appears, then unscrew it completely to release the driveshaft from the gearbox.

18 Make a final check that all components are disconnected or removed; the engine should now be retained by only its mounting bolts. Check that all electrical wires, control cables and breather hoses are neatly secured either to the frame or to the engine, as applicable, so that they cannot hinder the removal operation. Pull the carburettors (if still in place) carefully backwards to ensure that they are free from the engine; if the inlet stub mounting bolts have been removed it may be necessary to tap each one gently with a soft-faced mallet to break the seal with the cylinder head.

19 While it is possible to build a secure platform under the engine which will take its weight and allow it to be manoeuvred out of the frame, work is much easier if a trolley jack is used, as described below. Whichever method is to be used, enlist the aid of two assistants; the engine/gearbox unit is extremely bulky and quite heavy (approx 75 kg/165 lb). Its removal and refitting require care and patience if damage is not to be incurred.

20 Remove the retaining nuts from the mounting bolts. Carefully take the weight of the engine with the jack, ensure it is securely supported and tap out the lower rear mounting bolt, then unscrew both upper rear bolts. Do not lose the nuts.

21 With the assistants checking that the engine cannot fall, tap out or unscrew (as applicable) the front mounting bolts, lower the engine slightly and pull it forwards until the driveshaft is pulled clear of the gearbox shaft splines. Lower the engine fully and move it away from the machine.

22 If a mounting bolt is difficult to remove, apply a good quantity of penetrating fluid, allow time for it to work and release the bolt by rotating it before attempting to tap it out. Use the jack to adjust the engine's height while the bolts are removed, so that its weight does not stick them in place.

23 With the engine removed from the machine the coolant can now be drained and the radiator can be removed. Refer to Chapter 2.

24 Remove the nuts or bolts which retain the engine front mounting bracket to the crankcase and cylinder heads, then withdraw the bracket(s) and, on 500 models only, the fan cowling.

25 If these have not already been removed, disconnect the clutch cable and the tachometer drive cable.

26 Check carefully the final drive shaft gaiter; if split, torn or damaged this should be discarded and a new one purchased for refitting.

5.6 Disconnect battery terminals to prevent short circuits

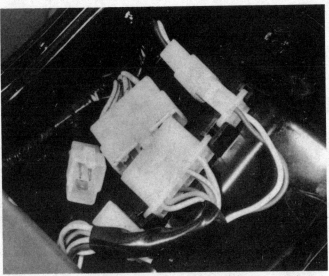

5.12 Disconnect all electrical leads at their connectors

5.13 Remove ignition HT coils where necessary and dismantle engine top mounting (where fitted)

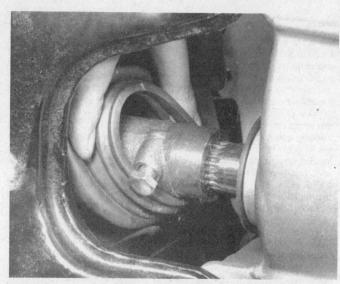

5.17 Where applicable, peel back rubber gaiter and remove drive-shaft pinch bolt

5.19a Engine's weight must be supported on platform or trolley jack ...

5.19b ... so that it can be slid forwards to disconnect driveshaft

5.24a Remove mounting nuts or bolts and withdraw fan cowling (where fitted) ...

5.24b ... and engine front mounting brackets(s)

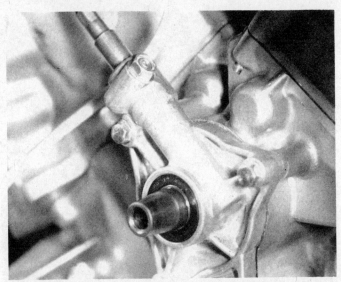

5.25 Remove pinch bolt to release tachometer drive cable

6 Dismantling the engine/gearbox unit: removing the cooling system components

1 Remove the bolts holding the thermostat bracket to the top of the crankcase. Loosen evenly and remove the two bolts which hold each flanged manifold to the cylinder head. Disconnect the pump bypass hose and remove the assembly. Note the O-ring sealing the face of each manifold. The transfer pipes are a push fit in the manifolds and thermostat housing, and may be detached. Note the O-ring on each end of the pipes. Remove the air spoiler plate (where fitted) from the top of the engine.

2 The chrome plated lower water pipe may be detached at this stage. It is held by two clamps, each of which is secured by two recessed socket screws. Note the sealing O-ring.

3 On 500 models only the fan must be removed. Unscrew the inspection cap from the engine front cover and apply a suitable spanner to the head of the primary drive gear retaining bolt to prevent rotation while the fan retaining bolt is unscrewed.

4 The fan must be pulled off the camshaft taper using a Honda service tool which is available under part numbers 07733-0010000, 07933-0010000, or 07933-2000001; if these are not available the machine's front wheel spindle will provide a substitute. Screw the extractor in until the fan is pushed from position. If the fan is very tightly in place and stubbornly refuses to move, do not continue tightening the bolt. When the bolt is tight, a few sharp blows on the bolt head with a hammer should release the tapered joint.

5 On all models, the water pump can be dismantled (except for the mechanical seal) with the engine/gearbox unit in the frame; the only preliminary work necessary is to drain the coolant (see Chapter 2) and to remove the chromed water pipe.

6 To dismantle the pump itself, remove the bolts holding the pump cover in place. The cover is located on two dowels and because of the presence of water, may be corroded in position. Tap the cover away from the engine rear cover using a rawhide mallet or a block of wood and hammer. The use of levers in an effort to remove a stubborn cover should be avoided because the risk of damage to the mating surfaces is high.

7 Remove the nut and copper washer from the end of the camshaft. The impeller is a tight push fit on the splined end of the shaft and should be eased from position using flat bladed levers. Great care must be taken not to damage the mating surfaces of the casing or the rear face of the impeller. Remove the small ceramic washer behind the impeller .

6.1a Remove thermostat assembly mounting bolts ...

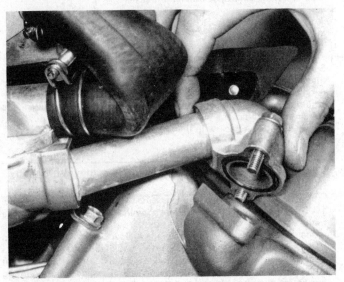

6.1b ... and note O-rings sealing coolant pipe flanged manifolds

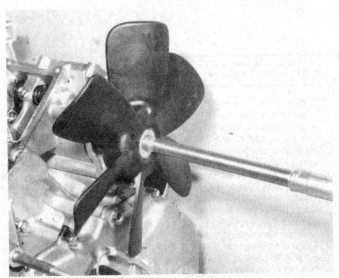

6.4 Fan (500 models only) can be pulled off camshaft using front wheel spindle

6.6 Note two locating dowels when removing water pump cover

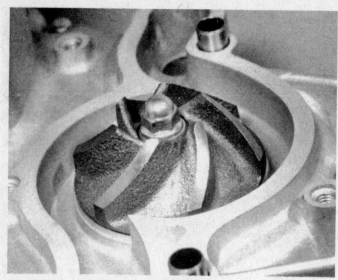

6.7a Remove dome nut and copper washer ...

6.7b ... to release water pump impeller ...

6.7c ... noting small spacer behind impeller

7 Dismantling the engine/gearbox unit: preliminaries

1   Before any dismantling work is undertaken, the external surfaces of the unit should be thoroughly cleaned and degreased. This will prevent the contamination of the engine internals, and will also make working a lot easier and cleaner. A high flash-point solvent, such as paraffin (kerosene) can be used, or better still, a proprietary engine degreaser such as Gunk. Use old paintbrushes and toothbrushes to work the solvent into the various recesses of the engine castings. Take care to exclude solvent or water from the electrical components and inlet and exhaust ports. The use of petrol (gasoline) as a cleaning medium should be avoided, because the vapour is explosive and can be toxic if used in a confined space.

2   When clean and dry, arrange the unit on the workbench, leaving a suitable clear area for working. Gather a selection of small containers and plastic bags so that parts can be grouped together in an easily identifiable manner. Some paper and a pen should be on hand to permit notes to be made and labels attached where necessary. A supply of clean rag is also required.

3   Before commencing work, read through the appropriate section so that some idea of the necessary procedure can be gained. When removing the various engine components great force is seldom required, unless specified. In many cases, a component's reluctance to be removed is indicative of an incorrect approach or removal method. If in any doubt, re-check with the text.

8 Dismantling the engine/gearbox unit: removing the cylinder heads

1   These components can be removed with the engine in or out of the frame. Note that if work is being carried out with the engine in the frame it is only necessary to remove the fuel tank if the rocker gear is to be dismantled. However if the cylinder heads are to be removed, the seat, side panels and fuel tank must be withdrawn, the carburettor inlet stubs must be released, the exhaust pipes and engine front mounting bracket(s) must be removed and the cooling system must be drained. Refer to Section 4. Also the air spoiler plate (where fitted) and the

thermostat/water pipe assembly must be removed. See Section 5. On early models only, the breather pipe must be disconnected from each cylinder head stub.

2   Each cylinder head should be dismantled separately, to prevent the interchange of identical components; the parts should be stored separately for the same reason. Before commencing, the remaining coolant which will still be in the water jacket must be drained by removing the small drain plug located at the front of each cylinder, just below the exhaust port. Tip the engine forward so that the draining coolant clears the crankcase.

3   Unscrew the two bolts securing the rocker cover and lift the cover away together with the sealing ring. Remove the spark plug. Before loosening the rocker arm carrier bracket the crankshaft should be rotated so that all the valves are closed. This will prevent any unwarranted stress being placed on the carrier by the partially open valves. To rotate the crankshaft, remove the inspection cap from the engine front cover and apply a spanner to the head of the primary drive gear retaining bolt.

4   Remove the two small bolts which pass through the cylinder head flange on the inner side of the head. Slacken them evenly, in a diagonal pattern, and then remove the rocker carrier bolts. The bolts serve also as the cylinder head holding bolts and must be loosened evenly to help avoid distortion. The rocker carrier is located on two hollow dowels. If necessary, ease the carrier from position using a flat bladed lever. Take special care not to damage the sealing surfaces of the cylinder head. With the rocker carrier removed the two pushrods may be withdrawn. Mark each pushrod clearly so that it may be refitted in its original position on subsequent reassembly.

5   The cylinder head is now free to be lifted from place. The use of a rawhide mallet may be required initially to release the cylinder head from the gasket and the two dowels on which it locates.

6   Note carefully the O-ring and the small oil feed jet which it surrounds, both of which fit into a stepped recess in the cylinder barrel mating surface. Remove and discard the O-ring and withdraw the jet. The jet should be stored carefully in a safe place. Remove and discard the head gasket.

8.2 Remove drain plug to drain coolant from cylinder water jacket

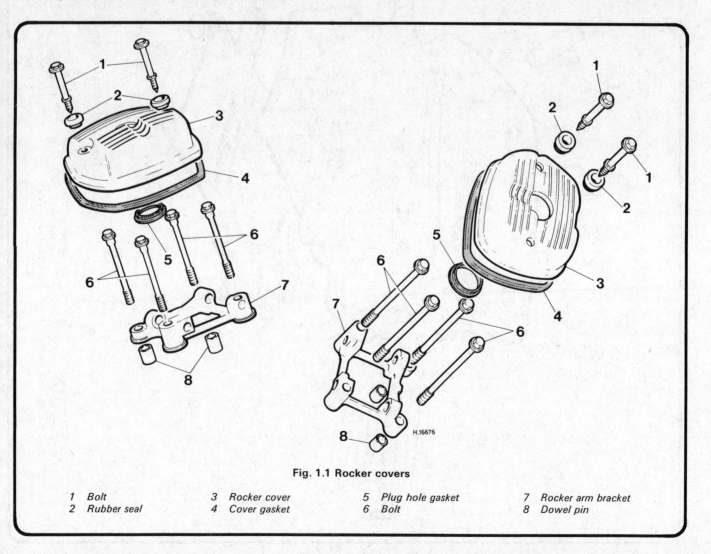

Fig. 1.1 Rocker covers

| 1 Bolt | 3 Rocker cover | 5 Plug hole gasket | 7 Rocker arm bracket |
| 2 Rubber seal | 4 Cover gasket | 6 Bolt | 8 Dowel pin |

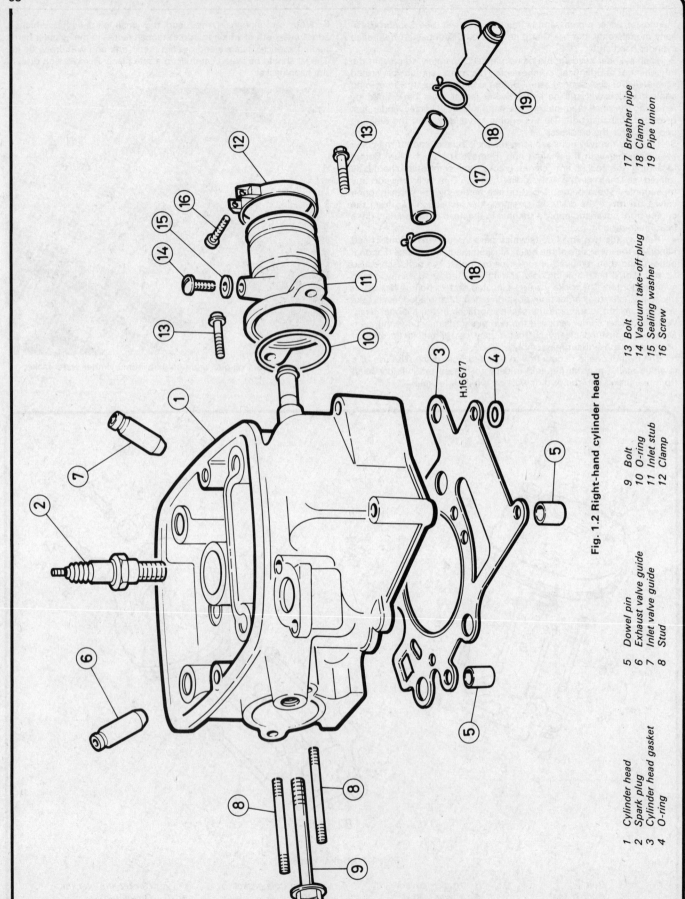

H.16677

Fig. 1.2 Right-hand cylinder head

1 Cylinder head
2 Spark plug
3 Cylinder head gasket
4 O-ring

5 Dowel pin
6 Exhaust valve guide
7 Inlet valve guide
8 Stud

9   Bolt
10 O-ring
11 Inlet stub
12 Clamp

13 Bolt
14 Vacuum take-off plug
15 Sealing washer
16 Screw

17 Breather pipe
18 Clamp
19 Pipe union

9 Dismantling the engine/gearbox unit: removing the clutch

1 The clutch may be removed with the engine in the frame or out of it, but in the former case the engine oil must be drained and the cable must be disconnected from the operating lever.

2 Unscrew its retaining bolts and withdraw the cover and its gasket. If the cover proves stubborn, apply a close-fitting open-ended spanner or a self-locking wrench to the operating lever and apply pressure as if attempting to operate the clutch; this will extract the cover with no risk of damage.

3 Unscrew evenly and remove the four bolts which retain the lifter plate in the centre of the clutch. Lift out the plate and bearing and the springs. The clutch is retained by a special nut which requires a peg spanner. A suitable tool can be fabricated from a short length of thick walled pipe, as shown in the accompanying illustration. Relieve the end of the pipe so that four pegs are formed which will engage securely with the special nut. This tool acts as a substitute for the Honda service tool part number 07716-0020201/2/3 which is applied with a handle 07716-0020500.

4 To prevent the clutch centre rotating lock the two parts of the clutch by refitting two or more clutch springs and bolts using suitable washers to take the place of the lifter plate. Remove the inspection cap from the engine front cover and apply a spanner to the primary drive gear retaining bolt. Holding the crankshaft with the spanner will lock the clutch and allow the clutch nut to be unscrewed.

5 Remove the refitted clutch springs and bolts and pull out the clutch centre, plates and pressure plate as one unit by grasping two of the four threaded columns into which the spring bolts screw. Before separating the plates and the clutch centre **carefully note** the sequence of plates to aid refitting, particularly that the plates are fitted in two separate sections, separated by a special double thickness plate. The clutch outer drum may now be withdrawn from the casing. Note that the drum is fitted with a central sleeve or a needle roller bearing; note carefully the arrangement of any thrust washers fitted, also the spacer fitted behind the clutch on CX500 C 1982, CX500 E-C and all GL500 models.

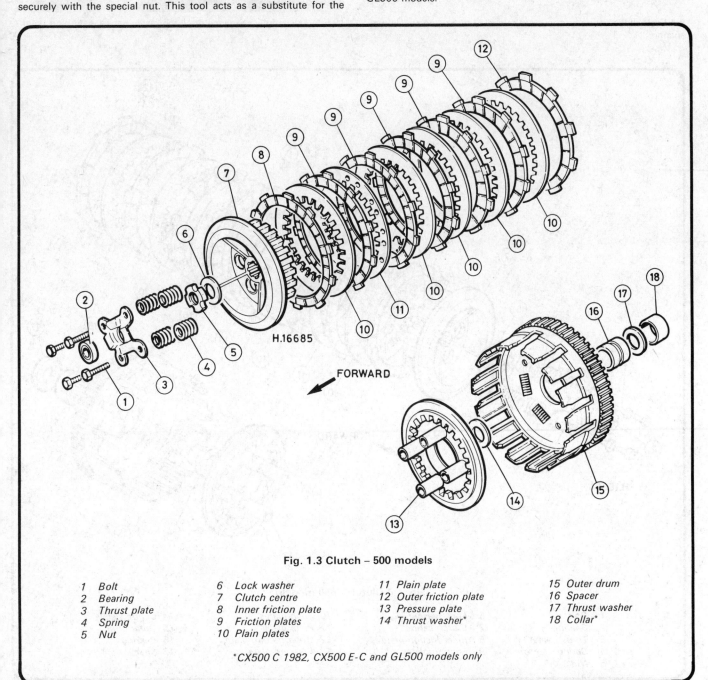

H.16685

FORWARD

Fig. 1.3 Clutch – 500 models

1 Bolt	6 Lock washer
2 Bearing	7 Clutch centre
3 Thrust plate	8 Inner friction plate
4 Spring	9 Friction plates
5 Nut	10 Plain plates
11 Plain plate	15 Outer drum
12 Outer friction plate	16 Spacer
13 Pressure plate	17 Thrust washer
14 Thrust washer*	18 Collar*

*CX500 C 1982, CX500 E-C and GL500 models only

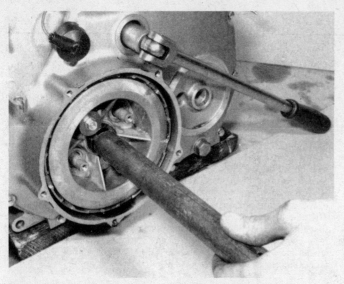

9.4 Lock clutch centre as shown and unscrew nut using special peg spanner

9.5 Note carefully sequence of clutch plates before disturbing them

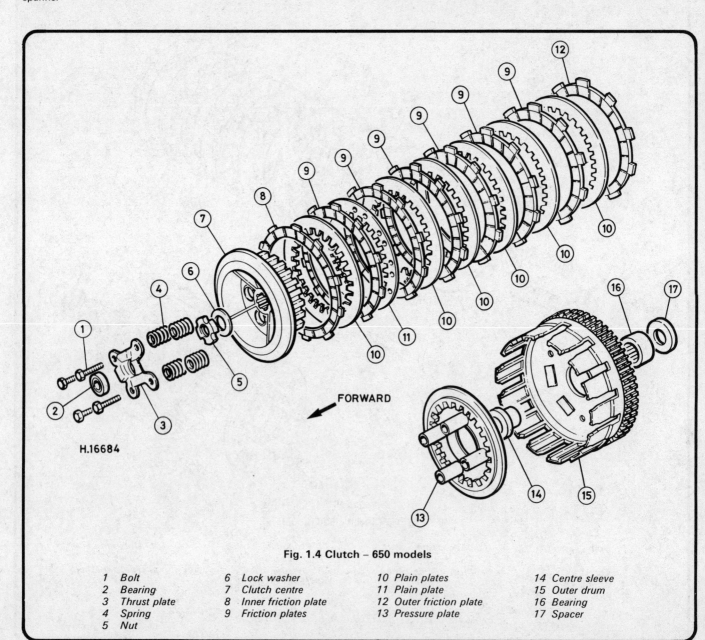

FORWARD

H.16684

Fig. 1.4 Clutch – 650 models

1 Bolt	6 Lock washer	10 Plain plates	14 Centre sleeve
2 Bearing	7 Clutch centre	11 Plain plate	15 Outer drum
3 Thrust plate	8 Inner friction plate	12 Outer friction plate	16 Bearing
4 Spring	9 Friction plates	13 Pressure plate	17 Spacer
5 Nut			

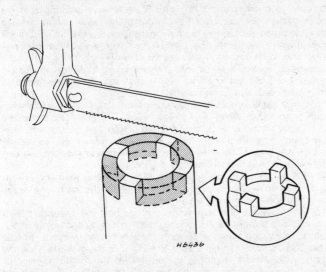

Fig. 1.5 Fabricated peg spanner

10 Dismantling the engine/gearbox unit: removing the engine front cover, primary drive gear and oil pump

1 While the components can be removed with the engine in or out of the frame, in the former case the engine oil and coolant must be first drained, the radiator, fan shroud and fan (500 models only), and the engine front mounting bracket must be removed, and the clutch cable and oil pressure switch wire must be disconnected. Release the wire from the clamp securing it.

2 Note that the front cover can be removed complete with the clutch cover, if required; there is no need to dismantle the clutch. The oil pressure switch also need not be unscrewed unless necessary.

3 Remove the cover retaining bolts, slackening them evenly in a diagonal sequence from the centre outwards and noting the presence of various cable and wiring clamps. Using a rawhide mallet break the joint between the cover and the gasket and then pull the cover off the location dowels. In addition to the small hollow locating dowels take careful note of the presence of two large oilway dowels and O-rings and the small oil feed jet and O-ring. These should be removed, the dowels and jet being stored safely and the O-rings discarded.

4 Remove its three retaining bolts, then pull the pump off the locating dowel and lift it away together with the drive chain. Withdraw the dowel and store it in a safe place.

5 To allow removal of the primary drive gear retaining bolt the crankshaft must be prevented from rotating. To achieve this a pinion locking plate should be fabricated, which bolts on the casing and engages with one or more teeth on the primary drive pinion. The tool is illustrated in the accompanying figure and replaces Honda tool 07924-4150000. It is recommended that this system of locking is adopted in preference to using a steel sprag or lever held by hand which is less controllable and may lead to damage of the teeth or casing. Unscrew the bolt and remove the oil pump drive sprocket.

6 On 650 models, the gear can be pulled off the crankshaft whenr required (see note below). On 500 models, pull off first the spring plate followed by the side plate and the two drive pins. Lift off the narrow auxiliary drive pinion. Mark the outer face of each pinion and the side plate to aid reassembly. Leave the primary drive gear in place until the alternator centre bolt has been loosened as described in Section 11.

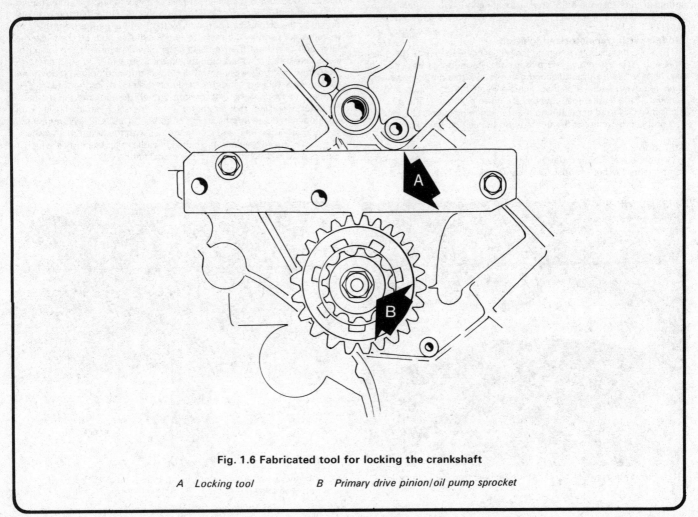

Fig. 1.6 Fabricated tool for locking the crankshaft

A Locking tool   *B Primary drive pinion/oil pump sprocket*

11 Dismantling the engine/gearbox unit: removing the engine rear cover, alternator and ignition components and the final output shaft

1   While the ignition pulser components can be reached without removing the engine from the frame this does require the removal of the swinging arm (see Chapter 5). If the rear cover is to be disturbed for any reason the engine must first be removed from the frame and the water pump cover and impeller must be withdrawn. Refer to Sections 5 and 6 of this Chapter.

2   If no marks were made by the manufacturer, mark the gearchange pedal and its shaft so that the pedal can be refitted in the same position, remove its retaining pinch bolt and pull the pedal off the shaft splines.

3   Remove its mounting bolts and withdraw the starter motor. On CX650 C, CX650 E-D and GL650 D2-E models, remove the circlip from the final output shaft.

4   Note that on all models fitted with CDI ignition (ie all CX500, CX500 C and CX500 D models except the CX500C 1982 models), the pulser advancer coils and cover need not be disturbed when removing the engine rear cover. For the full dismantling procedure, however, proceed as described below.

Models with CDI ignition

5   Remove the chromed pressed-steel cover from the rear of the engine rear cover. Disconnect the two leads from the advance pulser unit at the separate snap connectors. The unit is secured by two screws passing through elongated holes in the pulser stator. The provision of elongated holes allows the stator to be accurately aligned with the index mark scribed on the casing in the twelve o'clock position.

6   Unscrew the pulser rotor retaining bolt; there is no need to lock the crankshaft as a sharp tap on the end of a suitable spanner will be sufficient to release the bolt. Note its locating tang which engages in a slot in the alternator rotor centre as the pulser rotor is removed.

Models with transistorised ignition

7   Remove the six bolts retaining the pulser cover to expose the pulser coils and ATU assembly. The pulser coils are mounted on a circular backplate which can be detached once its two retaining screws have been removed. Mark a scribed line between the backplate and the material of the engine rear cover as an aid to accurate refitting.

8   The ATU is fixed in relation to the crankshaft by means of a locating peg and can be removed by unscrewing the central retaining bolt.

All models

9   With the water pump, starter motor, gearchange pedal and ignition components (where applicable) removed, slacken evenly, in a diagonal sequence, and then remove the screws securing the rear cover to the crankcase. Note the position of any cable clamps retained by various screws. Using a rawhide mallet separate the cover from the gasket with a few gentle taps, then lift the cover away. The gearchange shaft should be pushed inwards as the cover is withdrawn so that the arm inside remains meshed with the gear selector mechanism. To aid removal of the cover, the final output shaft may be grasped because this component can come away with the cover.

10   Check carefully that no small components such as thrust washers stick to the cover where they might be lost. The alternator stator and (on CDI ignition models only) the fixed pulsers may be removed from the casing after unscrewing the retaining bolts or screws. The various cables will have to be threaded through the casing aperture one at a time so that the block connectors do not snag.

11   Pull the final output shaft off the gearbox ouptut shaft end (or out of the rear cover). It fits over a splined collar on the shaft; remove the collar for safety and keep it with the shaft.

12   To remove the alternator rotor retaining bolt, the crankshaft must be locked to prevent rotation. If the clutch and primary drive are still in place select top gear and apply a self-locking wrench to the (temporarily refitted) final output shaft to lock the crankshaft via the transmission. If the wrench is braced against the shaft of the socket spanner (see accompanying photograph) the nut can be slackened easily by one person working alone. If the primary drive has been dismantled the fabricated tool described in the previous Section can be used to lock the primary drive gear. The use of a strap wrench is **not** recommended as it is unlikely that sufficient pressure can be exerted to allow the nut to be slackened without damaging the ignition trigger or the rotor.

13   When the bolt has been removed the rotor must be pulled off the crankshaft taper using a puller. It is recommended that this is done using only the Honda service tool, which is available under part numbers 07733-0020000/1, 07933-2000000, 07933-3950000 or 07933-3000000. The only alternative to this is to find a 20 mm (thread size) bolt with the correct pitch metric thread. Note that the required thread is the same size and pitch as the oil filter centre bolt; this can therefore be used as a pattern to find a suitable bolt. **Do not** use the filter bolt to extract the rotor; its large oilways render it too weak for such a task and it would probably shear if subjected to such stress.

14   Screw the tool or bolt into the rotor centre and tighten it down on to the crankshaft end. Tap it smartly on the end to jar the rotor free; if this fails to work at the first attempt tighten the tool and tap it again.

15   After releasing the alternator rotor from the taper, lift it from position complete with the starter clutch body and roller. Withdraw the starter motor idler gear and shaft together with the two thrust washers. The starter driven gear and the caged neeedle roller bearing on which it runs can be slid off the end of the crankshaft.

11.10 Fixed pulsers (CDI ignition) are marked to aid identification

11.13 Withdraw alternator rotor using only the correct service tool or a bolt (see text)

12 Dismantling the engine/gearbox unit: removing the gearchange external mechanism

1   Disconnect the spring at the upper end, which interconnects the gearchange pedal shaft with the selector claw arm. Withdraw the shaft from the crankcase, noting the thrust washer on its front end. Depress the selector claw arm against its spring pressure so that it is free of the selector pins and of the end of the selector drum. Grasp the selector claw arm and pull it away from the crankcase, complete with the shouldered sleeve and return spring.
2   Using a small screwdriver displace the outer end of the detent stopper arm return spring so that the tension on the arm is relaxed. Unscrew the stopper arm pivot bolt and remove the arm, washer, shouldered collar and the return spring.
3   Do not disturb the plates and selector pins from the end of the selector drum unless absolutely necessary; the gearbox can be dismantled with them in place. If removal is required, unscrew the single retaining bolt and remove the neutral indicator switch plate, the selector pin end plate, selector pins and spacer and then the detent cam plate. Note the way in which a tang on the switch plate engages with the pin end plate, also the shorter dowel pin which fits into the front face of the detent cam plate to align it correctly on the selector drum.

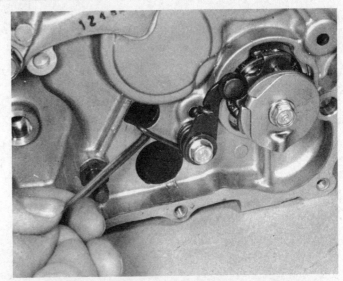

12.2 Use small screwdriver to displace outer end of detent stopper arm return spring

13 Dismantling the engine/gearbox unit: removing the cam chain and tensioner

1   These components can be removed only after the engine has been lifted out of the frame and the engine rear cover and alternator rotor have been withdrawn. Refer to Sections 5, 6 and 11 of this Chapter.
2   Before disturbing the tensioner components rotate the crankshaft until the left-hand piston is at TDC and the two camshaft drive sprocket bolts are in line with the index marks projecting from the crankcase next to the coolant passages.

Manual tensioner – all 500 models except the CX500 E-C and GL500 D-C
3   **Note:** since the cam chain tensioner components have been altered a number of times it is assumed here, for the sake of clarity, that the latest type is fitted; the removal procedure does not vary significantly. Refer to the accompanying illustrations for further details, if required.
4   Unhook the tensioner spring from the chain guide plate, then unscrew the four retaining bolts. Note that the top bolt also secures the chain guide blade upper end and the guide support plate; note carefully how the support plate fits over the mounting boss and against the crankcase wall. Withdraw the guide plate.
5   Unscrew the tensioner locking bolt; the tensioner unit is now free to be slid off the two pivot pins. Do not lose the collar that fits between the two sides of the tensioner arm, and through which the locking bolt passes. With the tensioner removed, there should be sufficient chain free play for the guide blade and support plate to be lifted away.

Automatic tensioner – CX500 E-C, GL500 D-C and all 650 models
6   Unscrew the three bolts securing the chain guide plate and withdraw the plate. Unlock the automatic tensioner by using a slim metal rod to press the steel ball back against spring pressure, until the tensioner pushrod can be pressed fully back into the tensioner body. Hold the pushrod by passing a suitable retaining pin through the hole in its right-hand end into the tensioner body. Unscrew the tensioner retaining bolt and withdraw the tensioner, followed by the tensioner blade.
7   Unscrew the two retaining bolts and lift away the guide blade, noting the two shouldered spacers.

All models
8   Remove the two bolts that pass through the camshaft sprocket into the mounting flange. The sprocket can now be removed together with the drive chain.

12.3 Do not disturb selector drum components unless necessary

13.6a Unlock automatic tensioner by pressing back steel ball as shown

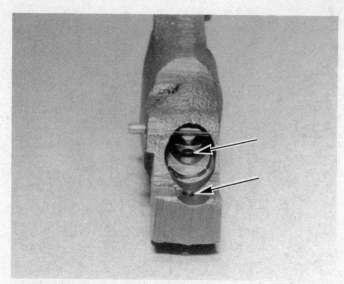

13.6b Hold pushrod by passing a pin through holes in pushrod and tensioner body (arrowed)

14 Dismantling the engine/gearbox: removing the camshaft and cam followers

1   These components can be removed only after the engine has been lifted out of the frame and the cylinder heads and cam chain have been withdrawn.

2   The camshaft sprocket mounting flange is retained on the end of the camshaft by a 27 mm nut deeply recessed in the mounting. Because of the position of the nut and the length of camshaft protruding, a box spanner will be required for nut removal. To prevent rotation of the camshaft when loosening the nut insert a suitable bar between one of the mounting flange lobes and the casing as shown in the accompanying photograph. Remove the nut and the special belville washer, noting that the second component is marked 'outside' on one face to aid correct assembly. Pull the sprocket mounting flange off the camshaft and displace the drive pin which lies in a hole drilled radially in the shaft.

3   The cam followers for each cylinder should be removed using the same procedure and stored individually to avoid the accidental interchange of similar components. In addition, the two followers in each set should be refitted on their shafts in the same sequence after removal, so that they may be refitted in their original positions on subsequent ressembly. To remove one set of cam followers unscrew the locating bolt which passes into the casing about half way along the length of the cam follower spindle. Screw a 6 mm bolt into the spindle threaded rear end, grip it with a large pair of pliers and withdraw the spindle from the casing. The spindle may be rotated by inserting a screwdriver in the slot provided, if difficulty is encountered in moving it initially. Lift out each follower and double coil spring from the casing. Repeat the procedure for the second set of cam followers.

4   The camshaft is removed from the front of the engine after detaching the forward bearing housing cum tachometer drive housing. After removal of the bolts ease the housing out of the front of the casing, taking care to keep it square to prevent tying. Withdraw the camshaft, noting the thrust washer on the rear end.

15 Dismantling the engine/gearbox unit: removing the gearbox components

1   Detach the oil separator plate from the front of the engine to gain access to the gearbox end cover. Slacken evenly, in a diagonal sequence, the bolts holding the end cover and then remove them. The gearbox components can be removed as a complete sub-assembly still attached to the end cover.

2   Rotate the selector drum until it can be seen that the rear end of the drum will clear the crankcase when the sub-assembly is withdrawn. Using a soft-nose mallet, tap the ends of the selector drum and of the output shaft, a little at a time, until the gearbox end cover is clear of the dowels upon which it locates. Lift the gearbox sub-assembly from position. If care is taken, the various selector components will remain in their correct positions relative to the gear clusters.

16 Dismantling the engine/gearbox unit: removing the pistons and connecting rods

1   To remove the connecting rods and pistons the engine must be removed from the frame and the cylinder heads, oil pump and gearbox must be removed. The left-hand cylinder big-end bearing cap is reached via the opening between the transmission and crankshaft chambers, the right-hand bearing cap being reached via the opening on the oil pump side.

2   **Note:** when removing the pistons and connecting rods either mark each component (left- or right-hand, as appropriate) or store them separately so that there is no chance of interchanging components on reassembly. This is vital to prevent the excessive wear that would result from mis-matching part worn components. Note that the pistons etc are removed **upwards** through the cylinder bores.

3   Rotate the crankshaft until a suitable spanner can be fitted to one nut on the forward (LH cylinder) big-end bearing cap. Slacken the nut and move the crankshaft again so that the second nut can be loosened. After removing the nuts separate the big-end cap from the connecting rod. If the cap shell becomes displaced refit it immediately so no confusion arises as to its correct position.

4   Rotate the crankshaft so that the up-moving crankpin pushes the piston upwards to the top of the stroke. Turn the crankshaft in the opposite direction so that the crankpin clears the lower end of the connecting rod. The connecting rod can now be pushed upwards so that the piston is displaced from the bore. Before doing this, however, it is worth checking to see whether there is a ring of carbon deposits at the top of the cylinder sleeve. If a ring exists it should be scraped away prior to displacing the piston. This will prevent breakage of the piston rings. After removal of the piston and connecting rod refit the big-end cap. Repeat the removal procedure for the right-hand connecting rod and piston.

5   If required, the pistons may be detached from their connecting rods. Displace both circlips from the piston using the end of a small screwdriver. The gudgeon pin is a light push fit and should prove easy to remove. If the gudgeon pin is a tight fit, support the piston on a block of soft wood, drilled with a hole slightly larger than the pin diameter. Using a suitable drift, drive the gudgeon pin out. Each piston is marked R or L on the crown to identify their correct positions. If the connecting rods are detached from the pistons, they too should be marked accordingly, to avoid confusion during reassembly.

14.2 Lock flange lug as shown to slacken retaining nut

15.2a Rotate selector drum so that it can pass through crankcase rear wall ...

15.2b ... as gearbox components are withdrawn

16.5 Displace circlips as shown – mark all components before separating them

17 Dismantling the engine/gearbox unit: removing the crankshaft

1   Loosen evenly and remove the bolts holding the rear main bearing housing to the engine casing. The housing is a tight press fit in the casing located on hollow dowels. Ideally the housing should be removed by pressing the crankshaft out from the forward end, using a special extractor, part No. 07935-4150000. This tool consists of a plate which bolts onto the front of the casing, and a pusher screw which engages with the crankshaft. If the correct tool is not available, the following method may be used; great care should be taken not to use excessive force or the crankshaft and rear main bearing may be damaged.

2   Wrap a length of tape around the splines at the rear end of the crankshaft. This will prevent the splines from damaging the main bearing surface. Using a rawhide mallet, tap the front of the crankshaft until the housing begins to move out away from the casing. Once the housing has moved out far enough, levers may be placed between the housing flange and the casing, and the housing eased out of place. Use only wide flat levers, placed carefully where the housing flange is thickest. If the housing proves stubborn, it is suggested that further attempts using this method are discontinued and the correct service tool is obtained.

3   Once the main bearing housing is clear of the casing, slide it off the crankshaft and then carefully lift the crankshaft from place. Note the two thrust washers (650 models only).

18 Examination and renovation: general

1   Before examining the parts of the dismantled engine unit for wear it is essential that they should be cleaned thoroughly. Use a petrol/paraffin mix or a high flash-point solvent to remove all traces of old oil and sludge which may have accumulated within the engine. Where petrol is included in the cleaning agent normal fire precautions should be taken and cleaning should be carried out in a well ventilated place.

2   Examine the crankcase castings for cracks or other signs of damage. If a crack is discovered it will require a specialist repair.

3   Examine carefully each part to determine the extent of wear, checking with the tolerance figures listed in the Specifications section of this Chapter or in the main text. If there is any doubt about the condition of a particular component, play safe and renew.

4   Use a clean lint-free rag for cleaning and drying the various components. This will obviate the risk of small particles obstructing the internal oilways, and causing the lubrication system to fail.

5   Various instruments for measuring wear are required, including a vernier gauge or external micrometer and a set of standard feeler gauges. The machine's manufacturer recommends the use of Plastigage for measuring radial clearance between working surfaces such as shell bearings and their journals. Plastigage consists of a fine strand of plastic material manufactured to an accurate diameter. A short length of Plastigage is placed between the two surfaces, the clearance of which is to be measured. The surfaces are assembled in their normal working positions and the securing nuts or bolts fastened to the correct torque loading; the surfaces are then separated. The amount of compression to which the gauge material is subjected and the resultant spreading indicates the clearance. This is measured directly, across the width of the Plastigage, using a pre-marked indicator supplied with the Plastigage kit. If Plastigage is not available both an internal and external micrometer will be required to check wear limits. Additionally, although not absolutely necessary, a dial gauge and mounting bracket is invaluable for accurate measurement of end float, and play between components of very low diameter bores – where a micrometer cannot reach. After some experience has been gained the state of wear of many components can be determined visually or by feel and thus a decision on their suitability for continued service can be made without resorting to direct measurement.

19 Bearings and oil seals – removal, examination and refitting

1   Wash each bearing thoroughly with petrol so that all old oil and any foreign matter has been removed. Allow the bearings to dry. Test the

bearing for roughness by spinning the outer race. Any roughness or snatching indicates wear or pitting in the race. The bearing should therefore be renewed. Bearings which are an interference fit in their housings should be tested in position as the outer races are designed to compress slightly when correctly positioned, giving the correct original tolerances. There should be no radial clearance on ball bearings though a small amount of side play is acceptable and on some bearings is evident even when new.

2   Removal of the input shaft bearing from the gearbox end cover plate should present no difficulties; the bearing may be drifted from place using a suitable drift of tubular construction. The input shaft rear bearing, fitted in the crankcase, may be drifted from place using a parallel punch passed through the oil hole in the rear wall. Care should be taken to keep the bearing square in the housing during this operation. The remainder of the bearings are fitted in blind holes and cannot be drifted from position. A special expanding puller is required to remove these bearings. If the correct type of puller is not available, it is suggested that the casings be returned to a Honda Service Agent who is suitably equipped.

3   New bearings may be drifted into place using a tubular drift of suitable size. Always fit the bearing with the manufacturer's serial marks outwards, ie, towards the drift, as it is this face of the bearing

which is specially hardened to take the force of the drift. Before refitting the input shaft rear bearing refit the oil guide plate in the housing. The bearing must be fitted with the sealed side to the rear.

4   It is recommended that the oil seals and O-rings be renewed whenever the engine is dismantled. This is particularly important with those seals and rings that are non-accessible so that major dismantling would be required for subsequent replacment. Seals which have previously given faultless service often begin to weep after reassembly due to damage during handling. This is particularly so with seals which are removed or replaced over splined shafts.

5   In most cases the oil seals are fitted between two separate cases, and can be removed easily after separation of the case. Other seals are a light drive fit in their housings and can be prised out of position using a screwdriver. Removal in this manner will invariably render the seal useless. For information relating to the mechanical water pump seal, see Chapter 2.

6   On refitting, ensure that the seals are the correct way round and drive them into place using a hammer and a tubular drift, such as a socket spanner, which bears only on the seal hard outer edge. Tap the seal in until it is flush with its surrounding crankcase area. Grease the seal lips to minimise damage as the shaft enters it.

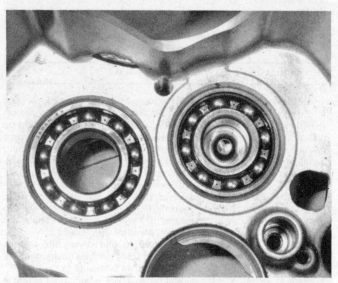

19.2a Input shaft rear bearing may be tapped out ...

19.2b ... using punch passed through hole in crankcase rear wall

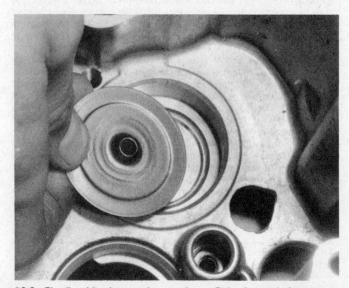

19.3a Fit oil guide plate as shown when refitting input shaft rear bearing ...

19.3b ... noting that bearing sealed surface faces to the rear

20 Cylinder heads and valves: dismantling, examination and renovation

1   Before examination and renovation of the cylinder heads can be carried out, the valves, springs and oil seals must be removed. Prior to this it is suggested that the carbon deposits in each combustion chamber are removed. Use a blunt scraper, which will not damage the surface, and then finish with a metal polish. Put an old sparking plug in place during cleaning, to prevent the fine particles of grit from clogging the plug hole threads.

2   A valve spring compressor is required to remove the valves. Compress the springs and remove the valve collets. Relax the springs and remove the valve spring collar, valve spring, valve spring seats and valve. Keep these components together in a set. Each valve must be replaced in its original location on subsequent reassembly. Prise off the oil seals and discard them. They should be renewed as a matter of course.

3   Clean the carbon from the inlet and exhaust ports and from the head of the valve.

4   Before attending to the valve seats, check the valve guide and stem wear. Valve clearance may be determined by subtracting the valve stem diameter from the internal diameter of the guide. If a valve or guide exceeds the stated wear limit it must be renewed. If the valve/guide clearance is excessive but the valve stem is still within the limits and is in good condition, the guide alone may be renewed. If an internal micrometer is not available for measuring the valve guide bore the extent of wear may be determined using a new valve and a dial gauge. Insert the new valve and measure the extent of lateral movement with the gauge at right angles to the stem. If the clearance is greater than that given in the Specifications, the guide must be renewed.

5   The valve guides may be drifted from position from the combustion chamber side of the cylinder head. Heating of the cylinder head before removal is not recommended by the manufacturers. Great care must be taken when carrying out this operation not to damage the cylinder head material. If inexperienced in this type of work, the advice of a Honda Service Agent should be sought. Before drifting a guide from place, remove any carbon deposits which may have built up on the guide end projecting into the port. Carbon deposits will impede the progress of the guide and may damage the cylinder head. If possible, use a double diameter drift. The smaller diameter should be close to that of the valve stem, and the larger diameter slightly smaller than that of the valve guide. Provided that care is exercised, a parallel shanked drift may be used as a substitute. A new guide may be fitted by reversing the removal procedure. A new guide will require reaming in order to bring the valve/guide clearance within the specified tolerances. It is probable also that the valve seat will require cutting in order that the valve seat/guide alignment is exact.

6   Check also that the valve stem is not bent, especially if the engine has been over-revved. If it is bent, the valve must be renewed.

7   The valves must be ground to provide a gas-tight seal, during normal overhaul, or after recutting the seat or renewing the valve.

8   Valve grinding is a simple, but laborious task. Smear grinding paste on the valve seat and attach a suction grinding tool to the valve head. Oil the valve stem. Rotate the valve in both directions, lifting it occasionally and turning it through 90°. Start with coarse paste if the seats are badly pitted and continue with fine paste until there is an unbroken matt grey ring on each seat and valve. After many re-grinds, the valve seat may become pocketed, when it should be re-cut. Wipe off very carefully all traces of grinding paste. If any remains in the engine, it will cause very rapid wear!

9   Valve sealing may be checked by installing the valve and spring and

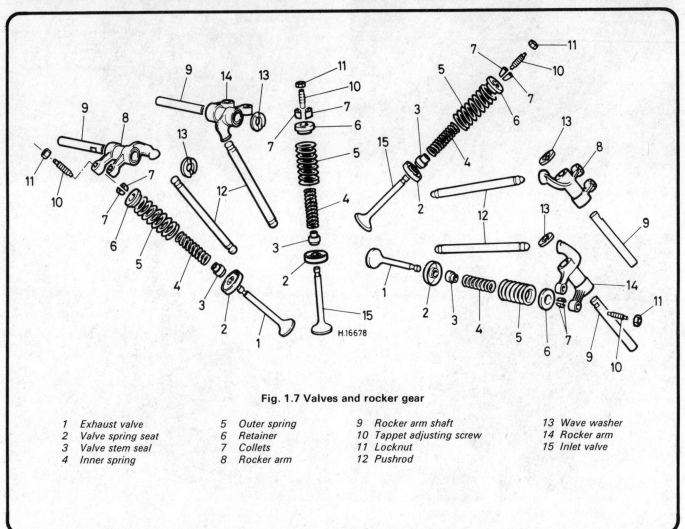

H.16678

Fig. 1.7 Valves and rocker gear

1 Exhaust valve	5 Outer spring	9 Rocker arm shaft	13 Wave washer
2 Valve spring seat	6 Retainer	10 Tappet adjusting screw	14 Rocker arm
3 Valve stem seal	7 Collets	11 Locknut	15 Inlet valve
4 Inner spring	8 Rocker arm	12 Pushrod	

pouring paraffin down the port. Check that none leaks past the valve seat.

10 If, after grinding, it is found that the width of the grey seating ring is greater than 2.0 mm (0.08 in) the valve seat must be recut using a special cutting tool. In order to restore the seat width, cutters of the correct diameter and cutting angle must be used, followed by a 45° tool. The correct seat width is within the range 1.1 – 1.3 mm (0.04 – 0.05 in). Because of the expense of purchasing three seat cutters and because of the accuracy with which cutting must be carried out, it is strongly recommended that the cylinder head be returned to a Honda Service Agent for attention.

11 Where deep pitting of the seat and valve is encountered, the seat should be recut as previously described. A pitted valve should be discarded and replaced by a new component. Regrinding of the valve face is not recommended with these valves due to the small head diameter and proportional dimensions.

12 Examine the condition of the valve collets and the groove on the valve stem in which they seat. If there is any sign of damage, new parts should be fitted. Check that the valve spring collar is not cracked. If the collets work loose or the collar splits whilst the engine is running, a valve could drop into the cylinder and cause extensive damage.

13 Check the free length of each of the valve springs. The springs have reached their serviceable limit when they have compressed to the limit readings given in the Specifications Section of this Chapter.

14 Reassemble the valve and valve springs by reversing the dismantling procedure. Ensure that all the springs are fitted with the close coils downwards towards the cylinder head. Fit new oil seals to each valve guide and oil both the valve stem and the valve guide, prior to reassembly. Take special care to ensure that the valve guide oil seal is not damaged when the valve is inserted. As a final check after assembly, give the end of each valve stem a sharp tap with a hammer, to make sure that the split collets have located correctly.

15 Check the cylinder head for straightness, especially if it has shown a tendency to leak oil at the cylinder head joint. If there is any evidence of warpage, provided it is not too great, the cylinder head may be machined flat. Most cases of cylinder head warpage can be traced to unequal tensioning of the cylinder head nuts and bolts by tightening them in incorrect sequence.

20.14a Fit spring seat and a new oil seal to guide ...

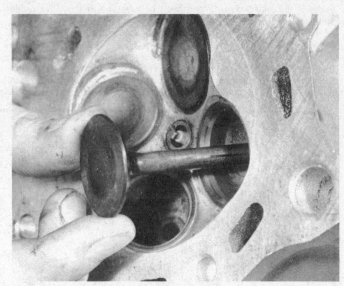

20.14b ... lubricate valve stem before refitting ...

20.14c ... and fit both springs with closer-pitched coils next to cylinder head

20.14d Refit spring top collar ...

20.14e ... and use a spring compressor to allow fitting of split collets

21 Rocker arms and spindles: examination and renovation

1   Dismantle each rocker carrier assembly separately, using an identical procedure, in order to prevent the accidental interchange of parts. Mark each spindle and rocker arm so that they may be replaced in their original positions. The spindles are a push fit and may be displaced using a drift, to free the arms. Note the wave washer fitted to each spindle between the carrier and the inner face of the arm.
2   Check the outside diameter of each rocker spindle and the inside diameter of each rocker arm bore. If the clearance is not within the service limit given, the relevant components must be renewed. Inspect the ball ends of the valve clearance adjuster screws for wear or damage. The hemispheres should have a smooth surface free of cracks or flaking metal. If damage is evident, the cupped ends of the pushrods may have suffered similarly.

22 Camshaft, cam followers and pushrods: examination and renovation

1   Check the clearance between each cam follower spindle and pair of followers in a manner similar to that used for the rocker arm and spindle inspection. Here again worn components should be renewed. Inspect the cam follower ends for wear or damage. If wear is evident, check for similar damage on the relevant mating component ie, the camshaft lobes or the pushrod lower ends.
2   Roll each pushrod on a flat surface as a check for straightness. If a bend is evident the pushrod in question should be renewed.
3   Measure and record the camshaft journal diameters. If either diameter is less than that given, the camshaft is in need of renewal. Measure the camshaft bearing inside diameters and compare the results with the specified service limits. The front bearing forms a detachable housing which also incorporates the tachometer drive gear. If wear is evident, the housing must be renewed. The rear journal of the camshaft runs directly in the main engine casing (crankcase). If wear develops here the complete casing must be renewed, because the bearing is not detachable. In view of the cost of replacing the casing, boring and sleeving the bearing back to the original size may be worth considering. This type of work is, of course, the domain of the specialist engineer.
4   Inspect the cam lobes for scoring or uneven wear and chipping of the hardened outer surfaces. If a cam lobe is badly worn or chipped it is probable that the rocker arm foot that runs on that lobe is also damaged and will require renewal. Measure the cam lobes at the highest point of lift (maximum diameter). Camshaft renewal will be necessary if one or more lobes does not come within the specified limits.

Fig. 1.8 Camshaft and followers

1   Bolt
2   Oil seal (Blanking plug – 650 models)
3   Camshaft bearing housing
4   Gasket
5   Camshaft
6   Drive pin
7   Thrust washer
8   Cam follower
9   Spring
10 Cam follower spindle
11 Cam follower
12 Bolt
13 Bolt
14 Washer
15 Tachometer drive gear
16 Thrust washer
17 Oil seal

H.16679

23 Cam chain, sprocket and tensioner: examination and renovation

1   Examine the drive chain for wear and loose or broken side plates. There are no specific figures available by which wear can be assessed, but some indication of wear can be obtained by the amount of movement left in the tensioner. If the tensioner arm was near its limit of travel and the guide blades are badly worn, the chain and blades should be renewed. The damage caused by a breaking chain can be imagined easily; therefore if in doubt, renew the chain.

2   The spring which controls the chain tension is unlikely to suffer damage or fatigue but its correct functioning is vital for accurate chain tensioning. Inspect the spring closely, particularly at the hook ends which may have become worn. Here again renew the spring if there is any doubt about its condition. On models with automatic tensioners, this will mean that the tensioner assembly must be renewed complete. Check that the plunger moves smoothly in the tensioner body.

3   Inspect the crankshaft sprockets for signs of wear or chipping of the teeth. Renew as required. The camshaft chain drive sprocket is integral with the crankshaft and therefore if wear is evident the crankshaft must be renewed. This unfortunate state of affairs is not likely to occur until the engine has covered a large mileage, provided that correct chain tension is maintained.

4   Check very carefully the condition of the chain guide and tensioner blades. If they are cracked, worn or damaged in any way they should be renewed. Always ensure, on models with manual tensioners, that only the very latest types of guide and tensioner blade are purchased.

5   The CX500 cam chain tensioner design has given rise to many problems and has been revised on several occasions. The problem lies in the unusual stresses placed on the chain by the camshaft configuration which resulted from the machine's unconventional layout and cylinder angle; the chain is under severe strain at one moment as the camshaft opens or closes the valves, the next moment it is almost too slack.

6   This irregular stress causes the guide blade to vibrate, either snapping itself near the top mounting or shearing off the top mounting bolt. In severe cases the crankcase boss would be damaged or even broken, necessitating crankcase renewal unless some ingenuity was brought to bear. Honda's temporary answer was to have dealers trim the guide blade top mounting down to bare metal in an attempt to provide a stronger mounting. Unfortunately this was not sufficient and various modifications were introduced including several progressively strengthened guide blades, a stronger top mounting bolt and the fitting of an extended guide plate in conjunction with a guide top mounting support plate.

7   Eventually the guide blade became so rigid that the stress was transferred to the tensioner blade, which then started to break. Both guide and tensioner blades were strengthened again.

8   The latest components can be identified initially by their strengthened mounting points and by the addition of rubber bosses on their rear faces to brace them against the rear main bearing cap. In view of the many different types that have been used, correct identification is only possible using the part numbers. If you are in any doubt at all about which items to fit, check first with a Honda Service Agent or with the importers before ordering replacement parts.

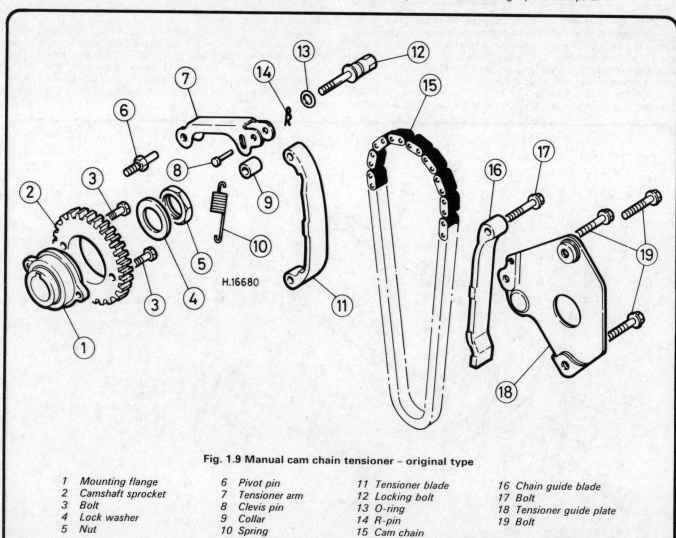

Fig. 1.9 Manual cam chain tensioner – original type

1 Mounting flange	6 Pivot pin	11 Tensioner blade	16 Chain guide blade
2 Camshaft sprocket	7 Tensioner arm	12 Locking bolt	17 Bolt
3 Bolt	8 Clevis pin	13 O-ring	18 Tensioner guide plate
4 Lock washer	9 Collar	14 R-pin	19 Bolt
5 Nut	10 Spring	15 Cam chain	

H.16680

9 Owners of any CX- or (US) GL500 model should treat the cam chain and tensioner components with extreme caution; if there is the slightest doubt about the condition of any of these components, particularly the chain tensioner spring, guide blade or tensioner blade, they should be renewed as a matter of course. Any machine fitted with the old type of guide plate should have it changed immediately for the new type in conjunction with the guide support plate and stronger bolt.

10 Note that at the time of writing the automatic tensioner fitted to CX500 E-C, GL500 D-C and all 650 models seems to be completely reliable. Owners should, however, take note of the above and pay particular attention to checking these components on their machines.

23.6 If early-type unmodified components such as these are found to be fitted, they must be replaced immediately by modified types

23.10 Automatic tensioners are usually reliable, but must still be checked carefully

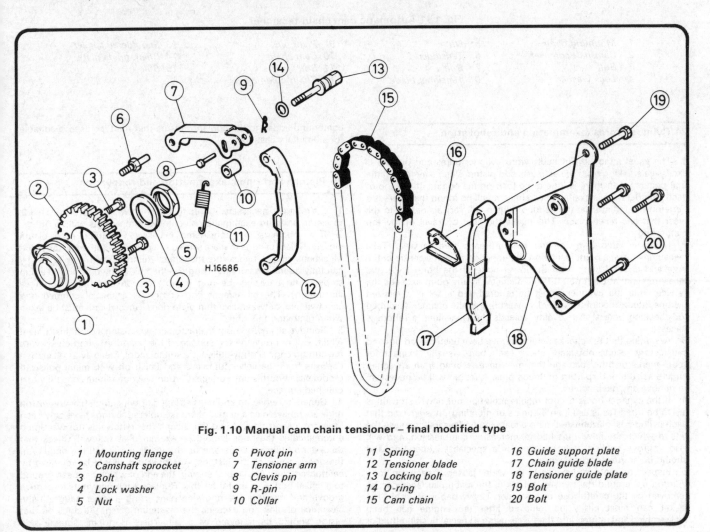

H.16686

Fig. 1.10 Manual cam chain tensioner – final modified type

1 Mounting flange	6 Pivot pin	11 Spring	16 Guide support plate
2 Camshaft sprocket	7 Tensioner arm	12 Tensioner blade	17 Chain guide blade
3 Bolt	8 Clevis pin	13 Locking bolt	18 Tensioner guide plate
4 Lock washer	9 R-pin	14 O-ring	19 Bolt
5 Nut	10 Collar	15 Cam chain	20 Bolt

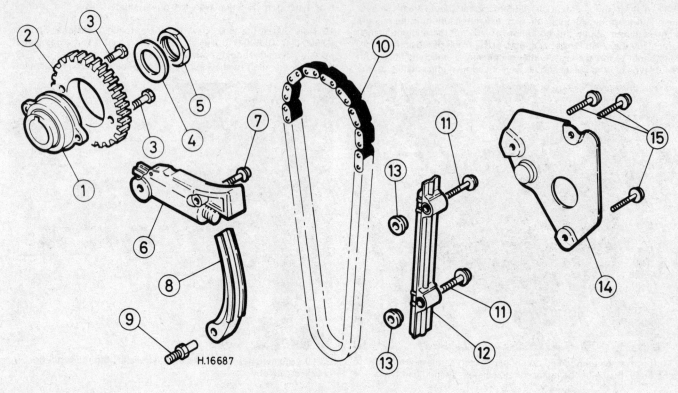

H.16687

Fig. 1.11 Automatic cam chain tensioner

1 Mounting flange	5 Nut	9 Pivot pin	13 Shouldered spacer
2 Camshaft sprocket	6 Tensioner	10 Cam chain	14 Chain guide plate
3 Bolt	7 Bolt	11 Bolt	15 Bolt
4 Lock washer	8 Tensioner blade	12 Chain guide blade	

24 Cylinder bores: examination and renovation

1　The usual indication of badly worn cylinder bores and pistons is excessive smoking from the exhausts and piston 'slap', a metallic rattle that occurs when there is little or no load on the engine. If the top of each cylinder bore is examined carefully, it will be found that there is a ridge on the thrust side, the depth of which will vary according to the wear that has taken place. This marks the limit of travel of the top piston ring.

2　Using an internal micrometer, measure each bore for wear. Take measurements at a point just below the upper ridge, at the centre of the bore and about 1 inch from the lower edge of the bore. Take two measurements at each point. If the diameter at any point exceeds the service limit, the cylinders must be rebored and a set of oversized pistons fitted. Check for ovality by measuring in the manner described for checking bore size. If ovality exceeds the wear limit, a rebore is necessary.

3　Assuming that the checks detailed above have been carried out and satisfactory results obtained, check each bore visually. Inspect for score marks or other damage that may have resulted from an earlier engine seizure or displaced gudgeon pins. A rebore will be necessary to remove any deep scores.

4　If the cylinder bores do not require attention but new piston rings are to be fitted (as is usual after engine dismantling), it is advised that each cylinder is 'glazebusted' before reassembly. This will increase the rate at which the new rings bed in, increasing compression, and will also improve cooling. 'Glazebusting' is a specialised operation and should be entrusted to an expert.

5　Inspect the coolant cavities in the water jacket of each cylinder. 'Furring up' around the transfer orifices in the mating surfaces can be removed by the careful use of a scraper. Heavy deposits within the jacket can most easily be removed after the engine has been reassembled and refitted into the frame, using a special flushing fluid or compound added to the coolant. Ensure that the type used is suitable for aluminium castings.

25 Pistons and rings: examination and renovation

1　If a rebore is necessary, ignore this Section where reference is made to piston and ring examination since new components will be fitted.

2　If a rebore is not considered necessary, examine each piston closely. Reject pistons that are scored or badly discoloured as the result of exhaust gases bypassing the rings. Remove each ring either by carefully opening each ring using the thumbs, or by placing three thin strips of tin between the ring being removed and the piston (see illustration). The oil scraper ring comprises a special crimped ring 'sandwiched' between two thin plain rings. Special care must be taken when removing this ring.

3　Remove all carbon from the piston crowns using a blunt instrument which will not damage the surface of the piston. A wood chisel with the cutting edge slightly dulled is a suitable tool. Clean away all carbon deposits from the valve cutaways and finish off with metal polish to produce a smooth shiny surface. Carbon will not adhere so readily to a polished surface.

4　Generally, when an engine is stripped down completely, the piston rings are renewed as a matter of course unless the rings have only been fitted for a short time. If ring life is such that renewal is not warranted automatically, the rings should be examined as follows. Check that there is no build-up of carbon on the inside surface of the rings or in the ring grooves of the pistons. Any build-up should be removed by careful scraping. An old broken ring, the end of which has been ground to a chisel profile, is useful for this. Replace each ring in its respective groove and measure the ring side play, using a feeler gauge. If the clearance on any ring exceeds the maximum given, the rings on that piston should be renewed as a set. There is no measurable side clearance on the oil ring as the crimped ring spring loads the plain rings.

5 Place each ring into the cylinder bore separately and measure the end gap. Push the ring down from the top of the bore, using the piston skirt, so that the ring remains square to the bore and is positioned about 1¹/₂ inches from the top. If the end gap exceeds the wear limit on any ring, the rings should be renewed as a set.

6 Check the outside diameter of each piston at the skirt by taking a measurement at 90° to the gudgeon pin at a point 7 – 10 mm (0.28 – 0.40 in) above the base of the piston skirt. If the piston is worn to below the service limit it must be renewed. Place each piston in its cylinder bore and using a feeler gauge, check that the clearance is within the specified limits.

7 The piston crowns will show whether the engine has been rebored on some previous occasion. All oversize pistons have the rebore size stamped on the crown. This information is essential when ordering replacement piston rings or in reboring.

8 The letter stamped on each piston ring indicates the top of the ring. It is essential that each ring is replaced so that it is the correct way up.

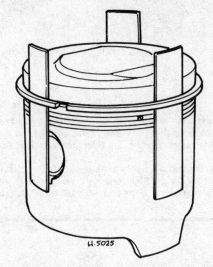

Fig. 1.12 Method of removing gummed piston rings

---

26 Connecting rods and big-end bearings: examination and renovation

1 Examine the connecting rods for signs of cracking or distortion, renewing any rod which is not in perfect condition. Check the connecting rod side clearance, using feeler gauges. If the clearance exceeds the service limit specified it will be necessary to renew the worn components. Connecting rod distortion, both bending and twisting, can only be measured using a great deal of special equipment and should therefore be checked only by an expert; otherwise the rods should be renewed if there is any doubt about their condition.

2 If the necessary equipment is available, the condition of the small-end assemblies can be checked by direct measurement, referring to the tolerances given in the Specifications Section of this Chapter. If the equipment is not available, it will suffice to ensure that the bearing surfaces in the connecting rod small-end bearing, in the piston bosses and over the entire gudgeon pin are smooth and unmarked by wear. The gudgeon pin should be a tight press fit in both connecting rod and piston, and there should be no free play discernible when the components are temporarily reassembled. If any wear is found, the component concerned should be renewed.

3 If a connecting rod is renewed at any time, it is essential that it is of the same weight group to minimise vibration. The weight group is indicated by the relevant letter marked on the machine surface across each rod and its cap.

4 Examine closely the big-end bearing shells. The bearing surface should be smooth and of even texture, with no sign of scoring or streaking on its surface. If any shell is in less than perfect condition the complete set should be renewed. In practice, it is advisable to renew the bearing shells during a major overhaul as a precautionary measure.

The shells are relatively cheap and it is false economy to reuse part worn components.

5 The crankshaft journals should be given a close visual examination, paying particular attention where damaged bearing shells were discovered. If the journals are scored or pitted in any way, a new crankshaft will be required. Note that undersized shells are not available, thus precluding re-grinding the crankshaft.

6 The standard connecting rod big-end eye inside diameter is given in the Specifications Section of this Chapter. To allow for manufacturing tolerances all connecting rods are divided into three size ranges; indicated by a number etched in the rod's machined surface. Similarly the standard crankpin diameter is subdivided into three groups; the group of each crankpin is indicated by its adjacent flywheel being marked with a letter painted on its inside edge. The various sizes are given in the Specifications Section of this Chapter.

7 Measure the crankpins using a micrometer, making a written note of each. Note that each crankpin should be checked in several different places and the smallest diameter noted. If significant ovality is found, renew the crankshaft. Compare the readings obtained with those given then mark the flywheel edges to indicate the diameter range. Disregard the existing flywheel marks if necessary. Note that on UK 500 models and US GL500 models two sets of dimensions are given for the crankpin and for the clearance between the shells and crankpin. Although it has not been possible to establish which is correct, both versions appearing with equal frequency in Honda's service manuals and parts catalogues, it is felt that the first version is the more accurate. Calculation reveals that the clearances are precisely correct with the various dimensions specified in this version, whereas in the second it may not always be possible to arrive at the required clearance given the crankpin diameter specified. **Note:** Owners of US CX500, CX500 C and CX500 D models should always refer to the first version as this is the only one specified for their machines.

8 Note that some very early CX500 models (up to engine number 2015993) suffered premature bearing failure because of incorrect shell selection at the factory. This resulted in tighter clearances than those specified; if the machines were carefully run in there was no problem, but enthusiastic use of the throttle soon caused failure. The bearing shell thicknesses were reduced to those specified in this Manual and the selection procedure was modified, again to that specified in this Manual. This procedure is identical for all models (the discrepancy noted above for UK 500 models and US GL500 models is of importance only if the crankpin or bearing clearance is being checked by measurement).

9 If necessary, the big-end eye diameter can be checked, with each connecting rod and cap assembled without shells and the cap nuts tightened to the correct torque setting, using a bore micrometer. It is unlikely that the diameter will have changed from that marked on the assemblies' machined faces because this area is not subjected to mechanical wear.

10 Using the sizings obtained, select the appropriate shell from the table shown below.

Connecting rod mark	Crankshaft mark	Bearing shell
1	A	F Pink
1	B	E Yellow
1	C	D Green
2	A	E Yellow
2	B	D Green
2	C	C Brown
3	A	D Green
3	B	C Brown
3	C	B Black

11 If the existing shells are to be checked for wear with a view to reusing them, this is best carried out using Plastigage to check the bearing shell/crankpin clearance as described in Section 18 of this Chapter; the only alternative is to calculate the clearance after careful measurement of each component, as described above. If the clearance exceeds the specified service limit the shells must be renewed.

12 Finally, carefully examine the condition of the connecting rod bolts and nuts, renewing any that are damaged or worn in the slightest way; these are very highly-stressed components. Note that it is usually considered good practice to renew bolts and nuts of this type whenever they are disturbed although the manufacturer makes no specific recommendation to this effect.

26.3 Letter indicates weight group (should be the same on both rods), number indicates size group

26.6 Inner letters are crankpin codes; outer letters are main bearing codes

27 Crankshaft and main bearings: examination and renovation

1   Examine the main bearings for visual signs of damage as described in the previous Section for the big-end bearings. The main bearing clearances can be determined only by direct measurement, using internal and external micrometers. This is because the bearings are of the bush type, rather than the shell type used for the big-ends. The correct clearances are given in the Specifications at the beginning of this Chapter.

2   Each main bearing is a tight interference fit in its housing. Although in practice it may be possible to drive an old bearing from position, when fitting a new bearing, the bearing must be pressed into position

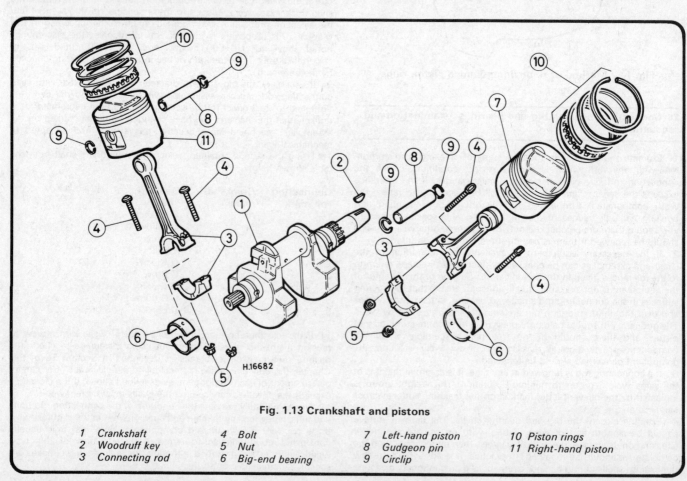

H.16682

Fig. 1.13 Crankshaft and pistons

1	Crankshaft	4	Bolt	7	Left-hand piston	10	Piston rings
2	Woodruff key	5	Nut	8	Gudgeon pin	11	Right-hand piston
3	Connecting rod	6	Big-end bearing	9	Circlip		

using a suitable fly-press and a correctly sized mandrel. Driving a new main bearing into place is to invite damage to the bearing itself and also the bearing housing. Because of the problems involved it is strongly recommended that the crankcase and crankshaft are returned to a Honda Service Agent for the work to be carried out.

3 If a fly-press and the necessary expertise for its use are available, selection of new main bearings should be carried out as follows.

4 Press out the main bearings from their respective housings so that the inside diameter of each housing may be determined and recorded. Note the main bearing journal code letters which are painted on the adjacent flywheel outside edge. Select the correct new main bearing bushes from the table shown below, noting that the housing inside diameters are divided into larger and smaller size groups.

Housing	Crankshaft mark	Bearing bush
Small group	A	C Brown
Small group	B	B Black
Large group	A	B Black
Large group	B	A Blue

5 Before pressing the new bearings into position in the housings, coat the outer surface of each with engine oil This will aid insertion. Ensure that the location tag projecting from each bearing aligns with the groove in the housing, before the pressing operation is started. The main bearings provided are sized to allow for the contraction which will result as they are pressed into the housings. In spite of this, the clearances **must be checked** to ensure all is well, before engine reassembly is commenced.

6 No official figures are available for service limits for crankshaft journal ovality or tapering. If wear is found the advice of an expert should be sought.

7 Check the security of the ball bearings which are used to plug the oilways in the crankshaft. They occasionally work loose, causing lubrication problems and subsequently bearing failure. A loose ball can be carefully caulked back into place after the surfaces have been cleaned and a locking agent applied.

8 The crankshaft should be cleaned using copious quantities of solvent and compressed air. Be very careful to check that all oilways are completely free from dirt and other foreign matter especially if the bearings have shown signs of failure due to lack of oil.

9 Examine the crankshaft closely. Any obvious signs of damage such as marked bearing surfaces, damaged threads, worn tapers and damaged chain sprockets will mean that it must be renewed. There are, however, light engineering firms advertising in the motorcycle press who can undertake major crankshaft repairs; in view of the expense of a new component it is worth trying such firms provided they are competent.

28 Clutch assembly and primary drive pinion: examination and renovation

1 Carefully clean all the clutch plates. Check the plain plate warpage by placing each plate on a flat surface and measuring with a feeler gauge. If the plates shown signs of 'blueing' or bad scoring they should be renewed.

2 Measure the thickness of each friction plate with a vernier gauge. The correct dimensions are given in the Specifications at the beginning of the Chapter. Note that one friction plate is slightly thicker than the remainder. This disc may be identified by the wider tangs. It is probable that all the friction plates will wear at a similar rate and should therefore be renewed as a complete set. Check that the tongues that locate in the outer drum and clutch centre are not worn; any serious burrs or indentations mean renewal. Very small burrs can be removed with a stone or a fine cut file. **Note:** Do not remove too much metal since the tongues will then be of unequal width and spacings, thus they will not take up the drive evenly and will wear the aluminium alloy clutch outer drum and centre.

3 Inspect the wave washer that lies between the two plates in the special 'sandwich' plate. If the washer is found to be breaking up, or has taken a permanent set, the plate should be renewed.

4 Examine the clutch outer drum and centre grooves, removing any small burrs with a blind edge file. Burrs left in the two components will prevent the clutch from disengaging smoothly, causing noisy gear selection.

5 Measure the free length of each clutch spring. If any spring has settled to the specified service limit or less, all the springs must be renewed as a set.

6 Measure the bearing surfaces of the clutch outer drum and centre sleeve, renewing any component which has worn to the specified limit or beyond, or shows any sign of wear or damage.

7 Inspect the clutch lifting mechanism in the clutch cover and also the ball bearing which fits into the centre of the clutch. Wear of the operating shaft, thrust piece and ball bearing will be self-evident; renew any worn or damaged components.

8 Examine the primary drive pinion and the larger pinion on the clutch outer drum for wear. Look for chipped, hooked or broken teeth. Wear can be expected to be low until a high mileage has been reached. Broken teeth may have resulted from a portion of displaced metal from another part of the engine finding its way between the meshing teeth of the pinion. Check the anti-backlash (narrower) pinion to ensure that the spring loading is still effective. If any wear or damage is found, the component must be renewed.

29 Gearbox components: examination and renovation

1 Before inspection of the gearbox components can take place the two clusters and the selector mechanism must be removed from the gearbox end cover. Commence by withdrawing the shaft which supports the two selector forks, and then lift the forks away. On 500 models only, do not lose the fork guide pins; these are a light push fit and may fall from place unnoticed. Repeat the operation with the second shaft holding the single fork. The selector drum may now be lifted from position. The two complete gear shaft assemblies should be removed simultaneously. The input shaft is a light drive fit in the ball bearing through which it passes and should be drifted from place using a rawhide mallet.

2 Examine the gearbox components very carefully, looking for signs of wear, chipped or broken teeth and worn dogs or splines.

3 If a pinion requires renewal, it is probable that the pinion with which it meshes will also be worn or damaged in a similar manner. Both pinions should be renewed at the same time to prevent problems due to uneven wear between a new component and a partially worn one. The pinions can be withdrawn after the removal of the various splined thrust washers and circlips. Carefully note the sequence of gears, washers and circlips to aid replacement. Their correct position is essential.

4 If the fit of any of the pinions on their shafts is suspect, measure the respective shaft diameters and pinion internal diameters and compare them with the dimensions specified. Renew where necessary.

5 Roll each selector fork shaft on a flat surface to check for straightness. If all is in order measure the shaft diameters and the selector fork bores and compare the results with those given. Also check that the claw end thickness of the forks is well within the recommended limits.

6 If the machine tends to jump out of gear it is most probably due to worn dogs on the gear pinions. Difficulty in selecting gears is usually caused by bent selector forks or worn fork guide channels in the selector drum. Visual indication of wear in the channels is usually most evident at abrupt changes of curvature.

7 Wear in the gear selector mechanism can only be rectified by direct replacement of the parts concerned. This applies equally to those components on the outside of the gearbox, such as the stop arms. If jumping out of gear or over-selection has been experienced, renew the stop arm and change the claw springs as a first step in eliminating the causes. If the pins in the end of the selector drum become worn they too can be renewed after removal of the end plate and neutral indicator switch plate. The cap is held by a bolt; apply locking fluid to the threads on reassembly.

8 Check the condition of the selector claw arm return spring. If the spring ears have become worn at the points where they rub against the stop bolt and arm, the spring should be renewed.

30 Final output shaft: examination and renovation

1 The shock absorber system comprises two cam-faced bosses loaded by a helical spring, which is retained under tension by an end cap and two split collets. Removal of the spring requires the use of a

special tool by which the spring may be compressed and the retaining collets displaced. This system of spring retention is also used on most types of rear suspension unit. If the special tool is not available, the type of clamp used for suspension unit dismantling could probably be used. Alternatively, a pair of scarf joint clamps as used by carpenter joiners can also be utilised.

2   Compress the main spring sufficiently to release the retaining collets and then release the tension slowly. Do not over-compress the spring as it may be damaged permanently. Examine the cam faces for excessive indentation or flaking. Although alteration of the cam profile due to wear will have little effect on the performance of the shock absorber it is probably wise to renew the two mating components if wear has progressed to a point where the case hardening has been worn through.

3   Measure the free length of the helical spring, renewing if it has settled to the specified service limit or less.

31 Starter clutch and drive gears: examination and renovation

1   The starter clutch assembly is housed in the rear of the alternator generator rotor. Check the condition of the three engagement rollers and the boss on the starter driven gear with which they engage. If necessary, the complete starter clutch can be removed from the rear of the rotor after unscrewing the three mounting screws. The screws are of the socket type but have a special star shaped recess in place of the more usual hexagonal recess. A special 'Torx' bit, (size T40), is therefore required for their removal.

2   Examine the condition of the starter drive gear and the driven gear. In the unlikely event of worn components, renewal is the only method of renovation. Note that Honda advise using new Torx screws to secure the starter clutch body to the alternator rotor; apply a locking compound to the threads and underside of the screw heads.

32 Engine reassembly: general

1   Before reassembly of the engine/gearbox unit is commenced, the various component parts should be cleaned thoroughly and placed on a sheet of clean paper, close to the working area.

2   Make sure all traces of old gaskets have been removed and that the mating surfaces are clean and undamaged. Great care should be taken when removing old gasket compound not to damage the mating surface. Most gasket compounds can be softened using a suitable solvent such as methylated spirits, acetone or cellulose thinner. The type of solvent required will depend on the type of compound used. Gasket compound of the non-hardening type can be removed using a soft brass-wire brush of the type used for cleaning suede shoes. A considerable amount of scrubbing can take place without fear of harming the mating surfaces. Some difficulty may be encountered when attempting to remove gaskets of the self-vulcanising type, the use of which is becoming widespread, particularly as cylinder head and base gaskets. The gasket should be pared from the mating surface using a scalpel or a small chisel with a finely honed edge. Do not, however, resort to scraping with a sharp instrument unless necessary.

3   Gather together all the necessary tools and have available an oil can filled with clean engine oil. Make sure that all new gaskets and oil seals are to hand, also all replacement parts required. Nothing is more frustrating than having to stop in the middle of a reassembly sequence because a vital gasket or replacement has been overlooked. As a general rule each moving engine component should be lubricated thoroughly as it is fitted into position.

4   Make sure that the reassembly area is clean and that there is adequate working space. Refer to the torque and clearance setting wherever they are given. Many of the smaller bolts are easily sheared if overtightened. Always use the correct size screwdriver bit for the cross-head screws and never an ordinary screwdriver or punch. If the existing screws show evidence of maltreatment in the past, it is advisable to renew them as a complete set.

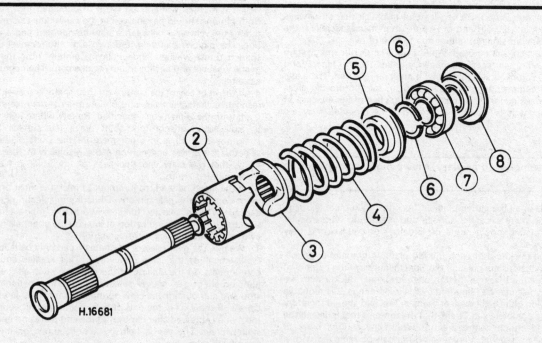

Fig. 1.14 Final output shaft

1 *Output shaft*	3 *Shock absorber cam*	5 *Spring collar*	7 *Bearing*
2 *Shock absorber cam*	4 *Spring*	6 *Collets*	8 *Oil seal*

33 Reassembling the engine/gearbox unit: refitting the crankshaft

1 Place the main crankcase on the workbench in an upright position. Lubricate thoroughly with clean engine oil the forward main bearing and journal. It is recommended that molybdenum disulphide grease be used in place of ordinary oil if new bearings have been fitted. On 650 models, ensure that the thrust washers are refitted correctly on each side of the front main bearing; note that the cutouts must engage with the crankcase lugs and that the oil grooves must face forwards on the front washer, to the rear on the rear washer.

2 Wrap the camshaft drive sprocket on the rear of the crankshaft with tape; this will prevent damage to the rear main bearing surface when the housing is installed. Insert the crankshaft through the rear of the casing, keeping it square so that the forward main bearing journal enters the bearing easily. Do not displace the thrust washers (650 models only).

3 Lubricate the rear main bearing journal with engine oil or molybdenum disulphide grease. Install the large hollow dowel and a new O-ring in the recess in the rear of the main engine casing. Slip the main bearing housing over the crankshaft end and rotate it so that it is roughly aligned in the correct position. The housing must be drifted into position very gently, using a rawhide mallet. Work round the edge of the housing so that it moves inwards squarely, and no tying occurs. To ensure that the bolt holes line up exactly when the housing is fully home, it is suggested that two or three extra long bolts are selected with the requisite thread, which may be temporarily screwed into the casing and used as guides for the housing.

4 When the housing is correctly positioned insert the retaining bolts through the flange. The bolts should be tightened evenly, in a diagonal sequence to 2.0 – 2.4 kgf m (14.5 – 17 lbf ft). During the tightening sequence check that the crankshaft is able to rotate freely.

5 The primary drive gear pinion may be refitted and secured at this stage, if required. See Section 36.

33.2 Lubricate bearings and insert crankshaft into front main bearing

33.3a Protect rear main bearing by wrapping camshaft drive sprocket with tape ...

33.3b ... and do not forget to refit locating dowel and new O-ring ...

33.3c ... before refitting rear main bearing housing

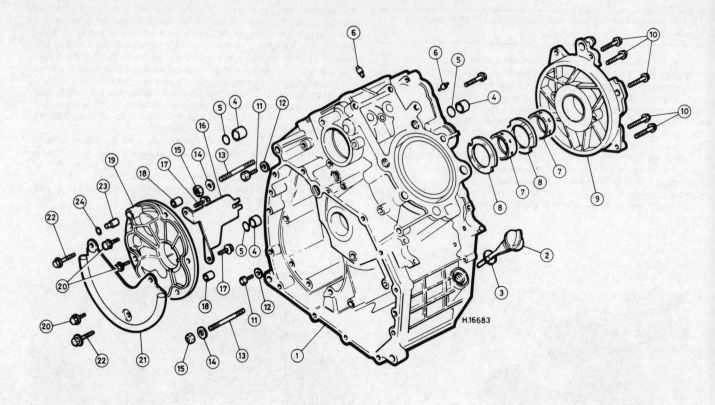

Fig. 1.15 Cylinder block

1 Cylinder block	9 Crankshaft rear main
2 Dip stick	bearing cap
3 O-ring	10 Bolt
4 Dowel	11 Bolt
5 O-ring	12 Sealing washer
6 Oil jet	13 Stud
7 Main bearing	14 Washer*
8 Crankshaft thrust washer –	15 Nut
650 only	16 Oil baffle plate*

17 Bolt*
18 Dowel
19 Gearbox front cover
20 Bolt
21 Oil separator plate
22 Bolt
23 Oil jet
24 O-ring

* 500 model only

34 Reassembling the engine/gearbox unit: refitting the pistons and connecting rods

1   Select one connecting rod and its original piston or a new piston of the correct type. Note that the two pistons are not interchangeable and are marked L or R on the crown to aid identification. Place the piston over the end of the rod so that the identification mark on the crown is on the opposite side of the rod to the small oil hole drilled in the rod flank. Insert the gudgeon pin and press it home. Clean engine oil or molybdenum disulphide grease should be applied to the small-end eye before the pin is inserted. Fit two new circlips ensuring that both locate correctly with the grooves in the piston bosses.
2   If not already in place, fit the piston rings as follows. Working from the bottom of the piston fit the first oil control ring side rail, the corrugated expander ring and the final side rail. The end gaps of each component in the oil control ring assembly should be spaced about $3/4$ inch apart. Now working from the top of the piston fit the 2nd compression ring and top compression ring. The upper rings must be fitted so that the letter mark, which is stamped on one side of each ring,

faces upwards. This is important to retain maximum compression. When fitting any piston into its cylinder bore, the rings should be arranged so that the end gaps are approximately 120° apart. Repeat the assembly procedure with the second piston and connecting rod.
3   Fit the big-end shells to the connecting rods and big-end caps. Ensure that the tongues locate correctly and the oil holes align with the oil holes in the rods. The bearings and journals should be lubricated thoroughly in the normal way. Lubricate both cylinder bores with a copious quantity of clean engine oil and then insert the right-hand piston and rod into the bore. A piston ring clamp should be used to compress the rings as they enter the bore. Fitting the piston without a clamp is possible but the risk of damage to the rings is great as there is no chamfered lead-in. The pistons should be fitted so that the oil hole in each connecting rod faces the top edge of the engine.
4   Engage the connecting rod lower end with the big-end journal and fit the cap. Ensure that the cap is fitted correctly so that the matching marks on the machined faces align. Fit the big-end nuts and tighten them evenly to the specified torque setting. Check that the crankshaft still rotates freely.
5   Repeat the assembly process for the left-hand piston.

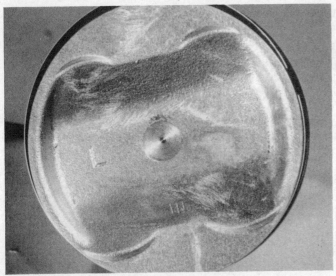

34.1 Each piston is identified by letter on crown. Ensure that they are correctly positioned

34.3a Tang on shell must engage with slot in cap or connecting rod

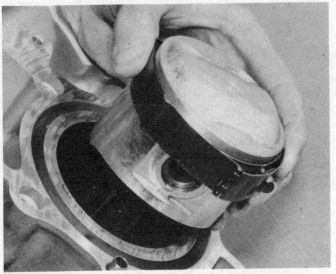

34.3b Ensure connecting rods and pistons are correctly aligned on refitting – note use of ring clamp

34.4a Ensure cap is correctly aligned with connecting rod

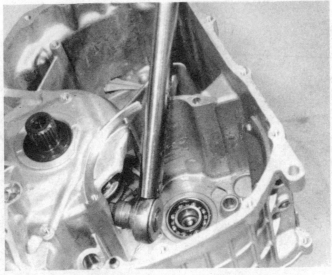

34.4b Tighten retaining nuts to specified torque setting

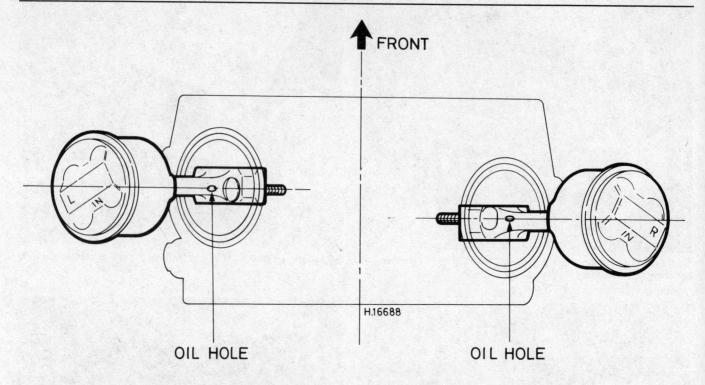

FRONT

OIL HOLE

H.16688

OIL HOLE

Fig. 1.16 Positioning guide for pistons and connecting rods

35 Reassembling the engine/gearbox unit: refitting the gearbox components

1 If the input or output shaft assemblies were dismantled for inspection they must be reassembled before refitting. Refer to the accompanying illustrations for the relative positions of the gears, washers and circlips. It is very important that the components are fitted in the correct order and the correct way round, or continual gearchange problems will arise. When the two shafts are complete, mesh them together and insert them into the end cover simultaneously, so that the output shaft enters its bearing and the input shaft passes through its bearing. Using a rawhide mallet, gently drift the input shaft fully home so that the gears are in line with each other.

2 Install the selector drum and then position the two output shaft selector forks so that they engage with the channels on the sliding pinions and the selector drum channels. Insert the selector fork shaft to locate the forks. Repeat the operation with the remaining fork and shaft.

3 Rotate the selector drum so that neutral gear is selected. The completed assembly may now be inserted into the gearbox chamber. Some care is required when carrying out this operation because the rear end of the selector drum must pass through the profiled aperture in the casing and the various shafts must engage with their bearings. With the components positioned satisfactorily push the end cover home to engage with the two hollow dowels.

4 Insert all the end cover bolts except those which secure the oil separator plate to the cover. Tighten the bolts evenly, in a diagonal sequence. The separatpr plate can now be fitted and the bolts tightened to the specified torque settings.

35.1 Fit both gearbox shafts simultaneously to gearbox cover

35.2a Install the selector drum, then refit ...

35.2b ... the two output shaft selector forks and their shaft ...

35.2c ... followed by the single input shaft fork and shaft

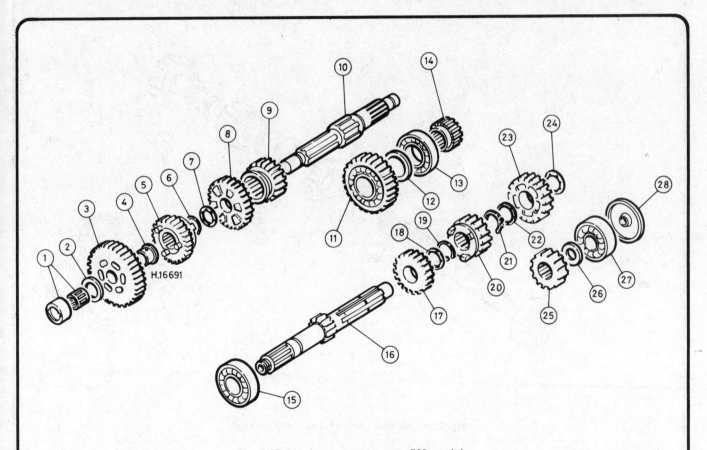

Fig. 1.17 Gearbox components – 500 models

1 Needle roller bearing	11 Output shaft 2nd gear	20 Input shaft 3rd gear
2 Thrust washer	12 Thrust washer	21 Circlip
3 Output shaft 1st gear	13 Bearing	22 Splined thrust washer
4 Bush	14 Output shaft collar	23 Input shaft 5th gear
5 Output shaft 4th gear	15 Bearing	24 Splined thrust washer
6 Circlip	16 Input shaft	25 Input shaft 2nd gear
7 Splined thrust washer	17 Input shaft 4th gear	26 Thrust washer
8 Output shaft 3rd gear	18 Splined thrust washer	27 Bearing
9 Output shaft 5th gear	19 Circlip	28 Oil guide plate
10 Output shaft		

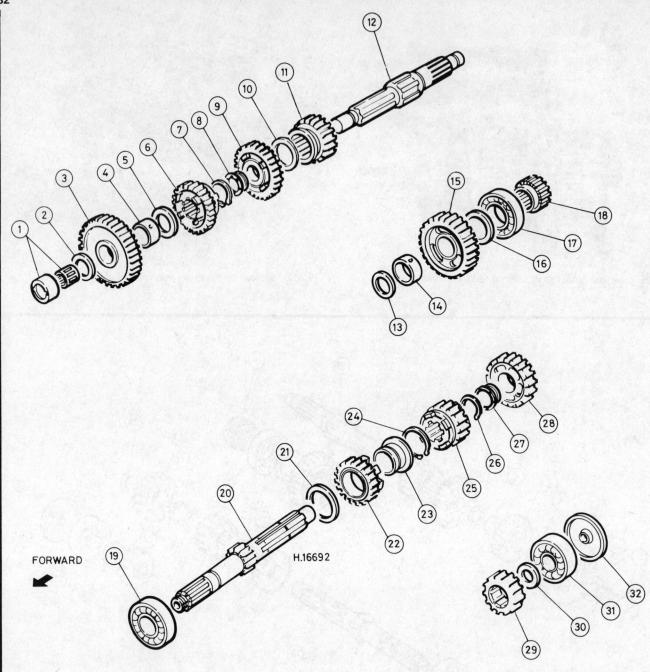

FORWARD

H.16692

Fig. 1.18 Gearbox components – 650 models

1	Needle roller bearing	9	Output shaft 3rd gear	17	Bearing	25	Input shaft 3rd gear
2	Thrust washer	10	Thrust washer	18	Output shaft collar	26	Circlip
3	Output shaft 1st gear	11	Output shaft 5th gear	19	Bearing	27	Bush
4	Bush	12	Output shaft	20	Input shaft	28	Input shaft 5th gear
5	Thrust washer	13	Thrust washer	21	Thrust washer	29	Input shaft 2nd gear
6	Output shaft 4th gear	14	Bush	22	Input shaft 4th gear	30	Thrust washer
7	Circlip	15	Output shaft 2nd gear	23	Bush	31	Bearing
8	Bush	16	Thrust washer	24	Circlip	32	Oil guide plate

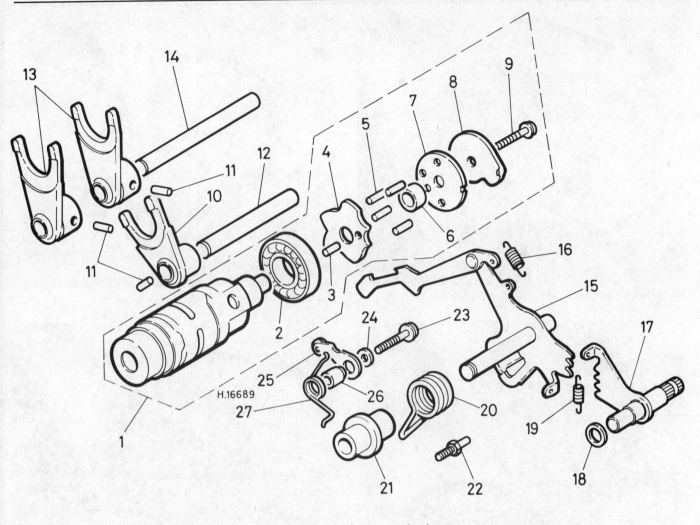

Fig. 1.19 Gear selector mechanism

1 Selector drum assembly	10 Selector fork	19 Spring
2 Bearing	11 Guide pin*	20 Return spring
3 Dowel pin	12 Selector fork shaft	21 Shouldered spacer
4 Selector drum cam	13 Selector fork	22 Spring anchor pin
5 Selector pins	14 Selector fork shaft	23 Bolt
6 Spacer	15 Selector claw arm	24 Washer
7 End plate	16 Spring	25 Stop arm
8 Neutral switch plate	17 Gearchange pedal shaft	26 Collar
9 Bolt	18 Thrust washer	27 Return spring

* Guide pins integral with forks on 650 models

36 Reassembling the engine/gearbox unit: refitting the primary drive pinion and oil pump

1   Refit the primary drive pinion on the splined front end of the crankshaft. On 500 models only, lubricate the auxiliary pinion and side plate and replace them on the shaft. All three components should have been marked during dismantling so that they may be refitted the correct way round. Insert the two drive pins through the side plate to locate with the primary drive pinion. Position the spring plate on the shaft and then slide the oil pump drive sprocket into place to locate the plate. Turn the plate so that two of the internally projecting tongues cover the drive pins. This will prevent the pins from working out of position during service.

2   Fit the drive pinion bolt and tighten it to a torque of 8.0 – 9.5 kgf m

(58 – 69 lbf ft). Prevent the crankshaft from rotating during tightening by using the locking tool fabricated during removal. The locking tool should be left in position until the clutch has been replaced.

3   Check that the oil pump locating dowel is positioned in the oil pump lower mounting hole in the main casing. Mesh the pump drive chain with the oil pump sprocket and offer up the assembly so that the chain may be meshed with the drive sprocket on the crankshaft. Push the oil pump into position so that the lower mounting locates with the dowel and then insert, but do not tighten fully, the mounting screws.

4   Position the oil pump so that the up and down movement of the chain measured in the middle of the upper run is 2.0 – 3.5 mm (0.08 – 0.14 in). Hold the oil pump in this position, tighten the crosshead screw or bolt on the right of the pump, then tighten the two remaining bolts, all to the specified torque setting. Recheck the chain tension.

36.1a Refit the primary drive pinion. Note (500 models only) the two drive pins

36.1b 500 models only – fit the auxiliary pinion ...

36.1c ... followed by the side plate, spring plate and pump drive sprocket

36.1d Place the spring so that the drive pin ends are covered before tightening the retaining bolt

36.4 Adjust oil pump drive chain free play before tightening pump mounting bolts

37 Reassembling the engine/gearbox unit: refitting the clutch and engine front cover

1   Thoroughly lubricate all clutch components before refitting. If new friction plates are to be fitted, soak them in clean engine oil before reassembly.

2   On early 500 mod ls, fit the thick thrust washer, followed by the centre sleeve and then the clutch outer drum, taking care to align the teeth of the anti-backlash gear so that the outer drum can be pushed fully into place. On CX500 C 1982, CX500 E-C and all GL500 models fit first the spacer to the input shaft, followed by the first thrust washer, the centre sleeve, the outer drum (aligning the teeth) and the second thrust washer. On all 650 models, fit first the thrust washer, then the outer drum; align its teeth by pressing the outer drum against those of the primary drive pinion, then push the centre sleeve into position with its shouldered end forwards.

3   On all models, it is easiest to reassemble the clutch by building up the plates as a unit. Use the notes made on dismantling to ensure that the plates are all fitted in their correct and original positions. If no notes were made refer to the accompanying illustrations and proceed as follows.

4   There are five distinct types of clutch plate fitted. The spring loaded double metal plate is easy to identify, as are the identical plain metal plates. Of the friction plates one is thicker, has wider projecting tongues and radial grooves; this is the Type B plate. Of the remaining, reasonably similar friction plates, one should be noticeably different by reason of the grooves; friction plates have either equally-spaced radial grooves or approximately eleven cross-grooves giving an unequal spacing at first glance. The 'different' plate is to be fitted against the pressure plate and is therefore assembled last. Unfortunately different models appear to have varying combinations of the same plates; therefore more precise identification is not possible. There should be one of one type of plate and five (500 models) or six (650 models) of the other.

5   Place the clutch centre forward face down on the work surface and fit the Type B friction plate, followed by a plain metal plate and one of the more common friction plates. Fit the double spring-loaded plate then alternately fit friction plates and metal plates to finish with the individual friction plate. Lastly, fit the pressure plate, passing the spring posts through the holes in the clutch centre.

6   Align the projecting tongues of the friction plates and insert the assembly into the clutch outer drum, rotating it to align the friction plate tongues with the outer drum slots and the clutch centre splines with those of the input shaft.

7   Fit the lock washer with the concave face (marked 'OUTSIDE') forwards, then fit the nut with its chamfered face to the rear (flat surface forwards). Lock the clutch by the method used on dismantling and tighten the nut to a torque setting of 8.0 – 10.0 kgf m (58 – 72 lbf ft).

8   Refit the clutch springs followed by the lifter plate and the spring bolts, which should be tightened fully. Install the ball bearing in the plate.

9   Remove the primary drive pinion locking tool. Check that the two small dowels are fitted to the mating surface at the front of the main engine casing. Insert the two large hollow dowels (oil transfer) into the casing recesses and fit a new O-ring to each. Install the oil jet and a new O-ring in the recess in the gearbox end cover. The jet should be fitted with the smaller diameter outermost. A new gasket can now be fitted to the casing mating surface and the engine front cover placed in position. On 650 models, align the projecting nozzle on the oil feed pipe with the crankcase oilway. Fit the casing bolts, noting that the bolt immediately above the oil pressure switch secures a cable clip. Tighten the bolts evenly, in a diagonal sequence.

10 Check that the clutch thrust piece is in place in the outer cover. The thrust piece should be lubricated with oil or grease. Fit the cover with a new gasket and tighten the securing bolts.

11 Apply a jointing compound to the threads of the oil pressure warning switch and then fit the switch. The switch should not be overtightened. Reconnect the switch lead and secure it by means of the cable clip.

37.2a Early 500 models – refit thrust washer and centre sleeve ...

37.2b ... then install clutch outer drum

37.5a Invert clutch centre and fit type B friction plate ...

37.5b ... followed by plain metal plate ...

37.5c ... one of the more numerous friction plates ...

37.5d ... and double spring-loaded plate

37.6 Align friction plate tongues with outer drum slots and clutch centre splines with input shaft to install assembly

37.7a Lock washer is of Belville type; fit as marked

37.7b Lock clutch centre to tighten retaining nut to specified torque setting

37.8 Refit clutch springs, bolts and lifter components

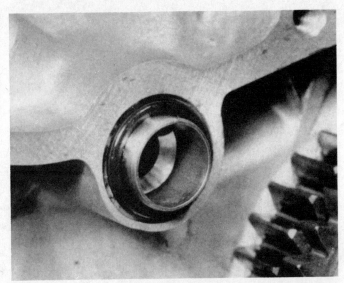

37.9a Do not forget to refit oil transfer dowels ...

37.9b ... and new O-rings to crankcase gasket surface and oil pump body

37.9c Fit oil metering jet as shown and renew sealing O-ring

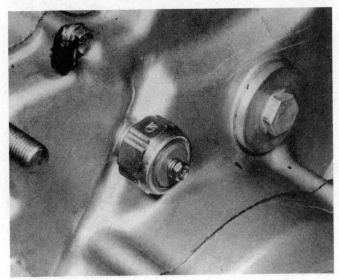

37.11 Do not overtighten oil pressure switch – front cover may crack

38 Reassembling the engine/gearbox unit: refitting the camshaft and cam followers

1 Lubricate the camshaft journals with molybdenum disulphide grease or engine oil. Slide the thrust washer onto the camshaft rear journal and then insert the shaft into the camshaft tunnel from the front of the engine.

2 Place a new gasket on the camshaft bearing housing and install the large hollow dowel (oil transfer) and a new O-ring in the engine casing. Lubricate the lip of the oil seal with grease (500 models only). This will help protect it when the seal passes over the shaft. Push the bearing housing over the end of the shaft and rotate the housing to bring the bolt holes in line with those in the casing. The housing should be drifted into place with a rawhide mallet, working round the periphery of the housing to prevent the boss tying in the main engine casing. During this procedure the camshaft should be rotated slowly to allow the tachometer worm gear on the camshaft to mesh correctly

with the driven pinion on the tachometer drive shaft. When the housing is fully home insert and tighten the screws. Check that the camshaft rotates freely.

3 Each pair of cam followers and spindles should be considered as a matched set and as such should be fitted separately. Position the two cam followers in the casing together with their end float control springs. As can be seen from the accompanying illustration, each follower is offset slightly about the centre line. The followers must be positioned so that the pushrod cup on each projects inwards towards the spindle support lug. The spring on each follower must be fitted on the centre lug side.

4 Insert the cam follower spindle from the rear of the engine with the slotted end outermost. Push the spindle fully home, aligning the followers as required so that they are located by the spindle. Place a screwdriver in the slotted spindle head and rotate the spindle until the relieved portion of the spindle within the central support lug is in line with the bolt hole. Insert and tighten the bolt.

5 Repeat the assembly procedure for the second cam follower set.

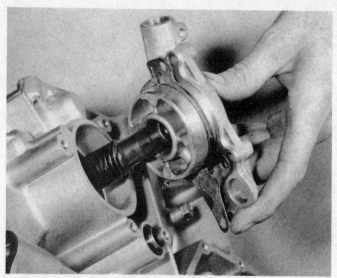

38.2 Always renew gasket and sealing O-ring when refitting camshaft front housing

38.4a Position cam followers and springs and insert spindle – slotted end to rear

38.4b Rotate spindle so that locking bolt can be fitted

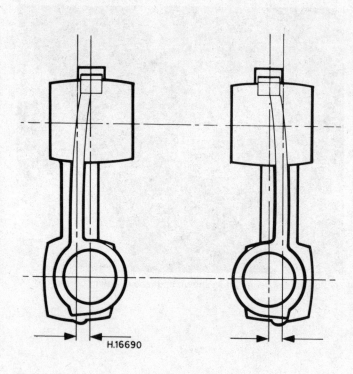

H.16690

Fig. 1.20 Cam follower off-set

39 Reassembling the engine/gearbox unit: refitting the gearchange external mechanism

1   Assemble the components that make up the detent stopper arm sub-assembly. The correct sequence is as follows; bolt, washer, stopper arm, shouldered collar and spring. Ensure that the narrow end of the collar engages with the arm. Install the unit in the casing so that the roller end of the arm locates wth the pins in the selector drum end. Tighten the bolt and check that the arm is free to pivot about the bolt. Grasp the inner end of the return spring with a pair of pliers and tension it in an anti-clockwise direction until the turned end can be engaged with the hole in the gearbox wall.

2   Check that the return spring and collar are correctly positioned on the spindle projecting from the selector claw arm. The spring must be fitted as shown in the accompanying photograph. Push the claw arm into position, simultaneously depressing the arm so that it clears the selector pins. Correctly positioned, the ears of the return spring should lie either side of the stop bolt screwed into the casing.

3   Place the thrust washer on the spindle end projecting from the rear

of the gearchange pedal shaft. Install the shaft in the crankcase so that the quadrant teeth mesh with those of the selector claw arm as can be seen in the photographs. Reconnect the tensioning spring which interconnects with the pedal shaft and selector claw arm.

40 Reassembling the engine/gearbox unit: refitting the cam chain and sprockets, and timing the valves

1   Check that the drive pin is correctly in place in the boss on the end of the camshaft. Place the camshaft sprocket flange over the shaft so

that the recess in the rear of the flange engages with the drive pin. Fit the belville washer and nut to secure the flange. The concave face of the washer is marked 'OUTSIDE'; the washer should be fitted accordingly. To prevent the camshaft rotating when tightening the nut, pass a suitable length of bar under the flange right-hand lug so that it engages in the recess in the casing. Tighten the nut to a torque setting of 8.0 – 10.0 kgf m (58 – 72 lbf ft).

2 The relative positions of the camshaft and crankshaft should now be set as follows before the cam chain is refitted: position the camshaft with the drive pin at about 2 o'clock. Adjust the position of the shaft until the two holes in the flange are exactly in line with the index marks projecting from the casing adjacent to the coolant transfer passage apertures. Rotate the crankshaft anti-clockwise until the left-hand piston is at TDC. In this position the keyway (or the key, if already fitted) on the crankshaft will be in line with the index mark cast into the top left-hand edge of the main bearing housing.

3 Mesh the HY-VO chain onto the camshaft sprocket and then, without allowing movement of the crankshaft or camshaft, fit the sprocket to the flange and mesh the lower run of the chain with the drive sprocket on the crankshaft. Insert the camshaft sprocket bolts and screw them in finger-tight. Check that the valve timing is still correct, referring to the accompanying diagram. When carrying out this check, push the chain inwards on the right-hand run to take up all slack. If the timing is correct, remove the two sprocket bolts and apply thread locking compound to their threads. Refit the bolts, tightening them to a torque setting of 1.6 – 2.0 kgf m (11.5 – 14.5 lbf ft).

Manual tensioner

4 Manoeuvre the guide blade into position ensuring that it engages correctly in its bottom mounting and that its rubber boss is braced against the bearing housing lug.

5 Connect the tensioner spring to the cam chain tensioner assembly. The shorter straight extension on the spring should be hooked onto the tensioner arm. Slide the tensioner onto the pivot pin projecting from the casing and engage the lower end of the blade with the pin to the right of the crankshaft. Install the tensioner adjuster bolt ensuring that the collar is not omitted. The collar should be positioned between the sides of the tensioner arm, to be located by the bolt as it is inserted. Do not tighten the bolt fully at this stage.

6 Fit the guide plate over the crankshaft and insert the support plate under the guide blade top mounting so that it fits over the crankcase boss and butts against the wall protrusion (ie pointing to 7 o'clock). Align the guide plate and fit the top mounting bolt, followed by the remaining three bolts. Hook the tensioner spring on to the guide plate and tighten the bolts to a torque setting of 0.8 – 1.2 kgf m (6 – 9 lbf ft).

7 Rotate the crankshaft anti-clockwise until the piston of the left-hand cylinder is at TDC, with the cam lobes for that cylinder away from the cam follower feet. Slacken the tensioner adjuster bolt to allow free movement of the arm and then tighten the bolt fully. The cam chain is now tensioned correctly. Place a new O-ring on the grooved bolt head.

Automatic tensioner

8 Refitting the two shouldered spacers so that their shoulders fit against the crankcase, manoeuvre the guide blade into position and tighten securely the two bolts. Fit the tensioner blade over its pivot.

9 Place the tensioner over its locating dowel then refit its mounting bolt and tighten it to a torque setting of 1.8 – 2.5 kgf m (13 – 18 lbf ft). **Note:** the tensioner bolt uses a special thread pitch and must **never** be replaced by any other type or the crankcase thread will be stripped.

10 Remove the temporary retaining pin from the pushrod end; the pushrod should immediately extend to tension the chain. Press the steel ball in and check that the pushrod moves smoothly and easily.

11 Refit the chain guide plate and tighten the three bolts to a torque setting of 0.8 – 1.2 kgf m (6 – 9 lbf ft).

All models

12 Rotate the crankshaft (anti-clockwise, seen from the rear) through one or two full turns, then check that the valve timing is still correct. Reset the timing if necessary.

39.2 Fit return spring and collar as shown to selector claw arm

39.3 Align teeth of gearchange pedal shaft with those of selector claw arm

40.1a Place sprocket flange on camshaft – drive pin should be in 2 o'clock position

40.1b Lock washer is of Belville type; fit as marked

40.1c Lock flange as shown to tighten retaining nut to specified torque setting

40.2a Align flange bolt holes with crankcase index marks

40.2b Align crankshaft keyway with rear main bearing housing index mark

40.3 Chain right-hand run must be pressed in when checking valve timing

40.4 Manual tensioner reassembly – manoeuvre guide blade into place ...

40.5 ... and refit tensioner arm and blade – note early un-modified types shown

40.8a Automatic tensioner reassembly – guide blade spacer shoulders must fit against crankcase

40.8b Insert guide blade and tighten bolts, then fit tensioner blade

40.9 Tighten tensioner mounting bolt to specified torque setting

40.11 Refit chain guide plate and tighten bolts

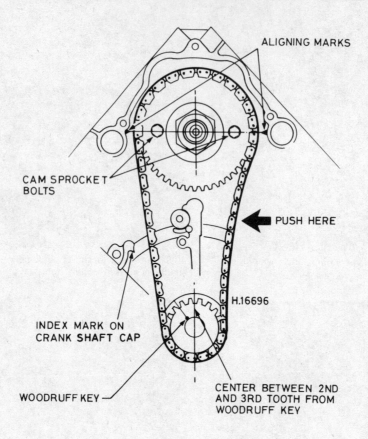

Fig. 1.21 Cam timing marks

41 Reassembling the engine/gearbox unit: refitting the alternator and engine rear cover

1   Check that the three roller assemblies are in place in the starter clutch to the rear of the alternator rotor. Insert the heavy washer and then install the starter driven pinion so that the boss enters the roller housing. To aid insertion of the boss rotate the driven pinion in an anti-clockwise direction.

2   Slide the pinion needle roller bearing onto the crankshaft and lubricate it with clean engine oil. Check that the Woodruff key is correctly positioned in the keyway in the tapered crankshaft end. Slide the rotor assembly onto the crankshaft so that the key engages with the keyway in the rotor internal bore. Insert and tighten the alternator rotor bolt to the specified torque setting.

3   Place the CDI pulser rotor against the end of the alternator rotor boss so that the projection locates with the recess. Insert and tighten the bolt.

4   Fit the starter intermediate pinion so that the smaller pinion engages with the driven pinion. One shim should be placed each side of the pinion on the spindle.

5   Before refitting the engine rear cover, lubricate the lips of the three oil seals with clean engine oil and check that the water pump mechanical seal is correctly fitted. Reference to this seal may be found in Chapter 2. If the CDI pulser (where fitted) or the alternator stator were disturbed, these too should be refitted.

6   When refitting the components, locking fluid should be applied to the bolt threads before they are inserted. When refitting the pulsers (CDI ignition models only), ensure that the pulser with the raised line on the outer face is installed nearest the inspection aperture. The raised line serves as the fixed index pointer used in checking the ignition timing.

7   Install the final output shaft splined collar on the end of the gearbox output shaft and then fit the shaft. Check that the two hollow dowels are in place in the crankcase rear mating surface. Fit the two large collars into the coolant transfer passages and fit a new O ring to

each. Place a new gasket on the mating surface and install the rear cover. Care should be taken to avoid damaging the oil seal lips as the shaft passes through the various seals. Fit the cover securing bolts and tighten them evenly, in a diagonal sequence to the specified torque settings. If the neutral indicator switch was removed it may now be refitted and the lead connected. Secure the lead by passing it through the guide channel integral with the pulser stator cover.

Models with CDI ignition

8   When refitting the advance pulser unit ensure that the line on the stator is lined up with the index mark before the screws are tightened or the advance operation will be incorrect. See the accompanying illustration. Reconnect the wires and fit the cover with a new gasket.

Models with transistorised ignition

9   To refit the ATU, locate the peg in the crankshaft end and replace the centre bolt, tightening it to a torque of 0.8 – 1.2 kgf m (6 – 9 lbf ft).

10   When refitting the pulser coil plate the ignition timing must be set statically. Remove the inspection plug to reveal the timing marks on the flywheel. Rotate the crankshaft until the FS mark for the right-hand cylinder (TR) aligns with the index mark. At this point the pulser coil plate can be fitted over the ATU and loosely secured with its two screws. Move the position of the plate so that the upper pulser coil (right-hand cylinder) alignment mark coincides with the tooth of the ATU cam and secure in place by tightening the two base plate screws. To check the ignition timing for the left-hand cylinder turn the crankshaft so that the FS mark for the left-hand cylinder (TL) aligns with the index mark. The lower pulser coil alignment mark should now coincide with the tooth of the ATU. If these marks do not align then the base plate screws must be slackened and the plate moved accordingly. It must be remembered, however, that by altering the position of the lower coil, the timing for the upper (right-hand cylinder) coil will automatically be changed. If there is a discrepancy in either of the units the misalignment can be shared between both cylinders.

11 To ensure the correct timing of the spark the air gap between the tooth of the ATU cam and the index mark on the pulser coil must be correctly adjusted. Remove the timing mark inspection plug from the top of the rear engine casing and turn the crankshaft until the FS mark for the left-hand cylinder (TL) aligns with the fixed pointer. At this point the tooth of the ATU cam should coincide with the mark on the lower pulser coil. Using feeler gauges check the air gap between the tooth and the coil mark. The correct gap should be between 0.45 – 0.65 mm (0.018 – 0.026 in). If adjustment is required slacken the screws retaining the left-hand coil to its backplate and move the coil position accordingly.

12 Turn the crankshaft until the FS mark for the right-hand cylinder (TR) is aligned and check the air gap between the tooth of the ATU and the upper pulser coil. Make any adjustment in the same way as for the left-hand cylinder.

13 Check the air gaps once again and install the pulser cover with its six bolts.

All models

14 Refit the starter motor. Refit the gearchange pedal, aligning it using the manufacturer's punch marks or the marks made on dismantling and tighten securely its pinch bolt.

15 On CX650 C, CX650 E-D and GL650 D2-E models, fit the circlip into its groove in the final output shaft.

41.1 Check that springs, plungers and rollers are in place in starter clutch

41.2a Fit starter driven pinion to rear of rotor and lubricate needle bearing before refitting

41.2b Lock crankshaft as shown when tightening rotor retaining nut

41.3 Align CDI pulser rotor lug with alternator rotor cutout before tightening bolt

41.4a Fit a thrust washer on each side of starter pinion ...

41.4b ... and fit pinion as shown

41.6a Use routing of electrical leads to position alternator stator correctly before tightening bolts.

41.6b CDI right-hand fixed pulser must be fitted next to rear cover inspection aperture

41.7a Fit splined collar to output shaft rear end ...

41.7b ... then refit final output shaft

41.9 Align ATU (transistorised ignition only) with locating peg before tightening bolt

41.14 Renew sealing O-ring before refitting starter motor

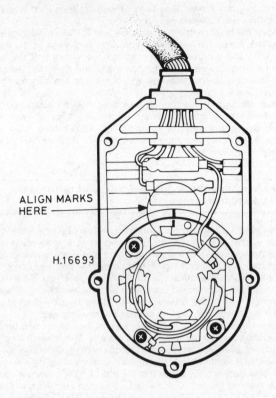

ALIGN MARKS
HERE

H.16693

Fig. 1.22 Alignment marks for advance pulser stator – CDI
ignition

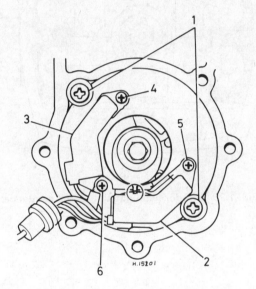

Fig. 1.23 Transistorised ignition pulser assembly

1   Pulser base-plate securing screws
2   Left-hand cylinder pulser coil
3   Right-hand cylinder pulser coil
4   Right-hand cylinder pulser coil adjusting screw
5   Left-hand cylinder pulser coil adjusting screw
6   Common adjusting screw

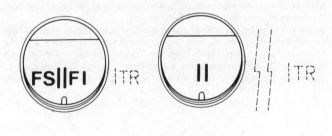

RIGHT–HAND CYLINDER

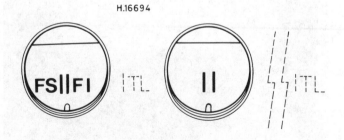

LEFT–HAND CYLINDER

Fig. 1.24 Alternator rotor ignition timing marks –
transistorised ignition system

42 Reassembling the engine/gearbox unit: refitting the cylinder heads and rocker gear

1   The procedure to be followed is identical for both cylinder heads and rocker gear assemblies. Complete the reassembly and installation of one complete assembly and then repeat the operation for the second unit. This will help prevent the accidental interchange of identical components.

2   Pour about 100cc of engine oil over the cam followers and camshaft lobes to ensure adequate lubrication when the engine is started. Install the two locating dowels and the oil control jet and O-ring in the cylinder mating surface. The control jet **must** be fitted with the narrower diameter innermost. Before fitting, check that the jet

orifices are clear. Place a new gasket over the dowels. The gasket should be thinly coated on both sides with a water and oil resistant liquid sealer. Place the cylinder head in position.

3   Grease the ball ends of both pushrods and then insert them through

the tunnel in the cylinder head so that the lower ends engage with the cups in the cam followers.

4   Rotate the crankshaft until the piston is at TDC on the compression stroke. That the piston is on the compression stroke can be determined by viewing the pushrods; both should be down. Install the rocker carrier locating dowels in the cylinder head and then refit the rocker carrier assembly so that the pushrods locate with the cupped ends of the rocker arms.

5   Insert and tighten the four cylinder head holding down bolts. The bolts should be tightened evenly, in the sequence shown in the accompanying illustration, to the specified torque setting.

6   Adjust the valve clearances as described in Routine Maintenance and refit the rocker covers.

43 Reassembling the engine/gearbox unit: refitting the cooling system components

1   Using ordinary household soap, apply a film of soapy water to the ceramic washer at the rear of the pump impeller, and the sealing face of the mechanical seal. Fit the narrow collar and the impeller onto the splined camshaft end and replace the copper washer and domed nut. The nut and washer must be of the type specified to ensure a water-tight seal. The domed nut should be tightened to a torque of 0.8 – 1.2 kgf m (6 – 9 lbf ft). After tightening, rotate the crankshaft to check that the impeller is not binding against the housing wall.

2   Place a new sealing ring in the groove in the pump cover face and position the cover on the two locating dowels. The cover bolts should be tightened evenly to the specified torque setting. Once again, rotate the crankshaft to check that the impeller is not fouling the casing.

3   On 500 models only, check that the taper on the camshaft end and internal taper in the fan boss are clean and then refit the fan. The fan securing bolt should be tightened to a torque of 2.0 – 2.5 kgf m (14.5 – 18.0 lbf ft).

4   Fit new O-rings to each end of the transfer pipes and to the flanges of the manifolds. Before fitting the various components apply a solution of soapy water to the rubber seals. This will aid fitting and help prevent seal damage.

5   Fit the left-hand manifold and tighten the two screws. Replace the thermostat carrier bracket and then install the left-hand pipe and the thermostat housing. Insert the thermostat and fit the cover. Do not omit the cover O-ring. Fit the bolts which pass through the bracket and cover, into the housing. Push the housing across to the left as the bolts

are tightened. This will ensure that the O-rings are firmly seated. Fit the right-hand pipe and manifold, again pushing the manifold across to the left as the bolts are tightened. Tighten all bolts to a torque setting of 0.8 – 1.2 kgf m (6 – 9 lbf ft).

6   Place a new O-ring on the end of the lower transfer pipe and fit the pipe into the passage mouth at the rear of the engine. Refit the pipe half clamps, tightening the upper socket screws first and then tightening the lower screws, until the pipe is gripped firmly, to a torque setting of 0.5 – 0.9 kgf m (3.5 – 6.5 lbf ft). Push the pipe into the union at the crankcase when tightening the screws.

7   Replace both cylinder drain plugs, ensuring that the sealing washers are not omitted and that they are in good condition. Refit the air spoiler baffle plate and connect the bypass hose to the water pump.

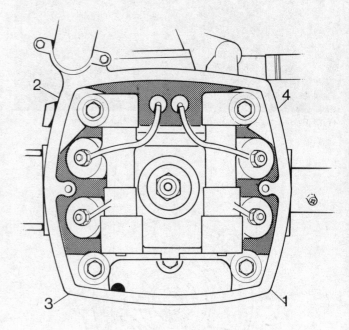

Fig. 1.25 Cylinder head nut tightening sequence

42.2a Check oil jet is clear and fit new sealing O-ring. Place new head gasket ...

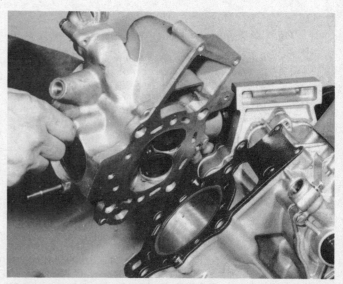

42.2b ... over locating dowels and refit cylinder head

42.3 Ensure pushrods are lubricated and correctly refitted in their original locations

42.4 Position piston at TDC on compression stroke before refitting rocker assembly

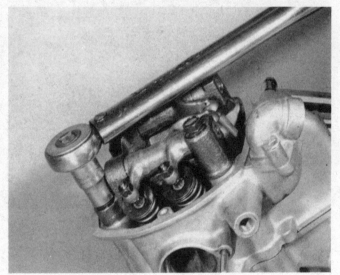

42.5 Tighten retaining bolts to specified torque setting

43.6 Tighten chromed coolant pipe clamps carefully, as described in text

44 Refitting the engine/gearbox unit in the frame

1   First check that all mounting bolts and nuts are completely clean and free from dirt or corrosion, and that the bolt shanks are well greased to aid refitting and to prevent corrosion. Use the specified grease to lubricate the final driveshaft splines.
2   If required, the engine front mounting bracket can now be refitted to the engine. Tighten the nuts to the specified torque setting and do not forget, where fitted, the cooling fan cowling. Similarly, refit the radiator if required, (refer to Chapter 2), and the tachometer drive and clutch cables.
3   Using the trolley jack in the same way as for removal, and with similar assistance, lift the engine into the frame until the driveshaft splines are engaged (rotate the rear wheel to help the splines engage fully), then move it to the rear and fit the rear mounting bolts. Fit (if not in place) the engine front mounting bracket and its mounting bolts. Ensure that the bolts are fitted from the left-hand side because the bolt heads are rounded to prevent chafing of the wiring leads. Fit, but do not tighten the nuts.
4   Refit the carburettors. See Chapter 3. Refit, where applicable, the engine top mounting plates and their bolts. On all 500 models and the

US GL650 models check that there is about 5 – 6 mm (0.20 – 0.24 in) of final output shaft spline showing in front of the universal joint, then refit and tighten to the specified torque setting the driveshaft pinch bolt. On CX650 C, CX650 E-D and GL650 D2-E models the shaft floats freely between two circlips; check that it is securely in place. There should be the same amount of final output shaft spline showing as described above for the earlier models. On all models, when the shaft is secure, refit the rubber gaiter over the gearbox or swinging arm flange, as appropriate.
5   With the engine/gearbox unit arranged securely and without strain on its mountings tighten all mounting bolts and nuts to the torque settings specified.
6   If this has not already been done, refit the radiator, check that the drain plug is securely fastened and all cooling system components are correctly refitted, then connect the expansion tank pipe to the radiator filler neck and refit the radiator shroud(s). Fill the system with coolant (see Chapter 2) and check for leaks. Do not forget to recheck the coolant level after the engine has first been started; accordingly do not fit the radiator cap until this has been done.
7   Ensuring they are correctly routed and secured neatly out of harm's way by any clamps or ties provided, reconnect to the main wiring loom all electrical components. Refit and tighten securely the HT coil rear

mounting bolt (where removed), and do not forget to connect the
starter motor lead and the battery earth lead. Check the spark plugs are
correctly gapped (see Routine Maintenance) and refit them. Connect
also the engine breather pipe(s) and the tachometer drive cable.
Connect the clutch cable and adjust it as described in Routine
Maintenance.

8   Refit the exhaust system. See Chapter 3. Refit the footrests (where
disturbed) and the brake pedal. If the rear brake was disturbed, adjust it
and the stoplamp rear switch as described in Routine Maintenance. On
CX650 C models refit the frame front cover.

9   Refit the battery and connect its leads again (see Chapter 7).

10 Working as described in Routine Maintenance, fit a new oil filter
element and refill the crankcase with oil. Note that a larger amount
than usual will be required if the engine has been dismantled. See
Chapter 3 Specifications.

11 On those machines so equipped, refit the fairing. See Chapter 5.

12 Make a final check that all components have been correctly refitted
and are correctly adjusted and working properly, where applicable.
Refit the fuel tank, the side panels and the seat, but remember that the
coolant and engine oil levels must be checked after the engine has
been run, also that the ignition timing has to be checked. It will be
necessary for example to remove the fuel tank again to check the
coolant level at the radiator.

44.4a Slide engine into place until driveshaft is engaged, then refit
pinch bolt (where fitted) ...

44.4b ... and rubber gaiter before tightening all mounting bolts ...

44.5 ... and retaining nuts to the specified torque settings

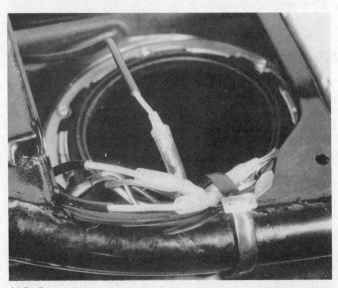

44.7a Route electrical leads carefully and secure with clamps or ties
provided

44.7b Ensure clutch cable is correctly routed

45 Starting and running the rebuilt engine

1 Start the engine using the usual procedure adopted for a cold engine. Do not be disillusioned if there is no sign of life initially. These machines are known to be difficult to start after they have been standing for a while, especially those models with vacuum fuel taps, where the fuel will take some time to work into the carburettors. Therefore a certain amount of perseverance may prove necessary to coax the engine into activity even if new parts have not been fitted. Should the engine persist in not starting, check that the spark plugs have not become fouled by the oil used during reassembly. Failing this go through the fault finding charts and work out what the problem is methodically.

2 When the engine does start, keep it running as slowly as possible to allow the oil to circulate. The oil warning light should go out almost immediately the engine has started, although in certain instances a very short delay can occur whilst the oilways fill and the pressure builds up. If the light does not go out, the engine should be stopped before damage can occur, and the cause determined. Open the choke as soon as the engine will run without it. During the initial running, a certain amount of smoke may be in evidence due to the oil used in the reassembly sequence being burnt away. The resulting smoke should gradually subside.

3 Check the engine for blowing gaskets and oil leaks. Before using the machine on the road, check that all the gears select properly, and that the controls function correctly.

4 When the machine has reached normal operating temperature, check the coolant level at the radiator filler neck. Top up if necessary (see Chapter 2) and fit the cap, then check the expansion tank level.

5 Referring to Chapter 4, check the ignition timing.

6 Finally, check the engine oil level and top up if necessary, as described in Routine Maintenance.

46 Taking the rebuilt machine on the road

1 Any rebuilt machine will need time to settle down, even if parts have been replaced in their original order. For this reason it is highly advisable to treat the machine gently for the first few miles to ensure oil has circulated throughout the lubrication system and that any new parts fitted have begun to bed down.

2 Even greater care is necessary if the engine has been rebored or if a new crankshaft has been fitted. In the case of a rebore, the engine will have to be run-in again, as if the machine were new. This means greater use of the gearbox and a restraining hand on the throttle until at least 500 miles have been covered. There is no point in keeping to any set speed limit; the main requirement is to keep a light loading on the engine and to gradually work up performance until the 500 mile mark is reached. Experience is the best guide since it is easy to tell when an engine is running freely.

3 If at any time a lubrication failure is suspected, stop the engine immediately, and investigate the cause. If any engine is run without oil, even for a short period, irreparable engine damage is inevitable.

4 When the engine has cooled down completely after the initial run, recheck the various settings, especially the valve clearances. During the run most of the engine components will have settled into their normal working locations. Check the various oil levels, particularly that of the engine as it may have dropped slightly now that the various passages and recesses have filled.

Chapter 2 Cooling system

Contents

Specifications

Note: *unless otherwise stated, information applies to all models*

Coolant

Type	Distilled water with corrosion-inhibited ethylene glycol antifreeze
Standard recommended mixture ratio	50% distilled water: 50% antifreeze
Boiling point of standard mixture:	
Unpressurised	107.7°C (226°F)
Pressurised – cap on	125.6°C (258°F)
Alternative mixture ratios for differing temperatures:	
Down to −32°C (−25°F)	55% distilled water : 45% antifreeze
Down to −37°C (−34°F)	50% distilled water : 50% antifreeze
Down to −44.5°C (−48°F)	45% distilled water : 55% antifreeze
Capacity – overall:	
500 models	2.00 lit (2.11 US qt/3.52 Imp pint)
650 models	2.08 lit (2.20 US qt/3.66 Imp pint)
Capacity of individual components – approximate:	
Radiator	1.40 lit (1.48 US qt/2.46 Imp pint)
Cylinder water jacket	0.40 lit (0.42 US qt/0.70 Imp pint)
Expansion tank – 500 models	0.20 lit (0.21 US qt/0.35 Imp pint)
Expansion tank – 650 models	0.38 lit (0.40 US qt/0.67 Imp pint)

Radiator

Cap valve opening pressure	0.9 kg/cm² (12.8 psi)
Tolerance	0.75 – 1.05 kg/cm² (10.7 – 14.9 psi)
Radiator core maximum test pressure	1.05 kg/cm² (14.9 psi)

Thermostat

Opens at	80 – 84°C (176 – 183°F)
Fully open at	93 – 97°C (199 – 207°F)
Valve minimum lift – fully open	8 mm @ 95°C (0.3150 in @ 203°F)

Fan motor thermostatic switch – 650 models only

Cuts in at coolant temperature of	98 – 102°C (208 – 216°F)
Cuts out at coolant temperature of	93 – 97°C (199 – 207°F)

Torque wrench settings

Component	kgf m	lbf ft
Water pump impeller nut	0.8 – 1.2	6.0 – 9.0
Water pump cover bolts:		
6 mm	0.8 – 1.2	6.0 – 9.0
8 mm	1.8 – 2.5	13.0 – 18.0
Water pipe manifold bolts	0.8 – 1.2	6.0 – 9.0
Chromed water pipe clamp Allen screws	0.5 – 0.9	3.5 – 6.5
Radiator drain plug	0.15 – 0.3	1.0 – 2.0
Cooling fan bolt – 500 models only	2.0 – 2.5	14.5 – 18.0
Fan cowling mounting nuts – 500 models only:		
All CX500, CX500 C and CX500 D models	3.5 – 4.5	25.0 – 32.5
CX500 E-C, all GL500 models	3.0 – 4.0	22.0 – 29.0

1 General description

The cooling system uses a water/antifreeze coolant to carry away excess energy produced in the form of heat. The cylinders are surrounded by a water jacket from which the heated coolant is circulated by thermo-syphonic action in conjunction with a water pump driven off the rear of the camshaft. The hot coolant passes upwards through a thermostat housing to the top of the radiator which is mounted on the frame downtubes to take advantage of maximum air flow. The coolant then passes downwards, through the radiator core, where it is cooled by the passing air, and then to the water pump and engine where the cycle is repeated. A fan is mounted behind the radiator to aid cooling; on 500 models it is mounted on, and driven by, the camshaft, but on 650 models a thermostatically-controlled electric fan is fitted. A wax pellet type thermostat is fitted in the system to prevent coolant flow through the radiator when the engine is cold, thereby accelerating the speed at which the engine reaches normal working temperature.

The complete system is partially sealed and pressurised, the pressure being controlled by a valve contained in the spring loaded radiator cap. By pressurising the coolant to approximately 13 psi, the boiling point is raised, preventing premature boiling in adverse conditions. The overflow pipe from the radiator is connected to an expansion tank into which excess coolant is discharged by pressure. The expelled coolant automatically returns to the radiator, to provide the correct level when the engine cools again.

2.4 Remove radiator shroud, if necessary, to reach drain plug

2 Cooling system: draining

1 **Warning:** to avoid the risk of personal injury such as scalding, the cooling system should be drained only when the engine and cooling system are **cold**. Note also that coolant will attack painted surfaces; wash away any spilled coolant immediately with fresh water.
2 Place the machine on the centre stand so that it rests on level ground. To gain access to the radiator cap the fuel tank and the seat must be removed. Refer to Chapter 3.
3 If the engine is cold, remove the radiator cap by pressing the cap downwards and rotating it in an anti-clockwise direction. If the engine is hot, having just been run, place a thick rag over the cap and turn it **slightly** until all the pressure has been allowed to disperse. A rag must be used to prevent escaping steam from causing scalds to the hand. If the cap were to be removed suddenly, the drop in pressure could allow the water to boil violently and be expelled under pressure from the filler neck. Apart from burning the skin the water/antifreeze mixture will damage paintwork. Where time and circumstances permit it is strongly recommended that a hot engine be allowed to cool before the cap is removed.
4 Place a receptacle below the front of the engine into which the coolant can be drained. The container must be of a capacity greater than the volume of coolant. Remove the radiator shroud or displace it far enough to permit the drain plug to be unscrewed. Allow the coolant to drain completely before refitting the drain plug. Note its specified torque setting. The coolant reservoir may be drained by pulling the lower hose off the union and removing the cap. To ensure all coolant is drained from the system, the drain plugs which screw into the front of the cylinders should be removed if possible.
5 The manufacturers recommend that the coolant be renewed at regular intervals. (See Routine Maintenance.) If the coolant being drained is to be re-used, ensure that it is drained into a clean non-metallic container.

3 Cooling system: flushing

1 After extended service the cooling system will slowly lose efficiency, due to the build-up of scale, deposits from the water and other foreign matter which will adhere to the internal surfaces of the radiator and water channels. This will be particularly so if distilled water has not been used at all times. Removal of the deposits can be carried out easily, using a suitable flushing agent in the following manner.
2 After allowing the cooling system to drain, replace the drain plugs (or lower hoses) and refill the system with clean water and a quantity of flushing agent. Any proprietary flushing agent in either liquid or dry form may be used, provided that it is recommended for use with aluminium engines. **Never** use a compound suitable for iron engines as it will react violently with the aluminium alloy. The manufacturer of the flushing agent will give instructions as to the quantity to be used.
3 Run the engine for ten minutes at operating temperature and drain the system. Repeat the procedure **twice** and then again using only clean cold water. Finally, refill the system as described in the next section.

4 Cooling system: filling

1 Before filling the system, always check that the drain plugs have been fitted and tightened and that the hose clips are tight.
2 The recommended coolant to be used in the system is made up of 50% corrosion-inhibited ethylene-glycol suitable for use in aluminium engines and 50% distilled water; this gives protection against the coolant freezing in temperatures of down to -37°C (-34°F). Other mixture ratios of the same ingredients for different temperatures are listed in the Specifications Section of this Chapter. To give adequate

protection against wind chill factor and other variables, the coolant should always be prepared for temperatures -5°C (-9°F) lower than the lowest anticipated.

3   Use only good quantity antifreeze of the type specified; never use alcohol-based antifreeze. In view of the small quantities necessary it is recommended that distilled water is used at all times. Against its extra cost can be set the fact that it will keep the system much cleaner and save the time and effort spent flushing the system that would otherwise be necessary. Tap water that is known to be soft, or rainwater caught in a non-metallic container and filtered before use, may be used in cases of real emergency only. Never use hard tap water; the risk of scale building up is too great.

4   So that a reserve is left for subsequent topping-up, make up approximately 2.5 litres (2.64 US qt/4.40 Imp pint) of coolant in a clean container. At the **standard** recommended mixture strength this will mean adding equal amounts of antifreeze and distilled water; do not forget to alter the ratio if different temperatures are expected.

5   Having checked the system as described in the subsequent Sections of this Chapter, add the new coolant via the radiator filler neck. Pour the coolant in slowly to reduce the amount of air which will be trapped; when the level is up to the base of the filler neck, fill the expansion tank to its upper level line. Refit the expansion tank filler cap.

6   Start the engine and allow it to idle until it has warmed up to normal operating temperature, with the temperature gauge needle giving its usual reading; the level in the radiator will drop as the coolant is distributed and the trapped air expelled. Add coolant as necessary. As soon as the thermostat opens, revealed by the sudden steady flow of coolant across the radiator and by a warm top hose, the level will drop again and more air will be expelled in the form of bubbles.

7   All trapped air must be expelled from the system before the radiator cap is refitted. When the level has stabilised for some time with the engine fully warmed up, and there are no more signs of air bubbles appearing, top the level up to the base of the filler neck and refit the radiator cap. Stop the engine, check that the expansion tank is topped-up to its upper level mark and refit the radiator shrouds, seat and side panel, as appropriate.

8   When the machine has been ridden for the first time after renewing the coolant and has cooled down, check the level again at the radiator cap to ensure that no further pockets of air have been expelled; top up if necessary. At all other times the coolant level should be checked at the expansion tank, as described in Routine Maintenance.

5   Radiator: removal, cleaning and examination

1   Drain the radiator as described in Section 2 of this Chapter. Where an Interstate fairing is fitted, it must be removed first. See Chapter 5.

2   On early CX500 models, the black plastic radiator guard is retained by two screws on each side, each screw being hidden by a rubber plug. Prise out the plugs and remove the screws. The guard can then be lifted away. On all later models, the guard is made of three parts; a central frame with a polished alloy cover on each side. Removal of the covers is quite straightforward. Remove the two screws recessed into the side of each cover and the two screws from the top edge. The radiator guard can now be lifted from the machine, taking care not to damage the grille on any of the cycle parts. Where applicable, remove the Interstate fairing mounting bracket (with horns) and the ignition HT coil.

3   Detach both main hoses from the radiator unions after slackening off the screw clips. Disconnect the expansion tank pipe from the union projecting from the rear of the filler neck and on 650 models only, disconnect the fan motor and thermostatic switch leads.

4   Remove the two radiator mounting bolts from the right-hand side of the machine. Support the radiator and remove the final single mounting bolt from the left-hand side of the radiator. Lift the radiator forward, until the mounting lugs clear the engine hanger, and then out to one side. Care should be taken to prevent the radiator matrix from fouling with any projecting cycle part.

5   Remove any obstructions from the exterior of the radiator core, using an air line. The conglomeration of moths, flies, and road dust usually collected in the radiator matrix severely reduces the cooling efficiency of the radiator.

6   The interior of the radiator can most easily be cleaned while the radiator is on the motorcycle, using the flushing procedure described in Section 3 of this Chapter. Additional flushing can be carried out by placing a hose in the filler neck and allowing the water to flow through for about ten minutes. Under no circumstances should the hose be connected to the filler neck mechanically as any sudden blockage in the radiator outlet would subject the radiator to the full pressure of the mains supply (about 50 psi). The radiator should not be tested to greater than 15 psi.

7   Bent fins can be straightened, if care is exercised, using two screwdrivers. Badly damaged fins cannot be repaired; a new radiator will have to be fitted if bent fins obstruct the air flow more than about 20%.

8   Generally, if the radiator is found to be leaking, repair is impracticable and a new component must be fitted. Very small leaks may sometimes be stopped by the addition of a special sealing agent in the coolant. If an agent of this type is used, follow the manufacturers instructions very carefully. Soldering, using soft solder, may be effective for caulking large leaks but this is a specialist repair best left to experts.

9   Inspect the three radiator mounting rubbers for perishing or compaction. Renew the rubbers if there is any doubt as to their condition. The radiator may suffer from the effect of vibration if the isolating characteristics of the rubber are reduced.

4.5 When refilling system, add coolant to base of radiator filler neck ...

4.6 ... check level at expansion tank at all other times

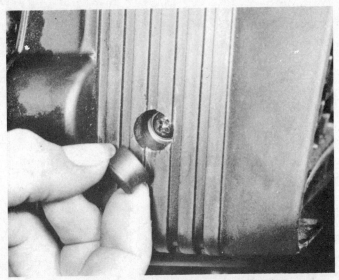

5.2a Radiator shroud mounting screws are covered by plugs on early models

5.2b On all other models remove screws at each side ...

5.2c ... and on top to release three-piece shroud

5.3a Slacken screw clips and release top hose ...

5.3b ... then bottom hose

5.4 Remove three mounting bolts (two shown) to release radiator

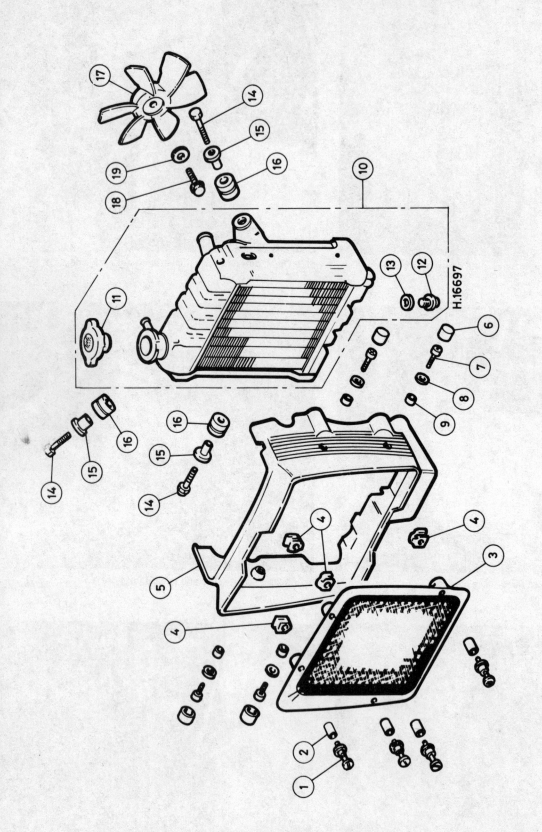

Fig. 2.1 Radiator and cooling fan – early CX500 models

H.16697

1 Screw
2 Spacer
3 Grille
4 Nut

5 Radiator guard
6 Plug
7 Screw
8 Washer

9   Collar
10 Radiator
11 Radiator pressure cap
12 Drain plug

13 Gasket
14 Bolt
15 Collar
16 Rubber grommet

17 Fan
18 Bolt
19 Washer

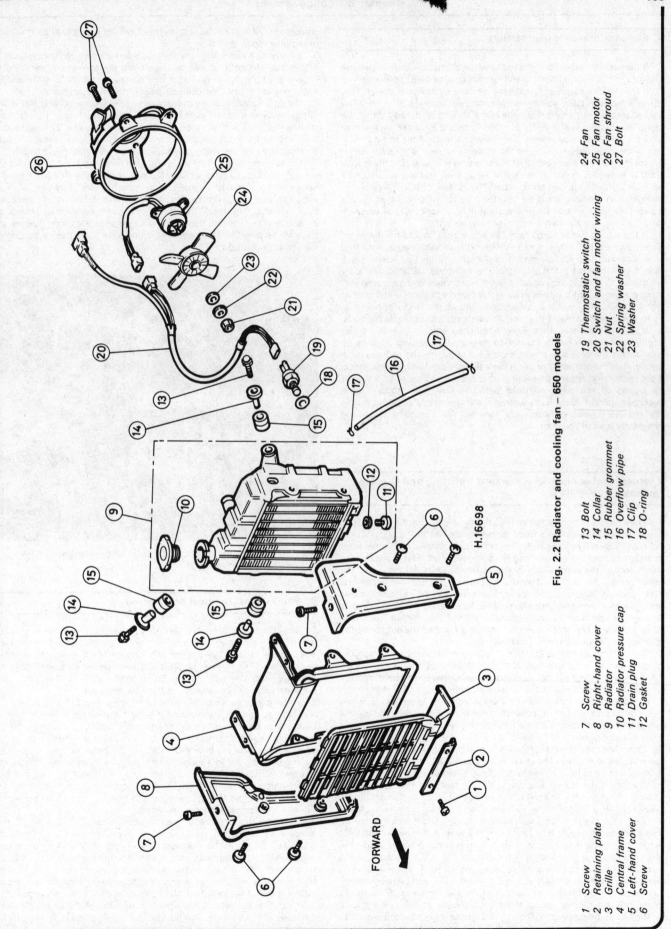

Fig. 2.2 Radiator and cooling fan – 650 models

1 Screw	19 Thermostatic switch
2 Retaining plate	20 Switch and fan motor wiring
3 Grille	21 Nut
4 Central frame	22 Spring washer
5 Left-hand cover	23 Washer
6 Screw	24 Fan
7 Screw	25 Fan motor
8 Right-hand cover	26 Fan shroud
9 Radiator	27 Bolt
10 Radiator pressure cap	
11 Drain plug	
12 Gasket	
13 Bolt	
14 Collar	
15 Rubber grommet	
16 Overflow pipe	
17 Clip	
18 O-ring	

H.16698

FORWARD

6 Radiator pressure cap: testing

1 If the valve or valve spring in the radiator cap becomes defective the pressure in the cooling system will be reduced, causing boiling over.
2 If the radiator cap is suspect, have it tested by a Honda dealer. This job requires specialist equipment and cannot be done at home. The only alternative is to try a new cap, but it should be noted that as the cap is very similar in size to those fitted to cars, a local car garage may have the necessary equipment; it is, therefore, worthwhile to find out if this is the case before going to the possibly unnecessary expense of substituting a new component. If the equipment is available, the cap is fitted to one end, using an adaptor if necessary, and a pressure of 13 psi (0.9 kg/cm²) applied to it, usually by means of a hand-operated plunger. The pressure must be held for a period of 6 seconds, during which time there should be no measurable loss; note that the pressure can, however, vary within a specified tolerance.
3 If the cap is found to be faulty in any way it should be renewed. Note however, that at the time of writing, the radiator cap is not listed as a separate item for all models and can only be purchased (as a Honda replacement part) as part of the complete radiator. The only alternative to this is to approach a good Honda service agent who will compare the cap with others in his stock (which fit other water cooled Honda machines) until a suitable cap is found. If this is not possible, try to find a second-hand but serviceable cap at a motorcycle breakers.
4 It should be noted that when tracing an elusive leak the entire cooling system can be pressurised to its normal operating pressure by connecting the test equipment described above to the radiator filler orifice. Remove the radiator cap, check the coolant level, topping it up if necessary, and apply a pressure of no more than 15 psi (1.05 kg/cm²) by means of the hand-operated plunger. Any leaks should soon become apparent. Most leaks will, however, be readily apparent due to the tell-tale traces of antifreeze left on the components in the immediate area of the leak.

7 Hoses and connections: removal, refitting and checking for leaks

1 The radiator is connected to the engine unit by two flexible hoses, there being an additional pipe between the water pump and the cylinder. The hoses should be inspected periodically and renewed if any sign of cracking or perishing is discovered. The most likely area for this is around the clips which secure each hose to its unions. Particular attention should be given if regular topping up has become necessary. The cooling system can be considered to be a semi-sealed arrangement, the only normal coolant loss being minute amounts through evaporation in the expansion tank. If significant quantities have vanished it must be leaking at some point and the source of the leak should be investigated promptly.
2 To disconnect the hoses, use a large pair of pliers to release the clamps and to slide them along the hose clear of the union spigot. Carefully work the hose off its spigots, noting that it may be necessary to slacken, or remove fully, the radiator mounting bolts to provide room to manoeuvre. The hoses can be worked off with relative ease when new, or when hot; do not, however attempt to disconnect the system when it is hot as there is a high risk of personal injury through contact with hot components or coolant.
3 **Warning:** the radiator hose unions, and coolant pipes are fragile; **do not use excessive force** when attempting to remove the hoses. If a hose proves stubborn, try to release it by rotating it on its unions before attempting to work it off. If all else fails, cut the hose with a sharp knife then slit it at each union so that it can be peeled off in two pieces. While expensive, this is preferable to buying a new radiator.
4 Serious leakage will be self-evident, though slight leakage can be more difficult to spot. It is likely that the leak will only be apparent when the engine is running and the system is under pressure, and even then the rate of escape may be such that the hot coolant evaporates as soon as it reaches the atmosphere, although traces of antifreeze should reveal the source of the leak in most cases. If not, it will be necessary to use testing equipment, as described in the previous Section, to pressurise the cooling system when cold, thereby enabling the source of the leak to be pinpointed. To this end it is best to entrust this work to a Honda service agent who will have access to the necessary

equipment if this is not available elsewhere, for example at a car garage or radiator repair agent.
5 In very rare cases the leak may be due to a broken head gasket in which case the coolant may be drawn into the engine and expelled as vapour in the exhaust gases. If this proves to be the case it will be necessary to remove the cylinder head(s) for investigation.
6 Other possible sources of leakage are the O-ring sealing the water pump body/crankcase joint, the mechanical seal and the O-rings sealing the coolant metal pipe unions. All these should be investigated and any leaks rectified by tightening the retaining screws, where applicable, or by renewing any seals which are worn or damaged.
7 On refitting hoses, first slide the clamps on to the hose and then work it on to each spigot in turn. **Do not** use lubricant of any type; the hose can be softened by soaking it in boiling water before refitting, although care is obviously required to prevent the risk of personal injury when doing this. When the hose is fitted, rotate it to settle it on its spigots and check that the two components being joined are securely fastened so that the hose is correctly fitted before its clamps are slid into position.

7.6 Coolant pipe unions are sealed by O-rings – renew whenever disturbed to prevent leaks

8 Thermostat: removal and testing

1 The thermostat is so designed that it remains in the closed position when it is in a normal cold condition. If the thermostat malfunctions, it will remain closed even when the engine reaches normal working temperature. The flow of coolant will be impeded so that it does not pass through the radiator for cooling and consequently the temperature will rise abnormally, causing boiling over.
2 If the performance of the thermostat is suspect, remove it from the machine as follows and test it for correct operation. Remove the fuel tank and drain the coolant as described in Section 2. Slacken off the screw clip which secures the radiator top hose at the thermostat union, and then displace the hoses. Disconnect the by-pass hose. Detach the air spoiler plate so that easier access may be made to the thermostat housing. Unscrew the bolts securing the thermostat bracket and also those which pass through the bracket into the thermostat housing. Lift off the bracket and the housing cover to gain access to the thermostat. The thermostat will lift from position.
3 Examine the thermostat visually before carrying out tests. If it remains in the open position at room temperature, it should be discarded. Suspend the thermostat by a piece of wire in a pan of cold water. Place a thermometer in the water so that the bulb is close to the thermostat. Heat the water, noting when the thermostat opens and the temperature at which the thermostat is fully open. If the performance is different from that specified, the thermostat should be renewed.
4 Heat the thermostat for about 5 minutes at 95°C (203°F) and measure the valve lift, which should be 8 mm (0.32 in). If the thermostat does not open fully it must be renewed.

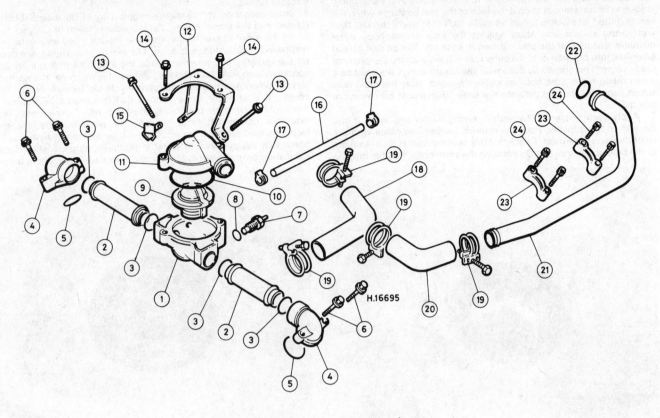

Fig. 2.3 Thermostat and water pipes

1 *Thermostat housing*	7 *Temperature gauge sender*	13 *Bolt*
2 *Coolant transfer pipe*	*unit*	14 *Bolt*
3 *O-ring*	8 *Sealing washer*	15 *Cable clip*
4 *Coolant pipe manifold*	9 *Thermostat*	16 *Bypass hose*
5 *O-ring*	10 *O-ring*	17 *Clip*
6 *Bolt*	11 *Thermostat cover*	18 *Radiator top hose*
	12 *Mounting bracket*	19 *Clamp*

20 *Radiator bottom hose*
21 *Chrome – plated coolant*
pipe
22 *O-ring*
23 *Clamp*
24 *Allen screw*

5   As an emergency measure only, if the thermostat is faulty it can be removed and the machine used without it. Take care when starting the engine from cold as the warm-up will take much longer than usual, and ensure that a new unit is fitted as soon as possible.

9   Water pump: removal, renovation and refitting

1   To prevent leakage of water or oil from the water pump to the crankcase and vice versa, two seals are fitted concentrically on the camshaft. The seal on the water pump side is of the mechanical type having a spring loaded annular face which bears against a ceramic washer mounted in the rear face of the impeller. The ceramic washer is mounted in a rubber bed to further prevent seepage. The second seal, which is mounted forward of the mechanical seal, is of the normal 'feathered' lip type used widely to prevent oil leakage along a shaft. Both seals are a drive fit in the engine rear cover.
2   Where work on the seals is required, the engine must be removed and the engine rear cover detached. The procedure for this is given in Chapter 1.
3   Removal of the seals from the engine rear casing is a straightforward operation. Prise out the oil seal from the front of the casing, using a flat bladed lever. Care should be taken not to bruise the periphery of the casing into which the seal fits. Select a tubular drift, the outside diameter of which is slightly smaller than that of the mechanical seal. Place the casing with the outside downwards, supported by blocks so that the mechanical seal can be driven from position. The ceramic washer in the rear of the impeller should be prised from position and the rubber seating displaced. Discard all the sealing components.
4   Replacement mechanical seals are a tight interference fit in the casing. To aid installation and prevent damage to the mechanical seal the casing **must** be heated prior to fitting of the seal. The casing should be heated in an oven to about 100°C (212°F). Do not use a blowtorch or other means of localised heating because distortion of the casing is likely if this approach is adopted. The mechanical seal is particularly susceptible to damage. Before installing the mechanical

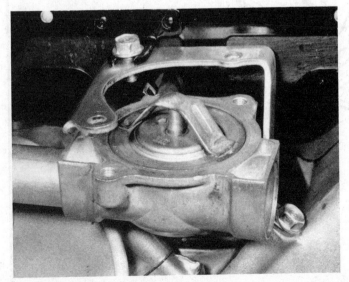

8.2 Thermostat is fitted as shown

seal apply a liquid jointing compound to the periphery of the seal body. Under no circumstances should the face of the seal be struck when the seal is drifted into place. Select a tubular drift that clears the seal face and spring shroud and abuts against the face of the body outer diameter. Ensure that the seal is driven in squarely. The oil seal should be drifted into position until the outer face is flush with the periphery of the housing. This should be done with the casing cold. Take great care when fitting either seal; both are easily damaged. Note the drain hole which runs into the gap between the seals. This should be cleared of any debris.

5   Apply soapy water to the washer seating rubber and insert it into the rear of the impeller. Push the ceramic washer into place with the smoother side outermost. Check that the ceramic washer face is square to the face of the impeller. Do not omit the small collar that fits into the impeller recess.

6   After renewing the seals the engine should be reassembled and refitted into the frame. Follow the procedures given in Chapter 1.

7   In the event of seal failure, the inter-mixing of oil and water in the engine will cause sludge to form. The extent of sludging will depend on the quantity of contaminant and the length of time during which contamination has taken place. Before the machine is returned to service, when new coolant and lubricant should be used as a matter of course, it is strongly recommended that the engine be flushed out thoroughly using a proprietary flushing oil. Furthermore, where extensive contamination is evident, an additional oil change is recommended approximately 500 miles after seal renewal. It should be noted that the presence of water in the lubricating oil will reduce its lubricating properties dramatically and may cause permanent damage.

9.3a Prise out oil seal from front of engine rear cover ...

9.3b ... to permit removal of pump mechanical seal

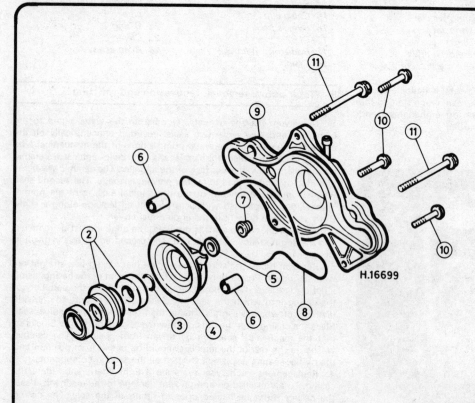

H.16699

Fig. 2.4 Water pump

1   Oil seal
2   Mechanical seal (carbon face)
3   Thrust washer
4   Water pump impeller
5   Sealing washer
6   Dowel pin
7   Nut
8   Sealing ring
9   Housing
10 Bolt
11 Bolt

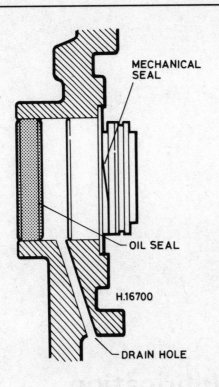

MECHANICAL
SEAL

OIL SEAL

H.16700

DRAIN HOLE

Fig. 2.5 Arrangement of oil seal and mechanical seal

10 Cooling fan: removal, examination and refitting

500 models

1   The cooling fan can be removed, as described in Section 6 of Chapter 1, after the radiator has been withdrawn. See Section 5 of this Chapter. The fan should be renewed if any of its blades are damaged. Refitting is described in Section 43 of Chapter 1.

650 models

2   The fan and motor assembly are automatically removed and refitted with the radiator. See Section 5 of this Chapter.
3   If the fan fails to work, connect a fully-charged 12 volt battery directly to the fan motor; for safety's sake this must only be done while the fan assembly is still attached to the radiator (on or off the machine) so that it is still enclosed by its protective shroud. If the fan motor works, the fault is in the thermostatic switch or in the main wiring loom; check for faults as described in Chapter 7.
4   If the fan motor does not work, it must be renewed. Remove the four shroud mounting bolts and withdraw the fan from the radiator, having released the motor lead from the wire retaining clip. Hold the fan and unscrew the retaining nut and washers. Remove the three mounting screws to separate the motor from the shroud.
5   On reassembly, install the motor with its 'TOP' mark facing to the top of the shroud, tighten the screws securely and refit the fan. Apply locking compound to the motor shaft threads before refitting the nut and washers. Tighten the nut securely and refit the fan assembly to the radiator.

11 Thermostatic switch and temperature gauge sender unit: removal, testing and refitting

1   These components should only be removed after the cooling system has been fully drained. See Section 2. The temperature gauge sender unit is screwed into the rear face of the thermostat housing, while the thermostatic switch is screwed into the radiator bottom rear face.
2   With the coolant drained, disconnect the unit wires and unscrew the unit. On reassembly, apply a suitable sealant such as Three-Bond No 1212 or equivalent to its threads, renew its sealing washer and tighten the unit securely. Do not overtighten.
3   The temperature gauge unit is part of the instruments and is removed as described in Chapter 5.
4   The testing of these components is described in the relevant Sections of Chapter 7.

Chapter 3 Fuel system and lubrication

Contents

Specifications

Note: *unless otherwise stated, information applies to all models*

Fuel tank capacity	Litre	US gal	Imp gal
All CX500 and CX500 D models:			
Overall	17.0	4.49	3.74
Reserve	3.5	0.93	0.77
CX500 C 1979, 1980:			
Overall	11.0	2.91	N/App
Reserve	2.5	0.66	N/App
CX500 C-B, CX500 C 1981, 1982:			
Overall	12.0	3.17	2.64
Reserve	2.5	0.66	0.55
CX650 C:			
Overall	12.4	3.28	N/App
Reserve	2.2	0.58	N/App
All GL500/650 models:			
Overall	17.6	4.65	3.87
Reserve	2.5	0.66	0.55
CX500 E-C, CX650 E-D:			
Overall	19.0	N/App	4.18
Reserve	2.5	N/App	0.55

Carburettor

Manufacturer	Keihin		
ID Number:			
Model	CX500 (UK)	CX500 1978	CX500 1979
Number	VB36A-A/C	VB26A-B	VB26A-B/C

Carburettor (continued)

Model / Number			
Model	CX500-A	CX500-B	CX500 C 1979
Number	VB36A-E/F	VB36A-F	VB27A
Model	CX500 C 1980	CX500 C-B	CX500 C 1981
Number	VB25A-B	VB37A-D	VB25A-C
Model	CX500 C 1982	CX500 D 1979	CX500 D 1980
Number	VB25A-D	VB23A	VB28A-B
Model	CX500 D 1981	CX500 E-C, GL500 D-C	GL500 and GL500 I 1981, 1982
Number	VB28A-C	VB1AA-A	VB29A-A
Model	CX650 E-D, GL650 D2-E	GL650, GL650 I	CX650 C
Number	VB2BA-A	VB2AA-A	VA2AC-A, VB2AB, VB2AC

Venturi diameter:
  CX500 C 1980, 1981, 1982, CX500 D 1980, 1981, GL500
  and GL500 I 1981, 1982 .... 34 mm (1.34 in)
  All other models .... 35 mm (1.38 in)
Primary main jet:
  CX650 E-D, GL650 D2-E 72
  All other models 78
Secondary main jet:
  CX500 C 1980, 1981, 1982, CX500 D 1980, 1981 115
  All other 500 models 112
  CX650 C 120
  All other 650 models 118
Slow jet – CX500 E-C, all GL500 models, all 650 models 45
Pilot mixture screw – turns out from fully closed:
  All US models 1978, 1979, GL650, GL650 I 2*
  CX500 (UK), CX500-A, CX500-B, CX500 C-B 2
  CX500 C 1980, 1981, 1982, CX500 D 1980, 1981 .... $1^{3}/_{4}$*
  CX500 E-C, GL500 D-C .... $1^{7}/_{8}$
  GL500 and GL500 I 1981, 1982 .... $1^{5}/_{8}$*
  CX650 E-D, GL650 D2-E .... $2^{3}/_{8}$
  CX650 C .... $2^{3}/_{8}$
  *Initial setting – to be used on renewal of pilot screw only, do not disturb
Float level .... 15.5 ± 1.0 mm (0.61 ± 0.04 in)
Idle speed .... 1100 ± 100 rpm
Fast idle speed:
  CX500 C 1980, CX500 D 1980 .... 1000 – 1500
  CX500 C 1981, 1982, CX500 D 1981 .... 1100 – 1500
  All other models .... 1500 – 2500
Maximum vacuum pressure difference between carburettors .... 40 mm Hg
Accelerator pump – US models from 1980 only:
  Pump rod/choke link arm clearance .... 0.1 – 0.3 mm (0.0039 – 0.0118 in)
  Choke link arm/carburettor stopper clearance .... 3.1 – 3.3 mm (0.1221 – 0.1299 in)

Engine lubrication system

Recommended oil .... Good quality SAE 10W/40 engine oil. API class SE or SF

	500 models	650 models
Capacity:		
At engine rebuild	3.0 lit (3.17 US qt/5.28 Imp pint)	3.6 lit (3.80 US qt/6.34 Imp pint)
At oil and filter change	2.6 lit (2.75 US qt/4.58 Imp pint)	3.1 lit (3.28 US qt/5.46 Imp pint)
At oil change	2.5 lit (2.64 US qt/4.40 Imp pint)	3.0 lit (3.17 US qt/5.28 Imp pint)

Relief valve opening pressure .... 5.0 – 6.0 kg/cm² (71 – 85 psi)
Pump delivery rate per minute:
  500 models – @ 3000 rpm .... 9.3 – 9.5 lit (2.46 – 2.51 US gal/2.05 – 2.09 Imp gal)
  650 models – @ 2500 rpm .... 14.0 – 14.2 lit (3.70 – 3.75 US gal/3.08 – 3.12 Imp gal)
Oil pump:
  Inner rotor/outer rotor maximum clearance .... 0.10 mm (0.0039 in)
  Outer rotor/pump body maximum clearance .... 0.35 mm (0.0138 in)
  Rotor maximum endfloat .... 0.10 mm (0.0039 in)

Final drive lubrication

Recommended oil grade .... Good quality hypoid gear oil API class GL-5
Viscosity:
  Above 5°C (41°F) .... SAE 90
  Below 5°C (41°F) .... SAE 80
Capacity .... 170 ± 10 cc (5.75 ± 0.34 US fl oz/5.98 ± 0.35 Imp fl oz)
Propeller shaft joint – 500 models only:
  Recommended grease .... Lithium-based multipurpose NLGI No 2 grease with molybdenum disulphide additive
  Quantity:
    On reassembly .... 45 cc (1.52 US fl oz/1.58 Imp fl oz)
    At routine maintenance .... 20 cc (0.67 US fl oz/0.70 Imp fl oz)

Torque wrench settings

Component	kgf m	lbf ft
Fuel tap filter bowl – where fitted	0.3 – 0.5	2.0 – 3.5
Fuel tap gland nut – all GL500 models, all 650 models	2.0 – 2.5	14.5 – 18.0
Air filter case mounting bolts	0.6 – 0.9	4.0 – 6.5
Throttle joint bolts – 650 models only	0.28 – 0.42	2.0 – 3.0
Carburettor/mounting bracket screws – 650 models only	0.28 – 0.42	2.0 – 3.0
Exhaust pipe/cylinder head nuts	0.8 – 1.4	6.0 – 10.0
Exhaust pipe/silencer clamp bolts	1.8 – 2.8	13.0 – 20.0
Balance chamber mounting bolts or nuts	2.4 – 3.0	17.0 – 22.0
Silencer mounting bolts:		
All GL models, CX650 C	4.5 – 6.0	32.5 – 43.0
CX500 E-C, CX650 E-D	3.0 – 4.0	22.0 – 29.0
Oil filter bolt	2.0 – 2.5	14.5 – 18.0
Engine oil drain plug	2.5 – 3.5	18.0 – 25.0
Oil pump mounting bolts	0.8 – 1.2	6.0 – 9.0
Oil pressure switch	1.8 – 2.3	13.0 – 16.0

1 General description

The fuel system comprises a tank from which fuel flows via a single tap to the left-hand carburettor and then to the right-hand carburettor through a transfer pipe. The tap, which incorporates a gauze filter, has a reserve position providing a small amount of additional fuel when the main source is exhausted. All models from 1981 on are fitted with a vacuum-controlled tap which is opened when the engine is running and therefore needs no 'Off' position although one is provided for emergency use or for long-term storage. For cold starting each carburettor has a butterfly valve choke, interconnected by a control rod and operated via a cable.

Two constant depression Keihin carburettors are fitted as standard, mounted on a shared bracket and interconnected by a control rod. The throttles are controlled by a push-pull two cable arrangement from a traditional twist-grip. To the rear of the carburettors, fitted within a plastic casing, is the corrugated paper air filter element.

US models from 1980 on are fitted with an accelerator pump which is mounted on the underside of the left-hand carburettor. Its purpose is to richen the mixture during acceleration, thus allowing the carburettor to be jetted for a weaker overall mixture to meet EPA emission requirements in the USA. The pump is operated by a spring rod connected to the throttle operating linkage. Once actuated by throttle opening, the pump feeds fuel into both carburettors.

The exhaust system comprises two individual pipes which lead into a large balance chamber under the rear of the engine/gearbox unit, and two silencers which lead the exhaust gases out of the chamber.

Lubrication is by the wet sump principle in which oil is delivered under pressure from the sump reservoir by a mechanical pump to the working parts of the engine. The oil pump, which is of the trochoid rotating vane type, is mounted to the left of the clutch and is driven by a chain from a sprocket on the crankshaft forward end. To protect the engine, oil is picked up from the sump through a gauze strainer and then passed through a full-flow paper filter element. The paper element can be removed and discarded at regular intervals. The engine oil is shared also with the primary drive and gearbox.

2 Fuel tank: removal and refitting

1 The fuel tank is secured at the rear by a single bolt passing through a lug projecting from the tank. The front of the tank is supported (except for the CX650 C and all CX500 C models) on two cups welded to the tank underside which rest on two rubber buffers, one of which is fixed either side of the frame top tube. On CX650 C and CX500 C models the front mounting consists of two bolts which pass through rubber mountings set in a flange at the front of the tank and into the frame.

2 First remove the seat, unscrew the tank rear mounting bolt (and the front mountings on CX500 C and CX650 C models), then lift the tank at the rear, check that the fuel tap is in the 'Off' position and disconnect the fuel feed pipe. Where applicable, disconnect also the fuel gauge sender unit wires, the vacuum pipe and the breather pipe. Lift the tank away.

3 The tank may be refitted by reversing the removal procedure. If difficulty is encountered in pushing the cups over the rubber buffers, a small amount of petrol or soap may be applied to give temporary lubrication. When fitting the rear bolt ensure that the rubber saddle is positioned correctly and that the collar is not omitted.

4 When connecting the pipes, ensure that their wire retaining clips are securely refitted over the tap spigots, although the main seal is effected by the interference fit of the pipes, the clips are an additional security measure to prevent fuel leaks. Do not forget to reconnect the fuel gauge wires (CX500 E-C, CX650 E-D). On models with vacuum taps, note that the **vacuum** pipe (from the carburettors) is fitted on the tap **front** union and the breather pipe is fitted to the rear union. If the pipes are reversed the tap will not open sufficiently to allow the engine to run properly, giving all the symptoms of a fuel blockage or a blocked tank breather.

3 Fuel tap and feed pipe: examination

Manual tap

1 Refer to Routine Maintenance for details of tap removal and refitting, and of filter cleaning. These taps cannot be dismantled; if one becomes blocked, attempt to clear it by applying a jet of compressed air through the feed pipe spigot when the tap is in the on or reserve position. If this fails to work, or if the tap is leaking, it must be renewed.

Vacuum tap

2 Refer to Routine Maintenance for details of tap removal and refitting and of filter cleaning.

3 To check the operation of the diaphragm remove the petrol tank from the machine, setting it up on wooden blocks for easy access to the tap. There is no need to drain the fuel or remove the tap for this test.

4 Place a clean container beneath the fuel delivery pipe (pipe on right-hand side of tap) and turn the tap to the 'On' position. No fuel should flow from the pipe. If it does this indicates failure of the diaphragm.

5 A further test can be undertaken by sucking gently on the end of the vacuum pipe (pipe terminating at the tap front union) to create a vacuum effect similar to that of the engine. The fuel tap should be positioned in the 'On' position. Fuel should flow freely from the delivery pipe. If a vacuum pump is available, the tap should open when a vacuum of 12 – 20 mm (0.5 – 0.8 in) of mercury is applied.

6 To inspect the diaphragm, drain the fuel from the tank and remove the fuel tap. Remove the screws retaining the diaphragm cover and lift the cover and gasket away. Remove the diaphragm and plunger valve for inspection. In the event of the diaphragm becoming holed or split, the plunger valve will close, thus blocking the fuel supply.

7 When reassembling the diaphragm assembly use new gaskets and seals. Do not use any of the silicone rubber gasket compounds available because they are attacked by fuel and will break up, thus causing small rubber-like particles to obstruct the carburettor jets.

Feed and vacuum pipes

8 Refer to Routine Maintenance.

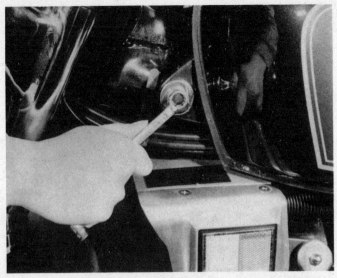

2.1a Fuel tank is secured by two bolts at front on Custom models ...

2.1b ... and by a single bolt at the rear on all others

Fig. 3.1 Vacuum fuel tap – all models 1981 on

1 Screw
2 Diaphragm cover
3 Spring
4 Spring seat
5 Diaphragm
6 Plunger
7 Diaphragm housing
8 Inner diaphragm
9 Filter bowl*
10 O-ring*
11 Filter gauze*
12 Nut
13 Tap body
14 O-ring
15 Filter

* 1982 on models only

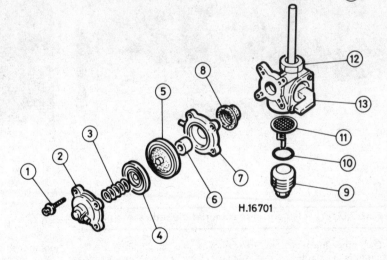

H.16701

4 Carburettors: removal and refitting

1 Remove the seat and fuel tank, following the procedure given in Section 2. Slacken the hose clips which secure the air filter hoses and inlet stubs to the carburettor mouths. Slide the clips away from the carburettors so that they will not become snagged during removal.

2 Prise the two air hoses backwards so that they clear the carburettor mouths and then pull the carburettors backwards out of the inlet stubs. Ease the carburettors out towards the left-hand side of the machine.

3 Before the carburettors can be lifted away, the two throttle control cables and the choke operating cable must be detached. Slacken the locknuts on the adjusters at the lower end of the cables, and screw the adjusters inwards to give as much slack as possible in the cables. Displace first one and then the other cable from the anchor bracket and then disconnect the inner cables from the pulley. The choke cable is secured to the carburettor by a small clamp held by a single screw. Slacken the screw to release the outer cable and disconnect the inner cable from the choke arm. With all the controls disconnected, pull the breather hoses through from below the carburettors and lift the assembly away from the machine.

4 Refitting is a straightforward reversal of the removal procedure. Ensure all breather pipes are routed correctly behind the engine/gearbox unit, re-connect the control cables before the carburettors are fully in place. When the carburettors are secured in position, adjust the cables as described in Routine Maintenance.

4.1 Slacken hose clips and displace air filter hoses ...

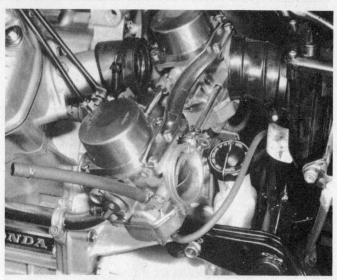

4.2 ... so that carburettors can be withdrawn far enough to disconnect ...

4.3a ... first the rear (opening) cable ...

4.3b ... and then the front throttle cable

5 Carburettors: dismantling, examination and reassembly

1 The procedure set out for removal and separation of the carburettors in paragraph 2 need not take place for normal examination and removal of the carburettor internal components such as the float assembly and jets. It is strongly advised that each carburettor be dismantled and reassembled separately, to prevent accidentally interchanging the components. Dismantle and examine each carburettor, following an identical procedure as described below.

2 If the carburettors are to be separated, displace the end of the light spring that interconnects the end of the choke link rod and the choke operating arm on the right-hand carburettor. Note very carefully the position of the spring before it is released. Remove the screws holding the lower mounting bar in place. These screws may be very tight, having been assembled using a locking fluid. Great care should be taken during removal because the screws are of a soft material and shear easily. On US 650 models, remove the split pin and withdraw the accelerator pump washer, spring and spring collar. On all 650 models, bend down the tabs securing the right-hand throttle joint bolt then

unscrew the bolt and remove the tab washer, balljoint seat and balljoint. Hold the joint pipe and turn the throttle link to separate the throttle linkage. Note very carefully the exact position of all components to ensure correct reassembly and do not lose the spring from inside the joint pipe. On all models remove its screws and withdraw the upper bar. Pull the carburettors apart; as this is done the choke rod will disengage from the choke arm, the throttle link will disengage from the spring loaded synchronisation screw and the fuel and (where fitted) the accelerator pump transfer pipe(s) will pull out of one or other of the carburettors. Do not lose the helical spring which lies concentrically between the ends of the throttle pivot rods (500 models only).

3 Invert each carburettor and remove the three screws which retain the float bowl. Lift the bowl from position. The sealing ring need not be displaced unless it has split or perished and leakage is likely to occur. The float assembly can be lifted from position after pushing the pivot pin out of the pivot posts. On early models the float needle will come away with the float assembly as it is retained by a spring clip hooked around the float tongue. On later models with plastic floats the needle must be lifted out.

4   Apply a small spanner to the needle jet holder hexagon (the jet assembly nearest the float needle seat), and using a screwdriver remove the secondary main jet. When removing any jet from a carburettor ensure that the screwdriver is of the correct size, fitting the slot closely. This will prevent damage to the jet and prevent burring, which may alter the orifice size. Unscrew the jet holder to free the needle jet. This jet projects into the venturi bore of the carburettor and may be displaced by inserting a finger through the carburettor mouth. Prise the rubber plug from the central jet pillar to gain access to the slow jet. On early models this jet is a tight press fit and must not be removed; on later models it is listed separately (see Specifications) and may be unscrewed for cleaning. From the third jet pillar unscrew the primary main jet and the primary nozzle.

5   The pilot screw (mixture adjusting screw) may be removed to aid cleaning of the internal passages. If this is done the pilot screw setting will have to be adjusted on reassembly as described in Section 8 or 9.

6   Lift the carburettor cap off the main body, after removing the two retaining screws. Pull out the piston very carefully, so that the piston needle does not get bent. The needle can be removed after lifting out the helical spring and removing the nylon plug and grub screw from the piston centre tube. If the piston is inverted the needle will fall out. Two air jets are obstructed by a small curved plate and gasket, which is retained by a single screw. These two jets and the two remaining jets (slow air jets) on the opposite side of the carburettor upper surface should not be removed.

7   It is not recommended that the 'butterfly' valves of either the throttle or choke be removed. The valves themselves are not subject to wear. If wear occurs on the operating pivots, a new carburettor will be required, as air will find its way along the pivot bearings, resulting in a weak mixture.

8   An air cut-off valve is fitted to the left-hand side of each carburettor. The valve is of the diaphragm type and automatically regulates the amount of air travelling through the pilot air system of the carburettor. When the engine is running at idling speeds, the valve remains open, allowing the correct quantity of air to enter the pilot system. If the throttle is closed when the engine has been running fast the high vacuum in the inlet manifold causes the valve to close, reducing the air flow and creating a rich mixture to prevent backfiring. Remove the valve cover, which is retained by two screws and lift the cover from position. Lift out the helical spring and valve diaphragm. Clean all the components in petrol.

9   On later (1980 on) US models dismantle the accelerator pump by unscrewing the three pump cover retaining screws and removing the cover and spring. This will expose the pump diaphragm and operating rod assembly. Remove the diaphragm and inspect it for signs of damage or deterioration. Check the rod is not bent and the spring is not broken. Check also the condition of the small gaiter at the carburettor body end of the operating rod and renew it if it is found to be split or perished. Note that it will be necessary to remove the float chamber to allow the gaiter to be withdrawn from its location in the carburettor

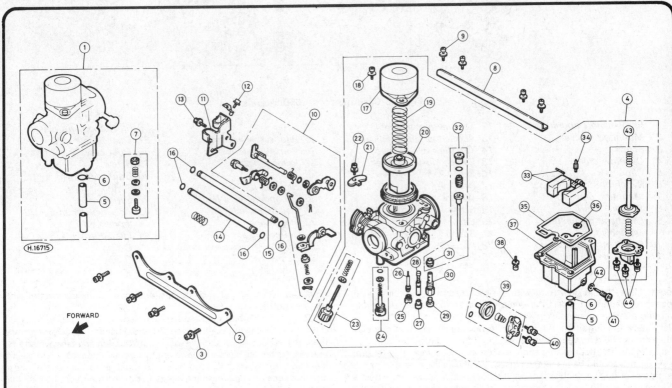

Fig. 3.2 Carburettor – 500 models

1 Right-hand carburettor	12 Screw	24 Pilot screw assembly	35 O-ring
2 Mounting bracket	13 Screw	25 Primary main jet	36 O-ring
3 Screw	14 Fuel transfer pipe	26 Primary nozzle	37 Float bowl
4 Left-hand carburettor assembly	15 Fuel transfer pipe	27 Plug	38 Screw
5 Overflow pipe	16 O-ring	28 Slow air jet – later models only	39 Air cut-off valve assembly
6 Clip	17 Cap	29 Secondary main jet	40 Screw
7 Synchronising screw set	18 Screw	30 Needle jet holder	41 Drain screw
8 Mounting bracket	19 Spring	31 Needle jet	42 O-ring
9 Screw	20 Piston	32 Jet needle assembly	43 Accelerator pump assembly – US models 1980 on
10 Operating linkage	21 Plate	33 Float and pivot pin	44 Screw
11 Cable adjuster bracket	22 Screw	34 Float needle valve	
	23 Idle adjusting screw		

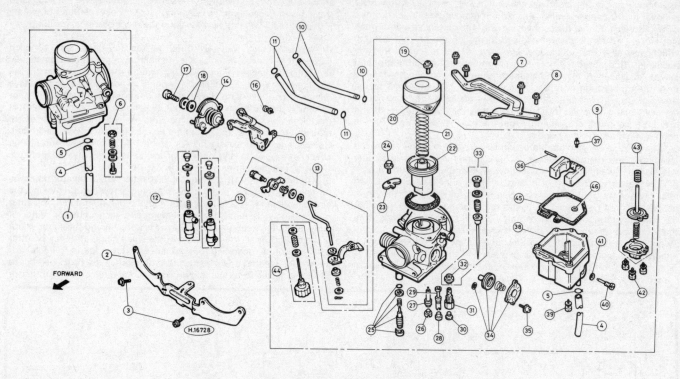

Fig. 3.3 Carburettors – 650 models

1	Right-hand carburettor	12	Throttle joint assembly	24	Screw
2	Mounting bracket	13	Accelerator pump linkage	25	Pilot screw assembly
3	Screw	14	Throttle cable pulley	26	Primary main jet
4	Overflow pipe	15	Cable adjuster bracket	27	Primary nozzle
5	Clip	16	Screw	28	Plug
6	Synchronising screw set	17	Bolt	29	Slow air jet
7	Mounting bracket	18	Washer	30	Secondary main jet
8	Screw	19	Screw	31	Needle jet holder
9	Left-hand carburettor	20	Cap	32	Needle jet
	assembly	21	Spring	33	Jet needle assembly
10	Fuel transfer pipe	22	Piston	34	Air cut-off valve assembly
11	Fuel transfer pipe	23	Plate	35	Screw

36	Float and pivot pin
37	Float needle valve
38	Float bowl
39	Screw
40	Drain screw
41	O-ring
42	Screw
43	Accelerator pump – US models only
44	Idle adjusting screw
45	O-ring
46	O-ring

body. Little should go wrong with the pump apart from a cracked or perished diaphragm, but if the pump system is completely dry it may require priming to expel air.

10 Check the condition of the floats. If they are damaged in any way, they should be renewed. Refer to Section 8 for details of the float level adjustment. The float needle and needle seating will wear after lengthy service and should be inspected closely. Wear usually takes the form of a groove or ridge, which will cause the float needle to seat imperfectly. Ideally, the needle and valve seat should be renewed as a pair. On this model of carburettor, however, the seat is not renewable and if the valve seat is faulty, the complete carburettor must be renewed. Furthermore the carburettor body is not supplied as a separate component but must be obtained as a fully jetted assembly. If faulty valve seating has caused persistent flooding of the float bowl, first attempt a cure by fitting a new needle only. If this fails to cure the problem, attempt to grind the seat in, using a light polishing compound (Brasso or Solvol Autosol). If the second procedure does not have the desired effect, the expense of a complete carburettor must be met.

11 After considerable service, the piston needle and the needle jet in which it slides, will wear, resulting in an increase in petrol consumption. Wear is caused by the passage of petrol and the two components rubbing together. It is advisable to renew the jet periodically in conjunction with the needle. The vacuum piston and carburettor cap also work as a pair. Examine the components for scoring and other damage, checking particularly that the piston does not have a 'tight spot' anywhere in its travel. Such a 'tight spot' could be caused by a bent needle so carry out the check with the needle

detached. Never interchange the piston or cap of one carburettor with that of another.

12 Before the carburettors are reassembled, each should be cleaned out thoroughly, using compressed air. Avoid using a piece of rag since there is always risk of particles of lint obstructing the airways and jet orifices. Never use a piece of wire or any pointed metal object to clear a blocked jet. It is only too easy to enlarge a jet under these circumstances and increase the rate of petrol consumption. If an air line is not available, a blast of air from a tyre pump will usually suffice.

13 Check the air cut-off valve diaphragm for splitting or other damage. Use compressed air to clean the by-pass channel and the valve plate seat. Inspect the small O-ring that fits in the by-pass orifice. Renew components as necessary.

14 Reassemble each carburettor, using the reversed dismantling procedure. Work must be carried out in absolute cleanliness. If possible, use new gaskets, O-rings can be re-used if there is no doubt as to their condition. When replacing the piston, note that it can be refitted in only one position. The groove in the piston side must locate with the projection in the main body.

15 Do not use excessive force when reassembling a carburettor since it is easy to shear a jet or some of the smaller screws. Furthermore, the carburettors are die cast in a zinc based alloy which itself does not have a high tensile strength.

16 Place the two completed carburettors together in preparation for refitting the mounting bars. When doing this attention should be paid to the following points as some of the refitting operations will need to be made more or less simultaneously. The fuel and accelerator pump

(US models only) transfer pipe(s), with an O-ring fitted to the groove in each end must enter the transfer ports in the carburettors. On 500 models, engage the synchronisation adjustment screw at the end of the throttle link rod with the fork in the operating pulley so that the spring and one washer lie against the upper side of the fork. The helical spring should be fitted concentrically between the end of the rod and the pulley rod. Engage the lightweight relief spring so that it interconnects the choke arm with the link rod. Refit the mounting bars one at a time.

17 On 650 models very carefully reassemble the throttle linkage following the reversal of the dismantling sequence. Tighten the mounting bar screws to the specified torque setting, then tighten the throttle joint bolt to the same torque setting and bend up against its flats an unused tab of the lock washer. Reassemble the accelerator pump collar, spring and washer, using a new split pin to secure them.

18 On all models, open and close the throttles a number of times and then check that the two throttle butterflies open and close simultaneously. If this is not the case, slacken the locknut on the synchronisation adjustment screw and turn it as necessary to bring the butterflies into alignment. Tighten the locknut. Whether or not adjustment is made at this juncture does not preclude the necessity for a synchronisation check to be made using a pair of vacuum gauges as described in Section 10.

19 With the throttles working properly and correctly adjusted, set the fast idle mechanism. Unscrew the throttle stop screw until both butterflies are fully closed, then check the clearance between the choke link arm and the throttle drum; this should be approximately 0.8 mm (0.032 in). If adjustment is necessary bend the forked end of the link arm open or closed until the clearance is correct. When the throttle stop screw is returned to its original setting this should give the correct fast idle speed (see Specifications).

20 On US models from 1980 on, the accelerator pump output is set up during manufacture but should be re-checked after the carburettors have been overhauled or the operating linkage has been disturbed. With the throttle stop screw slackened so that both butterflies are closed, check the clearance between the accelerator pump rod and the choke link arm. This should be 0.1 – 0.3 mm (0.004 – 0.012 in). Any necessary adjustment may be made by careful bending of the link arm.

21 At the other end of the arm, the pump stroke is limited by a second tang which stops against the projecting lug on the carburettor body. The specified gap here is 3.1 – 3.3 mm (0.12 – 0.13 in). Once again, adjustment can be made by bending the tang.

5.2a Before separating carburettors, note carefully how components of choke ...

5.2b ... and throttle linkages are fitted together

5.2c Do not lose small spring between pulley and rod – 500 models only

5.3a Remove retaining screws to release float bowl ...

5.3b ... and drive out pivot pin to remove float assembly

5.4a Unscrew secondary main jet ...

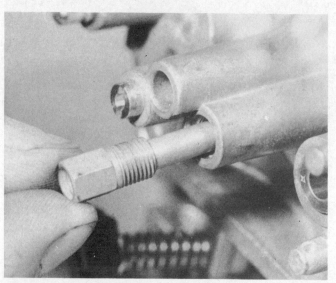
5.4b ... followed by jet holder ...

5.4c ... to permit removal of needle jet

5.4d Black rubber plug covers slow (pilot) jet – jet can be removed on later models only

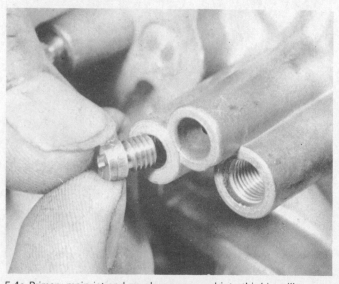
5.4e Primary main jet and nozzle are screwed into third jet pillar

5.6a Remove retaining screws to release carburettor cap and spring ...

5.6b ... then lift out piston assembly

5.6c Displace nylon plug from piston – note sealing O-ring ...

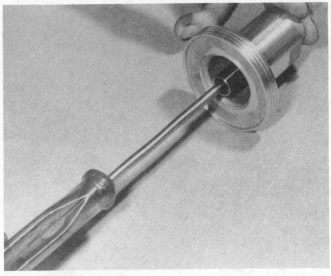

5.6d ... and unscrew retaining screw ...

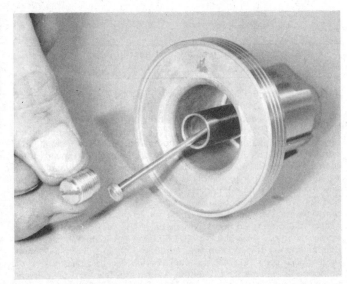

5.6e ... to release jet needle

5.6f Remove curved plate and gasket ...

5.6g ... to permit cleaning of air jets

5.8a Note spring behind air cut-off valve cover ...

5.8b ... and lift out valve diaphragm – renew if split or perished

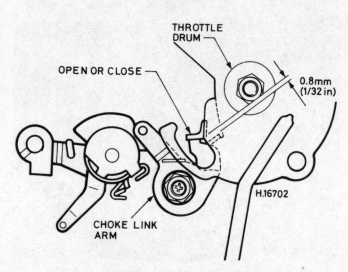

Fig. 3.4 Fast idle adjustment

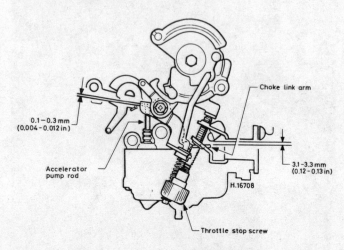

Fig. 3.5 Accelerator pump adjustment

6 Carburettor adjustment and exhaust emissions: general note

1   In some countries legal provision is made for controlling the types and levels of toxic emissions from motor vehicles.

2   In the USA exhaust emission legislation is administered by the Environmental Protection Agency (EPA) which has introduced stringent regulations relating to motor vehicles. The Federal law, entitled The Clean Air Act, specifically prohibits the removal (other than temporary) or modification of any component incorporated by the vehicle manufacturer to comply with the requirements of the law. The law extends the prohibition to any tampering which includes the addition of components, use of unsuitable replacement parts or maladjustment of components which allows the exhaust emissions to exceed the prescribed levels. Violations of the provisions of this law may result in penalties of up to $10 000 for each violation. It is strongly recommended that appropriate requirements are determined and understood prior to making any change to or adjustments of components in the fuel, ignition, crankcase breather or exhaust systems.

3 To help ensure compliance with the emission standards some manufacturers have fitted to the relevant systems fixed or pre-set adjustment screws as anti-tamper devices. In most cases this is restricted to plastic or metal limiter caps fitted to the carburettor pilot adjustment screws, which allow normal adjustment only within narrow limits. Occasionally the pilot screw may be recessed and sealed behind a small metal blanking plug, or locked in position with a thread-locking compound, which prevents normal adjustment.

4 It should be understood that none of the various methods of discouraging tampering actually prevents adjustment, nor, in itself, is re-adjustment an infringement of the current regulations. Maladjustment, however, which results in the emission levels exceeding those laid down, is a violation. It follows that no adjustments should be made unless the owner feels confident that he can make those adjustments in such a way that the resulting emissions comply with the limits. For all practical purposes a gas analyser will be required to monitor the exhaust gases during adjustment, together with EPA data of the permissible Hydrocarbon and CO levels. Obviously, the home mechanic is unlikely to have access to this type of equipment or the expertise required for its use, and, therefore, it will be necessary to place the machine in the hands of a competent motorcycle dealer who has the equipment and skill to check the exhaust gas content.

5 For those owners who feel competent to carry out correctly the various adjustments, specific information relating to the anti-tamper components fitted to the machines covered in this manual is given in the relevant Sections of this Chapter.

6 Note that if the machine is to be used for extended periods at high altitudes the pilot mixture screw settings will require careful re-adjustment, also if the machine is later returned to lower altitudes. This is to preserve its performance and to minimise exhaust emissions. See a local Honda dealer for details relevant to your model.

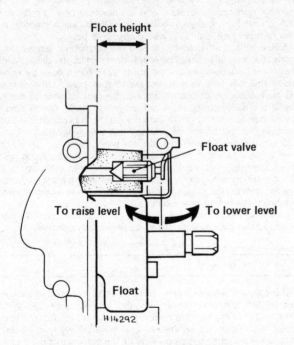

Float height

Float valve

To raise level   **To lower level**

Float

H14292

Fig. 3.6 Measuring the float height

7 Carburettors: checking the settings

1 The various jet sizes and needle position are predetermined by the manufacturer and should not require modification. Check with the Specifications list at the beginning of this Chapter if there is any doubt about the types fitted. If a change appears necessary it can often be attributed to a developing engine fault unconnected with the carburettor(s). Although carburettors do wear in service, this process occurs slowly over an extended length of time and hence wear of the carburettor is unlikely to cause sudden or extreme malfunction. If a

fault does occur check first other main systems, in which a fault may give similar symptoms, before proceeding with carburettor examination or modification.

2 Where non-standard items, such as exhaust systems, air filters or camshafts have been fitted to a machine, some alterations to carburation may be required. Arriving at the correct settings often requires trial and error, a method which demands skill borne of previous experience. In many cases the manufacturer of the non-standard equipment will be able to advise on correct carburation changes.

3 As a rough guide, up to $1/8$ throttle is controlled by the pilot jet, $1/8$ to $1/4$ by the throttle valve cutaway, $1/4$ to $3/4$ throttle by the needle position and from $3/4$ to full by the size of the main jet. These are only approximate divisions, which are by no means clear cut. There is a certain amount of overlap between the various stages. The above remarks apply only in part to constant depression carburettors which utilise a butterfly valve in place of the throttle valve. The first and fourth stages are controlled in a similar manner. The second stage is controlled by the by-pass valve which is uncovered as soon as the throttle valve (piston) is opened. During the third stage the fuel passing through the main jet is metered by the needle jet working in conjunction with the piston needle (jet needle).

4 If alterations to the carburation must be made, always err on the side of a slightly rich mixture. A weak mixture will cause the engine to overheat which may cause engine seizure. Reference to Routine Maintenance will show how, after some experience has been gained, the condition of the spark plug electrodes can be interpreted as a reliable guide to mixture strength.

8 Carburettors: adjustment

1 If the engine is not running properly, causing you to suspect the carburettors, bear in mind the points made in the previous Section when tracing the cause of the problem. Before touching the carburettors, ensure that the air filter is clean and that there are no leaks in the filter casing, air hoses or in the inlet stubs. Check the valve clearances (see Routine Maintenance) and, if a high mileage has been covered, carry out a compression test to check the engine top end. See Chapter 1. Check that the exhaust is in good condition and securely mounted with no leaks, check that the spark plugs are in good condition and correctly gapped. Although unlikely, the ignition timing should be checked (see Chapter 4) to ensure that the fault is not in the ignition system.

2 If the fault is still thought to be in the carburettors, remove them from the machine, dismantle them and clean them thoroughly. See Section 5. On reassembly check the float level to ensure that the basic mixture setting is correct.

3 The float level can only be checked with the carburettors removed from the machine and the float bowls removed. Hold the carburettors vertically with the air filter end upwards so that the float hangs from its pivot. Tilt the carburettors over to 15 – 45° from the vertical until the float is just resting on the needle so that the valve is just closed, then measure the distance between the raised edge of the float bowl mating surface on the main body and the lower surface of the float. The correct distance is 15.5 mm (0.61 in). On early models, make any adjustment by bending the float tang, which abuts against the float needle. Use a pair of electricians pliers to carry out the somewhat delicate adjustment. On later models with plastic floats, if the level is incorrect the float must be renewed. If this does not cure the fault then the needle must be renewed.

4 When the levels are correct, refit the float bowls, fit the carburettors to the machine, having checked the fast idle setting and accelerator pump setting (where applicable). Start the engine and allow it to warm up fully to normal operating temperature (a ten minute ride, or until the temperature gauge needle is in its normal position). Refer to Routine Maintenance and set the idle speed.

5 If attention is required to the pilot mixture screws, owners of US models from 1980 on should refer to the next Section. UK owners (and owners of early US models) proceed as follows.

6 When the engine is fully warmed up, stop it and screw in each mixture screw until it seats lightly and then unscrew it by the exact number of turns specified. This provides an initial setting which is exactly the same for each instrument and which may be used as a basis for the adjustment procedure. Start the engine and, if necessary,

re-adjust the tick-over speed to the normal setting of 1100 ± 100 rpm.

7   Select one carburettor for attention and turn the pilot screw first in one direction and then the other until the highest possible engine speed is obtained. Only very small movements should be made until an altered effect is noted. Now return the engine to the normal tick-over speed of 1100 ± 100 rpm.

8   Repeat this operation on the second carburettor and then readjust the tick-over. Carburettor adjustment is now complete, unless synchronisation is required. See Section 10.

8.6 Set pilot mixture screws carefully as described in text

9   Pilot mixture screw: adjustment – all US models from 1980 on

1   To meet EPA emission control regulations the pilot screw setting is preset at the factory and a limiter cap is fitted over the head of the screw to restrict adjustment to 7/8 of a turn. Any attempt to remove the limiter cap will inevitably break the pilot screw. The limiter cap is so designed that it cannot be turned to richen the mixture whilst the float bowl is attached. This is because the ear of the cap abuts against a lug cast in the float bowl. If it becomes necessary to remove the pilot screw for renewal or cleaning purposes access can be made as follows.

2   Remove the three screws and spring washers from the base of the float bowl and lift the bowl away. Screw the pilot screw in until it lightly touches its seat, counting the number of turns taken. The pilot screw can then be removed and examined for wear in the normal way.

3   When reassembling, screw the pilot screw in fully until it seats lightly and screw it out to the number of turns noted during dismantling.

4   In the event of a new pilot screw being fitted, set the screw position to the figure given in the specifications at the beginning of this Chapter. All new pilot screws should be fitted with limiter caps which are cemented to the screw head with Loctite 601 or an equivalent. See paragraph 8 below.

5   Start the engine and allow it to reach its normal operating temperature (take the machine on a journey of at least 10 minutes duration). By turning the throttle stop screw adjust the engine idle speed to 1100 ± 100 rpm.

6   Select one carburettor for attention and turn the pilot screw first in one direction and then the other until the highest possible engine speed is obtained. Using the throttle stop screw return the engine to the normal tick-over speed of 1100 ± 100 rpm. Screw in the pilot screw slowly until the speed is reduced by 100 rpm. If it is found that the pilot screw seats before the drop of 100 rpm is reached turn the screw out exactly 1 turn and then readjust the tick-over to the normal speed.

7   Repeat this operation on the second carburettor and then readjust the tick-over.

8   When the final setting has been reached, if a new limiter cap is to be fitted it must be positioned so that the pilot screw cannot be turned any further anti-clockwise to richen the mixture, ie with the cap ear bearing against the float bowl on that side. Glue the cap in place, being careful not to disturb the screw setting.

10   Carburettors: synchronisation

1   For the best possible performance it is imperative that the carburettors are working in perfect harmony with each other. If the carburettors are not synchronised, not only will one cylinder be doing less work, at any given throttle opening, but it will also in effect have to be carried by the other cylinder. This will reduce performance accordingly.

2   For synchronisation, it is essential to use a vacuum gauge set consisting of two separate gauges, one of which is connected to each carburettor by means of a special adaptor tube. The adaptor pipe screws into the outside lower end of each inlet stub, the orifice of which is normally blocked off by a crosshead screw plug. Most owners are unlikely to possess the necessary vacuum gauge set, which is somewhat expensive and is normally held by Honda service agents who will carry out the synchronisation operation for a nominal sum.

3   If the vacuum set is available to the owner, the adjustment necessary for synchronisation should be made as follows. First check all other systems and ensure that the carburettors are correctly adjusted, or it will be impossible to get the full benefit of the operation. Remove the blanking plugs from the inlet stubs and fit the adaptor pieces. Connect the gauge hoses to the adaptors.

4   Start the engine and allow it to reach normal working temperature before adjusting the tick-over to 1100 ± 100 rpm. Compare the readings of the two gauges. If the difference exceeds 4 cm Hg (1.6 in Hg) the carburettors require synchronisation. However, this is an extremely generous tolerance and in practice one should not be satisfied until both gauges produce **exactly** the same reading; this is by no means as difficult as it would appear to be.

5   Adjustment is effected by turning the screw that passes into the throttle link arm; this should be easily identified by the dab of white or yellow paint. Depending on the tools available it may be easier to remove the fuel tank to gain access to the screw. Either arrange a separate fuel supply or remove and refit the tank as often as necessary to refill the carburettors while adjustment is made. On models with vacuum taps, do not forget to plug the end of the vacuum pipe before attempting to run the engine with the fuel tank removed.

6   When making adjustments, do not press down on the adjusting screw while rotating it, hold the screw in exactly the same position while fastening the locknut and do not overtighten the locknut. Open and close the throttle quickly to settle the linkage after each adjustment is made, wait for the gauge reading to stabilise and note the effect of the adjustment before proceeding.

7   When the carburettors are correctly synchronised, stop the engine, disconnect the gauges and refit all disturbed components.

11   Air filter: general

The care and maintenance of the air filter element is described in Routine Maintenance. Never run the engine with the air filter disconnected or the element removed. Apart from the risk of increased engine wear due to unfiltered air being allowed to enter, the carburettors are jetted to compensate for the presence of the filter and a dangerously weak mixture will result if the filter is omitted.

12   Crankcase breather pipe: modifications – CX500-A model

1   A modified crankcase breather system has been introduced on the 1980 CX500-A. This modification affects models with engine numbers 2200013 to 2209731.

2   It was found that when the engine speed was increased above 7000 rpm, excess oil was being blown out of the breather tubes fixed to each cylinder head and accumulating in the air filter box. To rectify this, Honda have introduced a new crankcase breather system, whereby the timing inspection plug doubles as an engine breather.

3 In the event of the modified assembly not being fitted, the following parts can be obtained from an authorized Honda dealer.

Cylinder head plug (2 off) – 90547-415-305
Timing inspection/breather plug – 12361-415-610
O-ring – 91302-001-000
Breather tube – 95005-12460-20

4 Dismantle the previous breather assembly. Use the two plugs contained in the kit to blank off the cylinder head nozzles and secure with the wire clips used to retain the previous pipes. Fit the new inspection plug to the timing hole and run the breather pipe from the top of the plug to the pipe joint at the base of the air filter box.

13 Exhaust system: removal and refitting

1 Unlike the two-stroke engine, the exhaust gases of a four-stroke engine are usually not of an oily nature. The silencers are therefore not fitted with detachable baffles. If any fault develops in the silencer the complete unit should be renewed.

2 Removal of the exhaust system may be carried out as follows: slacken off the silencer and exhaust pipe clamp screws at the balance box joints. Loosen the screws as much as possible as the clamps are rounded over at the front edge to help secure the pipe ends. Each silencer may be removed individually after detaching the single support bolt which also secures the pillion footrest. Remove the two nuts that hold each exhaust pipe flange to the cylinder head and pull the flanges off the studs. Each exhaust pipe may be eased forward separately so that the pipe ends leave the exhaust port and balance box.

3 The balance box is supported by a single central bolt which passes through a bracket on the top of the box and a forked projection integral with the engine casing. Slacken the single bolt and then lift the box to the rear so it clears the fork. The box may now be lifted out from either side.

4 The exhaust system may be replaced by reversing the dismantling procedure. Fit a new gasket to each exhaust port to prevent leaks. The silencer and exhaust pipe junctions at the balance box are sealed by special sleeves. These should be renewed if leakage has occurred or if their condition is suspect. Do not tighten any of the securing bolts or nuts until the complete system has been assembled, then tighten them to their specified torque settings in sequence from the front of the machine. This method of assembly will prevent stresses being placed on the components which might cause fracture during service.

10.5 Carburettor synchronisation screw is identified by dab of paint

13.3 Exhaust balance box is mounted on engine by a single bolt

13.4a Always renew all gaskets on reassembly to prevent exhaust leaks

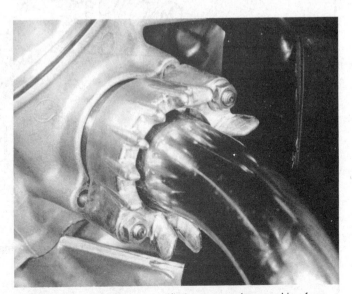

13.4b Tighten mountings to specified torque settings working from front to rear

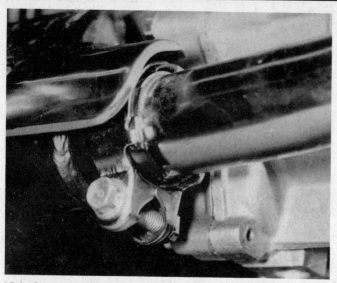

13.4c Check clamp bolts are securely fastened and gaskets intact

14 Engine lubrication system: general

1   Refer to Routine Maintenance for details of crankcase breather cleaning, changing the engine oil and checking its level, changing the filter element and cleaning the oil pump pick-up filter gauze and the filter bypass valve.

2   Refer to Chapter 1 for information on oil pump and pressure relief valve removal and refitting.

3   The oil pressure warning lamp circuit is checked as described in Chapter 7.

15 Oil pump: examination and renovation

1   After removal of the oil pump from the crankcase it should be dismantled as follows: detach the pick-up funnel/strainer unit after removing the two securing bolts. Note the O-ring; this should be discarded and a new item fitted on reassembly. Unscrew the bolt from the pump shaft and lift off the driven sprocket. Mark the outer face of the sprocket so that it may be refitted in the same position. Unscrew the pressure relief valve from the pump body; this unit should be put to one side for further dismantling. Remove the three screws which pass into the pump body and hold the end cover in place. Carefully ease the cover off the pump and then remove the end-float shim, pump shaft and drive pin. Before removing the pump rotors, check that the outer surface of the outer rotor has a small punch mark on it to aid identification. If no mark is evident, one should be made using a spot of paint. On reassembly the rotor must be fitted with the mark outermost again. Displace the two rotors.

2   Clean all the components thoroughly in petrol and allow them to dry. Check the pump castings for cracking or fracture and inspect the inner wall surface of the housing for scoring. Similarly, inspect the outer surface of the rotors. Using a feeler gauge, check the tip-to-tip clearance of the two rotors and the pump body/outer rotor clearance. If the clearances are greater than those given in the Specifications at the beginning of this Chapter, or if damage is evident, the complete pump unit must be renewed. The component parts of the oil pump proper are not available as individual spare parts. It should be noted that scoring of the rotors is only likely to occur as a result of the ingress of grit or metallic particles.

3   Reassemble the pump components in absolutely clean conditions; even a small particle of grit or metal may score the rotor or housing. Remember to fit the outer rotor in its original position. Lubricate the

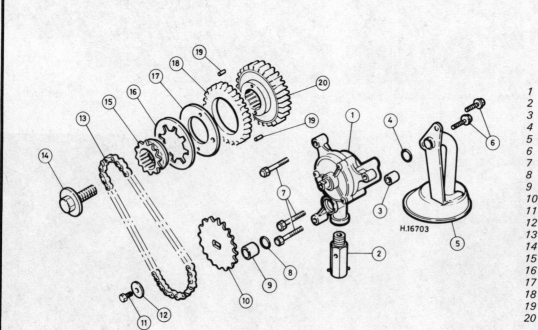

1   Oil pump
2   Pressure relief valve
3   Dowel
4   O-ring
5   Oil strainer
6   Bolt
7   Bolt
8   O-ring
9   Dowel
10 Pump driven sprocket
11 Bolt
12 Washer
13 Drive chain
14 Bolt
15 Pump drive sprocket
16 Spring plate
17 Side plate
18 Anti backlash pinion
19 Pin
20 Primary drive pinion

H.16703

Fig. 3.7 Oil pump and primary drive gear – 500 models

1   Oil pump
2   Bolt
3   Dowel
4   O-ring
5   Oil strainer body
6   Oil strainer
7   Bolt
8   Pressure relief valve
9   O-ring
10 Dowel
11 Pump driven sprocket
12 Bolt
13 Washer
14 Drive chain
15 Bolt
16 Pump drive gear
17 Primary drive pinion

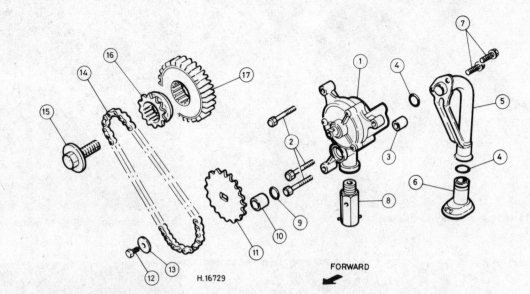

FORWARD

H.16729

Fig. 3.8 Oil pump and primary drive gear – 650 models

rotors thoroughly before fitting the cover and check that the pump shaft turns freely when the screws have been tightened. Final tightening of the relief valve and the sprocket bolt may be carried out after the pump has been installed in the casing.

16 Oil pressure relief valve: examination and renovation

The pressure relief valve may be dismantled for inspection and cleaning. Place the valve in the soft jaws of a vice with the threaded

portion downwards. The valve plunger is under tension from a spring and spring cap which are retained under load by a radially placed roll pin. Drive the pin out using a suitable drift. Before withdrawing the drift, depress the spring cap using a small screwdriver so that it and the spring are not ejected into the darkest recesses of the workshop. After removing the drift, slowly release the spring cap and spring and invert the valve body so that the valve plunger falls out. If the plunger has become stuck, it may be pushed out using a suitable rod inserted from the threaded end. If the valve plunger or the spring have become damaged, the complete valve unit should be renewed. Carefully clean all components before reassembly.

15.1a Mark sprocket outer face before removing retaining bolt

15.1b Pump cover is retained by three screws – check for rotor face identification marks before removing them

15.1c Note shim on pump driveshaft ...

15.1d ... and rotor drive pin – do not lose

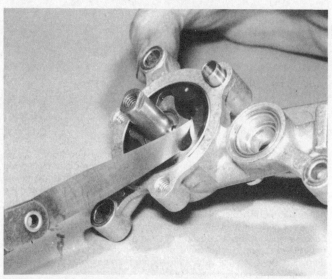

15.2a Measuring inner rotor/outer rotor clearance

15.2b Measuring outer rotor/pump body clearance

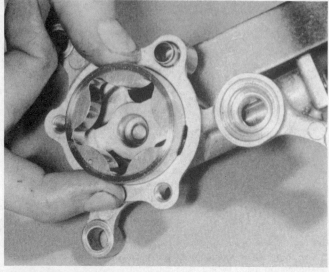

15.3 Rotor face identification marks must face outwards on reassembly

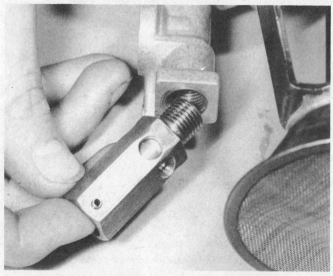

16.1 Oil pressure relief valve is screwed into pump body.

Chapter 4 Ignition system

Contents

Specifications

Note: *unless otherwise stated, information applies to all models*

Ignition system

Type:

All CX500, CX500 C (except CX500 C 1982) and
CX500 D models .... Capacitive Discharge Ignition (CDI)

All other models .... Transistorised electronic

Ignition timing – CDI system

Initial – FL or FR mark aligned .... 15° BTDC @ idle speed

Advance starts at .... 1750 – 2250 rpm

Full advance – two parallel lines aligned .... 37 ± 3° BTDC @ 5500 – 6000 rpm

Ignition timing – transistorised system

Initial – F1 mark (L/H or R/H) aligned .... 15° BTDC @ idle speed

Advance starts at .... 1500 ± 150 rpm

Full advance – two parallel lines aligned:

CX500 C 1982 .... 40 ± 1.5° BTDC @ 2780 rpm

CX500 E-C, all GL500 models .... 45 ± 1.5° BTDC @ 3000 rpm

All 650 models .... 40 ± 1.5° BTDC @ 3500 rpm

Ignition HT coil

Minimim spark gap .... 6 mm (0.24 in)

Pulser coil – transistorised system
Pulser coil/ATU rotor air gap .... 0.45 – 0.65 mm (0.018 – 0.026 in)

Spark plugs – standard recommendation

	NGK	ND
All UK 500 models, CX500 C 1982, GL500 and GL500 I 1982	DR8ES-L	X24ESR-U
All other US 500 models	D8EA	X24ES-U
All 650 models	DPR8EA-9	X24EPR-U9

Spark plug gap
500 models .... 0.6 – 0.7 mm (0.024 – 0.028 in)
650 models .... 0.8 – 0.9 mm (0.032 – 0.035 in)

1 General description

Early models are fitted with a full electronic ignition system of the CDI type in which two fixed pulser coils mounted on the inside of the engine rear cover are triggered by a small magnetic pick-up on the periphery of the alternator rotor; a source coil in the alternator stator provides power for the system. A CDI unit and two HT coils are the only external components of the system. The spark is advanced at higher engine speeds by a separate circuit controlled by two advance pulser coils which are triggered by a small rotor mounted on the rear end of the alternator rotor centre; these two coils are mounted on a stator plate, in a housing on the outside of the engine rear cover.

Later models are fitted with a fully-transistorised electronic ignition system in which two pulser or pick-up coils are mounted on a stator plate, in a housing on the outside of the engine rear cover, and are triggered by a raised tooth on the cam of a mechanical automatic timing unit (ATU) which is mounted on the rear end of the alternator rotor centre. The trigger pulse generated as the ATU tooth passes each coil's steel core signals a separate spark unit for each cylinder. The

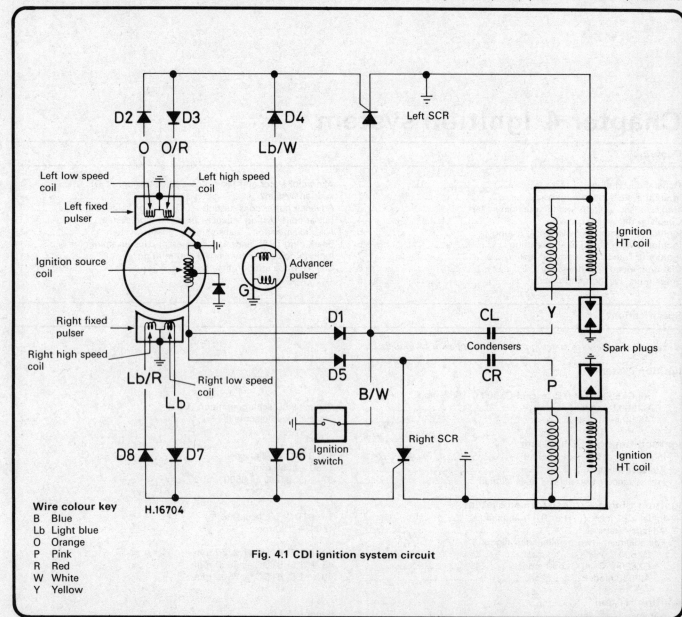

Wire colour key
B   Blue
Lb Light blue
O   Orange
P   Pink
R   Red
W   White
Y   Yellow

H.16704

Fig. 4.1 CDI ignition system circuit

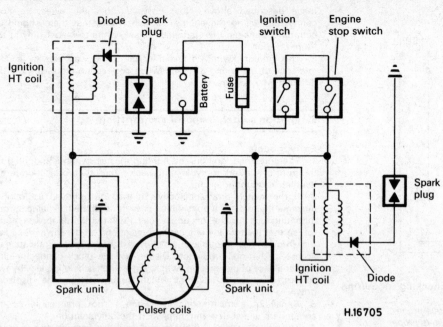

Fig. 4.2 Transistorised ignition system circuit

H.16705

spark units amplify the signal to activate the ignition HT coils. The components are completely separate for each cylinder.

Neither system requires any maintenance except for the spark plugs. See Routine Maintenance.

2   Ignition system: fault diagnosis

1   As no means of adjustment is available, any failure of the system can be traced to the failure of a system component or a simple wiring fault. Of the two possibilities, the latter is by far the most likely. In the event of failure, check the system in a logical fashion, as described below.

2   Remove the spark plugs giving them a quick visual check noting any obvious signs of flooding or oiling. Fit the plugs into the plug caps and rest them on the cylinder head so that the metal body of each plug is in good contact with the cylinder head metal. The electrode end of the plugs should be positioned so that sparking can be checked as the engine is spun over.

3   *Important note.* The energy levels in electronic systems can be very high. **On no account** should the ignition be switched on whilst the plugs or plug caps are being held. Shocks from the HT circuit can be most unpleasant. Secondly, it is vital that the plugs are soundly earthed when the system is checked for sparking. The system components **can be seriously damaged** if the HT circuit becomes isolated.

4   Having observed the above precautions, check that the kill switch is in the 'Run' position, turn the ignition switch to 'On' and turn the engine over. If the system is in good condition a regular, fat blue spark should be evident at the plug electrodes. If the spark appears thin or yellowish, or is non-existent, further investigation will be necessary, but first check the plugs by substituting new ones. Before proceeding further, turn the ignition off and remove the key as a safety measure.

5   Ignition faults can be divided into two categories, namely those where the ignition system has failed completely, and those which are due to a partial failure. The likely faults are listed below, starting with the most probable source of failure. Work through the list systematically, referring to the subsequent sections for full details of the necessary checks and tests. For models with transistorised ignition, refer to Section 3.

Total or partial ignition system failure:

a)   Loose, corroded or damaged wiring connections, broken or shorted wiring between any of the component parts of the ignition system
b)   Faulty main switch or engine kill switch
c)   Faulty ignition HT coil
d)   Faulty CDI unit or spark unit
e)   Faulty generator source coil (CDI system only)
f)   Faulty pulser coil

3   Transistorised ignition system: preliminary test

1   On later models a simple test can be used to check the system, but its usefulness is mitigated by the amount of preliminary dismantling required, and by the fact that it merely duplicates the much quicker test outlined in the previous Section.

2   Referring to the instructions given elsewhere in this Manual, remove the rear wheel and swinging arm. On some models it may also be necessary to remove the air filter casing and the inner rear mudguard. Remove the pulser cover.

3   Remove both spark plugs, connect each to its cap and place it so that each plug's metal body is firmly in contact with a good engine earth point. Check that the engine kill switch is in the 'Run' position and switch on the ignition.

4   Touch the blade of a screwdriver to each coil's steel core in turn; a healthy spark should appear at the respective spark plug's electrodes.

5   If sparks did not occur when carrying out the test as described in the previous Section, but do occur when carrying out this test, the fault can only be in the ATU, although there is a slight possibility of a wiring fault which should be checked as described later in this Chapter.

4   Checking the wiring

1   The wiring should be checked visually, noting any signs of corrosion around the various terminals and connectors. If the fault has developed in wet conditions it follows that water may have entered any of the connectors or switches, causing a short circuit. A temporary cure can be effected by spraying the relevant area with one of the proprietary de-watering aerosols such as WD40 or similar. A more permanent solution is to dismantle the switch or connector and coat the exposed parts with silicone grease to prevent the ingress of water. The exposed backs of connectors can be sealed off using a silicone rubber sealant.

2   Light corrosion can normally be cured by scraping or sanding the affected area, though in serious cases it may prove necessary to renew the switch or connector affected. Check the wiring for chafing or breakage, particularly where it passes close to part of the frame or its fittings. As a temporary measure damaged insulation can be repaired with PVC tape, but the wire concerned should be renewed at the earliest opportunity.

3   Using the wiring diagram at the end of the manual, check each wire for breakage or short circuits using a multimeter set on the resistance scale or a dry battery and bulb wired as shown in the accompanying illustration. In each case, there should be continuity between the ends of each wire.

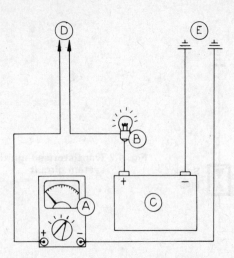

Fig. 4.3 Simple testing arrangement for checking the wiring

A   Multimeter   D   Positive probe
B   Bulb   E   Negative probe
C   Battery

5   Ignition and engine kill switches: testing

CDI ignition

1   The ignition system is controlled by the ignition switch or main switch which is bolted to the fork top yoke. The switch has several terminals and leads, of which two are involved in controlling the ignition system. These are the 'IG' terminal (black /white lead) and the 'E' terminal (green lead). The two terminals are connected when the switch is in the 'Off' position and prevent the ignition system from functioning by shorting the CDI unit to earth. When the switch is in the 'On' position the CDI/earth connection is broken and the system is allowed to function.
2   If the operation of the switch is suspect, reference should be made to the wiring diagram at the end of this book. The switch connections are also shown in diagrammatic form and indicate which terminals are connected in the various switch positions. The wiring from the switch can be traced back to the respective connectors where test connections can be made most conveniently.
3   The purpose of the test is to check whether the switch connections are being made and broken as indicated by the diagram. In the interests of safety the test must be made with the machine's battery disconnected, thus avoiding accidental damage to the CDI system or the owner. The test can be made with a multimeter set on the resistance scale, or with a simple dry battery and bulb arrangement as previously shown. Connect one probe lead to each terminal and note the reading or bulb indication in each switch position.
4   If the test indicates that the black/white lead is earthed irrespective of the switch position, trace and disconnect the ignition (black/white) and earth (green) leads from the ignition switch. Repeat the test with the switch isolated. If no change is apparent, the switch should be considered faulty and renewed. If the switch works normally when isolated, the fault must lie in the black/white lead between the switch and the CDI unit.
5   The kill switch, mounted on the right-hand handlebar, is tested the same way as its connections and functions are exactly as described above.
6   If either switch is found to be faulty it must be renewed. While each is a sealed unit and can only, officially, be repaired by renewing it as a complete assembly, there is nothing to be lost by attempting to repair it if tests have proven it faulty. Depending on the owner's skill, worn contacts may be reclaimed by building up with solder or in some cases, merely cleaning with WD40 or a similar water dispersant spray.

Transistorised ignition

7   The switches for models equipped with this system are tested (and repaired or renewed, as necessary) in substantially the same way as

those described above. Note however, that the wires are now connected first to the engine kill switch and that there should be continuity between their terminals (black/white to black) only in the 'Run' position.
8   The ignition switch is tested in the same way, but by careful reference to the wiring diagram switch connection tables to note the different wire colours.

6   Ignition switch: removal and refitting

Early models

1   The combined ignition and lighting master switch is mounted on the front of the fork top yoke, and passes through a projection in the handlebar mounting clamp.
2   If the switch proves defective, it may be removed as follows: unscrew the two small bolts which pass through the headlamp shroud mounting lugs into the fork upper yoke. Push the shroud forward at the top so that the two lower locating projections on the lower yoke free the shroud. The shroud can be moved forward sufficiently to gain access to the ignition switch. Disconnect the block connector plug from the base of the ignition switch. The switch is held in place by two screws, after the removal of which the switch can be displaced downwards.
3   Repair of a malfunctioning switch is not practicable as the component is a sealed unit; renewal if the only solution.
4   The switch may be refitted by reversing the dismantling sequence. Tighten the bolts to the specified torque setting. Remember that a new switch will also require a new set of keys.

Later models – 1981 on

5   Disconnect the instrument drive cables from the console and move the cables clear. Remove the two bolts which retain the instrument console to its mounting plate. This will allow the console to be removed upwards just enough to reveal the two bolts retaining the ignition switch to the yoke. Where fitted, take care not to damage the front fork air pressure balance pipe when slackening the bolts. Disconnect the block connector from the base of the switch and remove the switch. If necessary the headlamp unit can be moved slightly from position by removing the single bolt on each side, to allow removal of the switch.
6   The only part of the switch that can be dismantled is the wiring connector at its base. Position the ignition key between the 'On' and 'P' positions of the lock and with the use of a screwdriver push the plastic ears of the connector out of location through the three rectangular slots, around the base of the switch. The connector can now be removed for inspection.
7   Reassembly is a direct reversal of the dismantling procedure. If the headlamp unit position has been disturbed it must be refitted so that the punch marks on each side coincide with those on the support brackets.

7   Ignition HT coils: location and testing

All models

1   Two separate ignition coils are fitted, each of which supplies a different cylinder. The coils are mounted below the fuel tank, each side of the frame top tube.
2   If a weak spark, poor starting or misfiring causes the performance of the coils to be suspected, they should be tested on a spark-gap tester by a Honda Service Agent or an auto-electrician who will have the appropriate equipment.
3   It is unlikely that the coils will fail simultaneously. If intermittent firing occurs on one cylinder, the coils may be swopped over by interchanging the low tension terminal leads and the HT leads. If the fault then moves from one cylinder to the other, it can be taken that the coil is faulty.
4   The coils are a sealed unit and therefore if a failure occurs, repair is impracticable. The faulty item should be replaced by a new component.

Transistorised ignition models

5   On these later models the HT coils can also be tested by measuring the resistance of their windings. Note that this will serve to give some

indication of a coil's condition, but will not be as effective as the spark-gap test mentioned above. If any of these tests show a coil to be defective, it must be renewed; if however they show it to be in good condition but the fault persists, then the spark-gap test must be carried out to check.

6 Remove the seat and fuel tank to gain access to the ignition coil mountings. Remove the bolts retaining the coils to the frame and disconnect the wiring connectors. Using a multimeter set on the resistance scale inspect the primary windings as follows. Checking one coil at a time, connect one probe of the tester to one contact in the plastic block connector and the other probe to the remaining contact. The resistance measured should be 2-3 ohms. Carry out the same test on the primary windings of the second coil.

7 Before checking the secondary windings check carefully the coil body. If an S mark is moulded into it, the winding resistance can be checked as described in paragraphs 8 and 9 below; if no S mark is found the test is as described in paragraph 10. In both cases, when checking the resistance of the secondary windings it is necessary to use either a multimeter of the Sanwa SP10-D or Kowa type because the correct readings quoted by Honda can only be obtained with these testers. Part numbers for the testers are, Sanwa – 07308-0020000 and Kowa – TH-5H-1.

8 Set the tester to the K ohms scale and connect the negative probe of the tester to the black and white wire contact in the plastic block connector and the positive probe to the high tension terminal. The reading obtained on the Sanwa tester should be within 200-350 K ohms and 50-200 K ohms on the Kowa tester.

9 A single diode is fitted to the secondary windings to stop the high tension current from flowing back up the feed wire and into the secondary windings of the other coil. In this event both cylinders would fire at the same time. To test the correct operation of the diode, transpose the probes of the tester used for the secondary winding test. An infinitely high resistance should register on the tester scale, indicating that the diode is in working order.

10 Connect two 12 volt batteries as shown in the accompanying illustration and check that 23-25 volts is available before starting. Set the tester to the required milli-amps range (Sanwa 25 mA, Kowa 100 mA) and connect it as shown. A reading of 3 mA approximately should be obtained on the Sanwa tester; on the Kowa tester the needle should swing slightly. Reverse the tester polarity to test the diode (see above); there should be no continuity.

8 CDI unit: location and testing

1 The CDI unit is located below the dualseat where it is mounted on top of the rear mudguard. This unit, which contains the various transistorised components that control the ignition system, is sealed for life, the components being encapsulated in a resin. To remove the unit, withdraw the seat, unplug the wiring connectors and unbolt the unit from its mounting.

2 To test the unit, Honda advise against the use of any test meter other than the Sanwa Electric Tester Type SP-10D (Honda part number 07308-0020000) or the Kowa Electric Tester (TH-5H), because the use of other instruments may result in inaccurate readings.

3 Most owners will find that they either do not possess a multimeter, in which case the CDI unit will have to be checked by a Honda Service Agent, or own a meter which is not of the specified make or model. In the latter case, a good indication of the unit's condition can be gleaned in spite of inaccuracies in the readings. If necessary, the CDI unit can be taken to a Honda Service Agent or auto-electrician specialist for confirmation of its condition.

4 The test details are given in the accompanying illustration in the form of two tables of meter probe connections with the expected reading in each instance. If an ordinary multimeter is used, the resistance range may be determined by trial and error. The diagram illustrates the CDI unit connections referred to in the table. For owners not possessing a test meter the unit or the complete machine can be taken to a Honda Service Agent for testing. When using the tables, identify the table to be used by the finish of the unit ie, plated (galvanised) or black-painted. The models given are as precise as possible, but not conclusive.

5 If the CDI unit is found to be faulty as a result of these tests, it must be renewed. No repairs are possible. It might be worthwile trying to obtain a good secondhand unit from a motorcycle breaker, in view of the cost of a new part.

7.1 Ignition HT coils are mounted each side of frame top tube

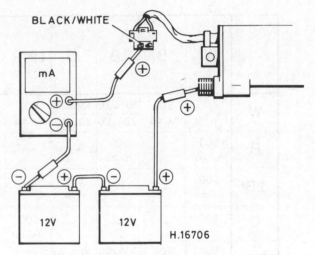

Fig. 4.4 Transistorised ignition system – testing non S-marked HT coil secondary windings

8.1 Location of CDI unit

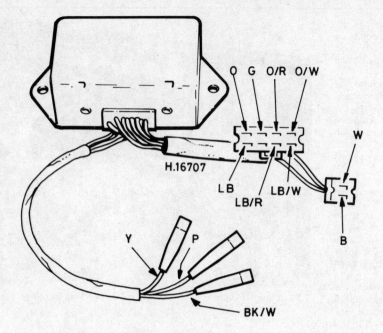

Fig. 4.5 CDI unit test connections

See test table Fig. 4.6 for early models up to 1979 or Fig. 4.7 for later models 1979 on

Note: *the results shown in both tables relate to two recommended testers. The upper figure is the expected result when using a Sanwa tester set on the x K ohm scale and the lower, when using a Kowa tester set on the x 100 ohm scale.*

Wire colour key

B	Blue	P	Pink
BK	Black	R	Red
G	Green	W	White
LB	Light blue	Y	Yellow
O	Orange		

	W	B	LB	O	LB/R	O/R	G	LB/W	O/W	P	Y	BK/W
W		500 ~ ∞	10~20 / 20~60	10~20 / 20~60	500 ~ ∞	500 ~ ∞	3 ~ 8	10~20 / 20~60	10~20 / 20~60	500 ~ ∞	500 ~ ∞	500 ~ ∞
B	500 ~ ∞		500 ~ ∞	←	←	←	←	←	←	←	←	←
LB	500 ~ ∞	←		500 ~ ∞	←	←	←	←	←	←	←	←
O	500 ~ ∞	←	←		500 ~ ∞	←	←	←	←	←	←	←
LB/R	500 ~ ∞	500 ~ ∞	10~20 / 20~60	10~20 / 20~60		500 ~ ∞	3~8 / 5~20	10~20 / 20~100	10~20 / 20~100	500 ~ ∞	←	←
O/R	500 ~ ∞	500 ~ ∞	10~20 / 20~60	10~20 / 20~60	500 ~ ∞		3~8 / 5~20	10~20 / 20~100	10~20 / 20~100	500 ~ ∞	←	←
G	500 ~ ∞	500 ~ ∞	3~8 / 5~20	3~8 / 5~20	500 ~ ∞	500 ~ ∞		3~8 / 5~20	3~8 / 5~20	500 ~ ∞	←	←
LB/W	500 ~ ∞	←	←	←	←	←	←		500 ~ ∞	←	←	←
O/W	500 ~ ∞	←	←	←	←	←	←	←		500 ~ ∞	←	←
P	500 ~ ∞	∞ ⌒	←	←	500 ~ ∞	500 ~ ∞	∞ ⌒	←	←		∞ ⌒	∞ ⌒
Y	500 ~ ∞	∞ ⌒	←	←	500 ~ ∞	500 ~ ∞	∞ ⌒	←	500 ~ ∞	500 ~ ∞		∞ ⌒
BK/W	500 ~ ∞	10~20 / 20~60	500 ~ ∞	←	←	←	←	←	←	←	←	

H.12583

Fig. 4.6 Testing the CDI unit – plated type (early models up to 1979)

PROBE − \ PROBE +	W	B	LB	O	LB/R	O/R	G	LB/W	O/W	P	Y	BK/W
W		500 ~∞	10~20 / 20~60	10~20 / 20~60	500 ~∞	500 ~∞	3~8 / 3~20	10~20 / 20~60	10~20 / 20~60	500 ~∞	500 ~∞	500 ~∞
B	500 ~∞		500 ~∞	500 ~∞	500 ~∞	500 ~∞	500 ~∞	500 ~∞	500 ~∞	500 ~∞	500 ~∞	500 ~∞
LB	500 ~∞	500 ~∞		500 ~∞	500 ~∞	500 ~∞	500 ~∞	500 ~∞	500 ~∞	500 ~∞	500 ~∞	500 ~∞
O	500 ~∞	500 ~∞	500 ~∞		500 ~∞	500 ~∞	500 ~∞	500 ~∞	500 ~∞	500 ~∞	500 ~∞	500 ~∞
LB/R	500 ~∞	500 ~∞	10~20 / 20~60	10~20 / 20~60		500 ~∞	3~8 / 5~20	10~20 / 20~100	10~20 / 20~100	500 ~∞	500 ~∞	500 ~∞
O/R	500 ~∞	500 ~∞	10~20 / 20~60	10~20 / 20~60	500 ~∞		3~8 / 5~20	10~20 / 20~100	10~20 / 20~100	500 ~∞	500 ~∞	500 ~∞
G	500 ~∞	500 ~∞	3~8 / 5~20	3~8 / 5~20	500 ~∞	500 ~∞		3~8 / 5~20	3~8 / 5~20	500 ~∞	500 ~∞	500 ~∞
LB/W	500 ~∞	500 ~∞	500 ~∞	500 ~∞	500 ~∞	500 ~∞	500 ~∞		500 ~∞	500 ~∞	500 ~∞	500 ~∞
O/W	500 ~∞	500 ~∞	500 ~∞	500 ~∞	500 ~∞	500 ~∞	500 ~∞	500 ~∞		500 ~∞	500 ~∞	500 ~∞
P	500 ~∞	∞ (diode)	∞ (diode)	∞ (diode)	500 ~∞	500 ~∞	∞ (diode)	∞ (diode)	∞ (diode)		∞ (diode)	500 ~∞
Y	500 ~∞	∞ (diode)	∞ (diode)	∞ (diode)	500 ~∞	500 ~∞	∞ (diode)	∞ (diode)	∞ (diode)	500 ~∞		500 ~∞
BK/W	500 ~∞	20~100 / 100~500	500 ~∞	500 ~∞	500 ~∞	500 ~∞	500 ~∞	500 ~∞	500 ~∞	∞ (diode) / 500~∞	∞ (diode) / 500~∞	

H.12582

Fig. 4.7 Testing the CDI unit – black painted type (later models 1979 on)

9 Spark units: location and testing

1 The spark units are two identical rectangular units mounted in an aluminium alloy casing and mounted by one bolt. They are to be found mounted side by side under the seat on the CX500 C 1982 model and behind the left-hand side panel to the rear of the battery on all other models except the CX650 C. In all of the above cases, the units can be disconnected at their block connectors and unbolted.

2 On the CX650 C model, remove the left-hand side panel and the battery, then remove the three bolts securing the battery carrier to the air filter casing, and withdraw the carrier to expose the units. They can then be removed as described above.

3 To test each unit, first disconnect the pulser coil four-pin block connector. Start by testing the left-hand cylinder spark unit. Use the diagram accompanying this text for guidance. Connect a multimeter set to the voltage measuring function, between the blue with yellow tube wire of the six pin block connector, and earth the other side to the frame. Run another wire from the blue with white tube wire from the pulser coil's four-pin connector to earth on the frame. Turn the ignition on and check the reading on the multimeter. The needle should change from 12 to 0 volts each time the white-tubed wire is earthed. Turn the ignition off and disconnect the apparatus. To test the right-hand spark unit, connect the voltmeter between the yellow wire of the six-pin block connector and earth. Run a wire from the yellow with white tube wire of the pulser coil's four-pin connector to earth on the frame. Turn the ignition on and check that the voltmeter behaves in the same way as before.

4 If a fault is apparent in either of the spark units, that unit should be renewed because repair is impracticable.

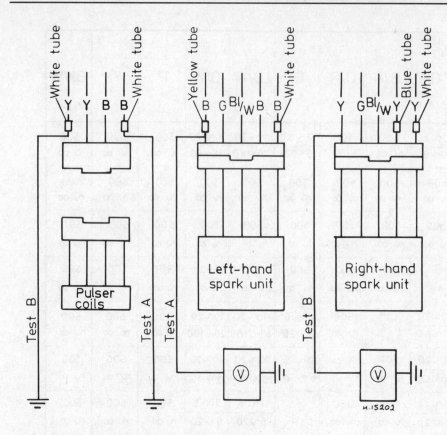

Fig. 4.8 Voltmeter connections for spark unit tests

Test A – Left-hand spark unit test

Test B – Right-hand spark unit test

Y	Yellow	G Green
B	Blue	Bl Black
W	White	

9.1 Location of spark units (transistorised ignition models) except CX500 C and CX650 C

to the top of the rear mudguard. Measure the resistance across the following pairs of wires on the larger male connector plug and compare the readings with the following specified resistance figures.

Wire colour	Resistance (Ohms)
White to Blue	77 – 95
Green to White	387 – 473*
Orange to Green	95 – 116
Light blue to Green	95 – 116
Orange/Red to Green	81 – 99
Light blue/Red to Green	81 – 99

Note: on some models, unspecified by the manufacturer, the resistance range for this test is 315 – 385 ohms.

If the resistance of one or more pairs of wires is found to be outside the range given in the table, there is evidence that the components are not serviceable. The alternator stator, the two pulser units and the alternator rotor are available only as a set and as a result the replacement cost is high. It is wise to have the system double-checked by an expert before renewing the components.

3   Apparent failure of the ignition source coil or the pulser may be a result of breakage of wires external to the units themselves. This may be verified by a visual inspection after removal of the components from the engine casing. New wires may be fitted and then joined at a suitable point, using a soldering iron.

10 Alternator stator and fixed pulser coils: testing – CDI ignition system

1   The ignition source coil and the fixed pulsers can be tested whilst still in place on the machine for breakdown of internal insulation and connections. An ohmmeter or a multi-meter set to the x10 ohm resistance position is required to carry out this test.

2   Remove the seat, disconnect the battery and disconnect the large block connector and two smaller block connectors which are clipped

11 Advance pulser coils: testing – CDI ignition system

1   The advance pulser coils may be tested in place on the machine in a manner similar to that described for the fixed pulsers. The two wires from the coils are connected by two separate snap connectors close to the unit and hidden behind the plated cover at the rear of the engine rear cover. In addition to this, the wires are connected at one of the smaller block connectors on the rear mudguard. By virtue of good accessibility, make the initial resistance test at the block connector. Check that the resistance between the Light blue/White lead and a suitable earth point (the Green wire) is within the range 185 – 225 ohms. Repeat the test between the Orange/White wire and earth; the

result should be within the same specified range. If either reading is incorrect, further inspection should be made.

2 Removal of the chromed pressed-steel cover which encloses the advance pulser at the rear of the engine is not impossible, merely difficult due to lack of working space. If care is taken, the bolts holding the plate in position can be removed and the cover slipped from place. Access may now be made to the two separate snap connectors which connect the pulser leads that terminate at the block connector at which the initial test was made. Disconnect the snap connectors and repeat the resistance test between each lead and the stator ring of the advance pulser unit. If the readings are both correct, it is possible that the initial readings were not reliable, being caused by poor conductivity at the snap connectors.

3 If the final readings taken are not within the specified range, the unit must be renewed. Check first that the fault is not due to a break or short in the wiring which might be repaired.

11.1 Advance pulser coils (CDI ignition system) are located on engine rear cover

12 Pulser coils: testing – transistorised ignition system

1 Remove the seat (or the left-hand side panel) to gain access to the block connector. Disconnect it and carry out the following resistance test on that half of the connector which carries the wires from the pulser coil assembly. Connect a multimeter set on the resistance scale across the wire contacts for the right-hand cylinder, these being the two yellow wires. The coil resistance should be 530 ± 50 ohm at around 20°C (68°F). Carry out the same test across the two blue wires for the left-hand cylinder pulser coil.

2 If either unit does not produce the expected result it is faulty. This means that the complete pulser stator assembly must be renewed. Check first that the problem is not due to a wiring fault which might be repaired.

13 ATU: examination – transistorised ignition system

1 The unit comprises spring loaded balance weights, which move outward against the spring tension as centrifugal force increases. The balance weights must move freely on their pivots and be rust-free. The tension springs must also be in good condition. Keep the pivots lubricated and make sure the balance weights move easily, without binding. Most problems arise as a result of condensation, within the engine, which causes the unit to rust and balance weight movements to be restricted.

2 If the unit is severely worn or if greasing does not cure sticking, it must be renewed. When reassembling the unit after lubrication, align

the rotor tooth with the backplate cutout. Use a light high melting-point grease for lubrication.

14 Spark plug (HT) leads and suppressor caps: examination

1 Erratic running faults and problems with the engine suddenly cutting out in wet weather can often be attributed to leakage from the high tension leads and spark plug caps. If this fault is present, it will often be possible to see tiny sparks around the leads and caps at night. One cause of this problem is the accumulation of mud and road grime around the leads, and the first thing to check is that the leads and caps are clean. It is possible to cure the problem by cleaning the components and sealing them with an aerosol ignition sealer, which will leave an insulating coating on both components.

2 Water dispersant sprays are also highly recommended where the system has become swamped with water. Both these products are easily obtainable at most garages and accessory shops. Occasionally, the suppressor caps or the leads themselves may break down internally. If this is suspected, the components should be renewed.

3 Where the HT leads are permanently attached to the ignition coil it is recommended that the renewal of the HT leads is entrusted to an auto-electrician who will have the expertise to solder on new leads without damaging the coil windings.

4 When renewing the suppressor caps, be careful to purchase ones that are suitable for use with resistor spark plugs (where fitted).

15 Ignition components: removal and refitting

1 All components of the ignition system that are outside the engine unit are removed and refitted as described in the relevant Sections of this Chapter.

2 The ignition source coil (CDI system), alternator and fixed pulser coils (CDI system) are mounted inside the engine rear cover. If they are to be renewed, the engine must be removed from the frame and the rear cover must be detached. See Chapter 1.

3 The advance pulser coils and rotor (CDI system) or the pulser coils and ATU (transistorised system), however, represent a different story. Depending on the owner's ingenuity it may be possible to reach them with the engine in the frame. Honda's recommendation is to remove the rear wheel and swinging arm to enable them to be reached.

4 For instructions on removing and refitting all these components, refer to the relevant Sections of Chapter 1. Note particularly the need to align the fixed pulser and/or pulser stator (as applicable) with fixed timing marks on reassembly; also note the procedure necessary to set the pulser coil air gap (transistorised ignition).

16 Checking the ignition timing

1 Since no provision exists for adjusting the ignition timing (with the minor exceptions mentioned below), and since no ignition component is subject to mechanical wear there is no need for regular checks; only if investigating a fault such as a loss of power or a misfire should the timing be checked.

2 The ignition timing can be checked only whilst the engine is running using a stroboscopic lamp; therefore a suitable timing lamp will be required. The inexpensive neon lamps should be adequate in theory, but in practice may produce a pulse of such low intensity that the timing mark remains indistinct. If possible, one of the more precise xenon tube lamps should be employed powered by an external source of the appropriate voltage. Do not use the machine's own battery as an incorrect reading may result from stray impulses within the machine's electrical system.

3 Remove the inspection plug from the right-hand side of the engine rear cover. The set of timing marks for each cylinder, and the fixed timing pointer, can now be identified. Refer to the appropriate accompanying diagram; for the left-hand cylinder on CDI models the TL line represents TDC, the FL line the initial timing mark, and the two adjacent unmarked parallel lines the full advance mark. Similar marks are given for the right-hand cylinder. On transistorised ignition models the TDC marks are the same as on CDI machines, but the initial timing mark is made up of two lines; the FS mark is for static timing (see

Chapter 1) and the FI mark is the dynamic (engine running) mark. The full advance marks are the same.

4   To check the ignition timing connect a timing lamp to the right-hand cylinder HT lead as directed by the lamp's manufacturer. Start the engine and when warm set the idle speed at 1100 ± 100 rpm. Point the lamp at the rotor and check that the initial timing mark is aligned with the fixed index pointer. At a slightly increased engine speed (see Specifications) the timing will start to advance and the mark on the flywheel will go out of alignment with the index pointer. Increase the engine speed to that specified and check that the full advance marks align with the index pointer.

5   Stop the engine and connect the stroboscope up to the left-hand cylinder HT lead. Check the timing as shown for the right-hand cylinder.

6   Because of the nature of the ignition system it is unlikely that the timing will be incorrect. For this reason no actual means of adjusting the timing is provided. A small amount of adjustment can, however, be made by altering the position of the pulser coil baseplate in relation to the rotor (CDI ignition) or ATU (transistorised ignition). This will of course require considerable dismantling and the need for the air gap to be rechecked. It is also worth checking that the alternator rotor is securely fastened and correctly located on the crankshaft by its Woodruff key; the same applies to the rotor or ATU (as applicable). Check also that all coils are securely fastened.

7   On models with CDI ignition, any deviation from the correct ignition timing can only be a result of failure of one or more of the controlling components: ie the ignition source coil, timing pulsers or the CDI unit.

8   On models with transistorised ignition, unless some component is improperly fastened or not correctly located, the only cause of incorrect timing can be the ATU. If it cannot be cured by careful dismantling, lubrication and reassembly, it must be renewed.

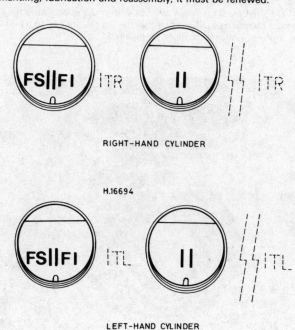

Fig 4.10 Ignition timing marks – Transistorised ignition

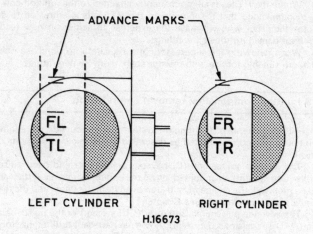

Fig. 4.9 Ignition timing marks – CDI ignition

Chapter 5 Frame and forks

Contents

Specifications

Note: *unless otherwise stated, information applies to all models*

Front forks

	Main spring	Top spring
Travel:		
All CX500, CX500 C and CX500 D models .... 139.5 mm (5.5 in)		
All GL models, CX500 E-C, CX650 E-D .... 150.0 mm (5.9 in)		
CX650 C .... 160.0 mm (6.3 in)		
Fork spring free length:		
CX500 D 1981	565.2 mm (22.2519 in)	N/App
Service limit	556.6 mm (21.9133 in)	N/App
CX500 C 1981, 1982	503.1 mm (19.8071 in)	100.7 mm (3.9646 in)
Service limit	495.1 mm (19.4921 in)	96.7 mm (3.8071 in)
CX500 E-C	458.9 mm (18.0669 in)	N/App
Service limit	449.7 mm (17.7047 in)	N/App
CX500-B, CX500 C-B	N/Av	N/Av
All other CX500, CX500 C and CX500 D models	461.7 mm (18.1771 in)	98.2 mm (3.8661 in)
Service limit	449.5 mm (17.6968 in)	90.1 mm (3.5472 in)
All GL500, GL500 D and GL500 I models	508.1 mm (20.0039 in)	100.7 mm (3.9646 in)
Service limit	493.0 mm (19.4094 in)	97.7 mm (3.8465 in)
GL650, GL650 D2-E, GL650 I	466.9 mm (18.3819 in)	123.6 mm (4.8661 in)
Service limit	451.8 mm (17.7874 in)	120.6 mm (4.7480 in)
CX650 E-D	480.5 mm (18.9173 in)	N/App
Service limit	470.9 mm (18.5393 in)	N/App
CX650 C	498.3 mm (19.6181 in)	N/App
Service limit	483.4 mm (19.0315 in)	N/App
Stanchion maximum warpage	0.2 mm (0.0079 in)	

Stanchion OD:

CX500 C 1981, 1982	34.950 – 34.975 mm (1.3760 – 1.3770 in)
Service limit	34.900 mm (1.3740 in)
CX500 D 1981	32.950 – 32.975 mm (1.2972 – 1.2982 in)
Service limit	32.900 mm (1.2953 in)
Recommended fork oil	Automatic Transmission Fluid (ATF)

Fork oil capacity – per leg:

	Left-hand	Right-hand
CX500 (UK), CX500-A, CX500-B, CX500 1978, 1979, CX500 C-B, CX500 C and CX500 D 1979, 1980	135cc (4.6 US fl oz/ 4.8 Imp fl oz)	as left
CX500 C 1981, 1982	220cc (7.4 US fl oz/ 7.7 Imp fl oz)	as left
CX500 D 1981	185cc (6.3 US fl oz/ 6.5 Imp fl oz)	as left
CX500 E-C	265cc (9.0 US fl oz/ 9.3 Imp fl oz)	250cc (8.5 US fl oz/ 8.8 Imp fl oz)
All GL500, GL500 D and GL500 I models	210cc (7.1 US fl oz/ 7.4 Imp fl oz)	as left
CX650 C	480cc (16.2 US fl oz/ 16.9 Imp fl oz)	as left
CX650 E-D	290cc (9.8 US fl oz/ 10.2 Imp fl oz)	275cc (9.3 US fl oz/ 9.7 Imp fl oz)
GL650, GL650 D2-E, GL650 I	275cc (9.3 US fl oz/ 9.7 Imp fl oz)	as left

Fork air pressure – forks cold and fully extended:

CX500 C 1981, 1982, CX500 D 1981	10 – 16 psi (0.7 – 1.1 kg/cm²)
CX500 E-C, all GL500, GL500 D and GL500 I models	11 – 17 psi (0.8 – 1.2 kg/cm²)
CX650 C, CX650 E-D	0 – 6 psi (0 – 0.4 kg/cm²)
GL650, GL650 D2-E, GL650 I	6 – 17 psi (0.4 – 1.2 kg/cm²)
All other models	N/App

Rear suspension

Travel:

All CX500, CX500 C and CX500 D models	85 mm (3.4 in)
CX650 C, all GL500, GL500 D and GL500 I models	120 mm (4.7 in)
CX500 E-C, all other 650 models	110 mm (4.3 in)

Suspension unit spring free length:

All CX500, CX500 C and CX500 D models	249.3 mm (9.8149 in)
Service limit	245.5 mm (9.6653 in)
CX650 C	241.9 mm (9.5236 in)
Service limit	234.6 mm (9.2362 in)
All other models	N/App

Suspension unit air pressure:

CX500 E-C, GL500 1981, 1982, CX650 E-D, GL650	0 – 71 psi (0 – 5.0 kg/cm²)
GL500 D-C, GL500 I 1981, 1982, GL650 D2-E, GL650 I	14 – 71 psi (1.0 – 5.0 kg/cm²)

Torque wrench settings

Component	kgf m	lbf ft
Steering stem nut	9.0 – 12.0	65.0 – 87.0
Steering stem clamp bolt – where fitted	1.8 – 2.5	13.0 – 18.0
Handlebar clamp bolts:		
CX500 (UK)	1.8 – 2.5	13.0 – 18.0
All other CX500, CX500 C and CX500 D models	2.5 – 3.0	18.0 – 22.0
CX500 E-C, all GL500 models, all 650 models	2.5 – 3.5	18.0 – 25.0
Fork top yoke pinch bolts:		
All CX500, CX500 C and CX500 D models, CX650 C	0.9 – 1.3	6.5 – 9.5
CX500 E-C, CX650 E-D, all GL models	0.9 – 1.5	6.5 – 11.0
Fork bottom yoke pinch bolts:		
All CX500, CX500 C and CX500 D models	1.8 – 2.5	13.0 – 18.0
CX500 E-C, GL500 D-C, CX650 E-D, all GL650 models	3.0 – 4.0	22.0 – 29.0
CX650 C	4.5 – 5.5	32.5 – 40.0
Fork top plug:		
All models without air-assisted forks	4.5 – 5.5	32.5 – 40.0
All models with air-assisted forks	1.5 – 3.0	11.0 – 22.0
Air hose connections – CX500 C 1981, 1982, CX500 D 1981, all GL500 models:		
Right-hand fitting (separate)	1.5 – 2.0	11.0 – 14.5
Left-hand fitting (hose)	0.4 – 0.7	3.0 – 5.0
Hose connector	0.4 – 0.7	3.0 – 5.0
Fork brace bolts – 650 models	1.8 – 2.8	13.0 – 20.0
Anti-dive case mounting Allen screws – CX500 E-C, CX650 E-D ..	0.6 – 0.9	4.0 – 6.5
Fork drain plug	0.6 – 0.9	4.0 – 6.5
Damper rod Allen screw	1.5 – 2.5	11.0 – 18.0
Rider's footrest mounting bolts	3.0 – 4.0	22.0 – 29.0
Pillion footrest mounting bolts – CX500 E-C, all GL500 models, all 650 models	4.5 – 6.0	32.5 – 43.0

Component	kgf m	lbf ft
Side stand mountings – CX500 E-C, all GL500 models, all 650 models:		
Pivot bolt	1.0 – 2.0	7.0 – 14.5
Locknut	3.0 – 4.0	22.0 – 29.0
Centre stand pivot bolt – CX500 E-C, all GL500 models, all 650 models	3.0 – 4.0	22.0 – 29.0
Ignition switch mounting bolts	1.0 – 1.4	7.0 – 10.0
Rear mudguard 8 mm mounting bolt – CX650 C, all GL models	3.0 – 4.0	22.0 – 29.0
Swinging arm pivot shaft:		
All CX500, CX500 C and CX500 D models	0.8 – 1.2	6.0 – 9.0
CX500 E-C, GL500 D-C, CX650 E-D, GL650 D2-E	1.7 – 2.1	12.0 – 15.0
GL500 and GL500 I 1981, 1982	0.9 – 1.2	6.5 – 9.0
CX650 C – right-hand	9.0 – 12.0	65.0 – 87.0
CX650 C – left-hand (preload)	2.0	14.5
CX650 C – left-hand (normal)	1.0 – 1.4	7.0 – 10.0
Swinging arm pivot shaft locknut:		
All CX500, CX500 C and CX500 D models	8.0 – 12.0	58.0 – 87.0
CX500 E-C, CX650 E-D, all GL models – with service tool	8.2 – 10.8	59.0 – 78.0
CX500 E-C, CX650 E-D, all GL models – normal	9.0 – 12.0	65.0 – 87.0
CX650 C – with service tool	9.3 – 11.5	67.0 – 83.0
CX650 C – normal	10.0 – 13.0	72.0 – 94.0
Rear suspension unit mountings – upper and lower:		
Except Pro-Link models	3.0 – 4.0	22.0 – 29.0
Pro-Link models	4.5 – 5.5	32.5 – 40.0
Final driveshaft pinch bolt:		
All CX500 models except CX500 E-C	1.8 – 2.5	13.0 – 18.0
CX500 E-C, all GL500 models, all 650 models	1.8 – 2.8	13.0 – 20.0
Final drive gear case/swinging arm mounting nuts:		
Pro-Link models	4.5 – 7.0	32.5 – 50.5
CX650 C	3.0 – 3.5	22.0 – 25.0
All other models	4.5 – 6.0	32.5 – 43.0
Rear suspension unit air hose unions – Pro-Link models	0.4 – 0.7	3.0 – 5.0
Suspension linkage pivot bolt retaining nuts – Pro-Link models	4.5 – 5.5	32.5 – 40.0

1 General description

The frame is based around a tubular steel spine with heavy gussetting and some use of fabricated, pressed steel components. The engine/gearbox unit is supported at the rear on pressed brackets which extend rearwards and serve also as the swinging arm support points. The front of the engine is suspended on a large U-shaped fabricated bracket which serves as a mounting for the radiator and which is detachable to aid engine removal.

The telescopic forks are conventional in design, being coil-sprung with hydraulic damping, but later models (see Specifications) have air valves, linked on some models, which can be used to vary the air pressure in the fork. This alters the spring rate to suit the rider's needs. These later forks also have two Teflon coated bushes which can be renewed to offset wear. CX500 E-C and CX650 E-D models also have a brake-operated anti-dive mechanism on the left-hand fork leg.

On all CX500, CX500 C and CX500 D models and the CX650 C, rear suspension is of the swinging arm type using oil-filled suspension units to control the damping action. The units are adjustable so that the spring rating can be changed within certain limits to match the load carried. The right-hand swinging arm member serves also as a torque tube to which the final drive casing is attached, and through which the final driveshaft passes.

While CX500 E-C, CX650 E-D and all GL models use the same swinging arm, they are fitted with a single air/oil suspension unit which acts on the swinging arm via a variable rate linkage (Honda Pro-Link system).

2 Front fork legs: removal

1 On models fitted with Interstate fairings, owners may wish to remove the fairing to prevent any risk of damage.
2 As described in Chapter 6, remove the front wheel from the machine and ensure that the machine is well supported so that it cannot fall.

3 Remove the brake caliper(s) from the fork legs if still fitted, wedge a piece of wood between the pads to prevent the risk of damage should the brake lever be applied inadvertently, and tie them out of harm's way. Do not twist or distort the brake hoses.
4 Remove the front mudguard, and on 650 models only, the fork brace.
5 On CX500 E-C and CX650 E-D models, tilt the headlamp fairing forwards.
6 On CX500 models, the headlamp shroud must be detached and moved forwards slightly. The shroud is held by two bolts at the top which pass into the fork upper yoke and is located at the bottom by two projections on the lower yoke.
7 On CX500 C 1981, 1982, CX500 D 1981 and all GL500 models it is first necessary to release the air pressure and disconnect the air balance pipe. In order to make access to the pipe connections the instrument console must be raised slightly. Unscrew the knurled speedometer and tachometer cable retaining rings and withdraw them from the instrument heads. Remove the two bolts, visible from the top of the console, to allow the instruments to be raised from position on their mounting bracket. Remove the air valve dust cap from the right-hand fork leg and release the air pressure by depressing the centre stem of the valve. Disconnect the balance pipe from first the left-hand leg and then the right-hand leg; this will obviate any chance of the pipe twisting along its length. Remove the cover (two bolts) from the bottom yoke.
8 On all other models with air-assisted forks, remove the valve caps and depress the valve cores to release the air pressure.
9 It is a good idea to slacken the fork top plugs and damper rod Allen screws at this stage, while the fork legs are still held in the frame and pressure can be applied (particularly to the Allen screw) with minimum risk of damage. See Section 3.
10 Slacken fully the pinch bolts securing each leg in the fork top and bottom yokes. The forks legs can now be eased downwards, out of position. If the clamps prove to be excessively tight, they may be gently sprung, using a large screwdriver. This must be done with great care, in order to prevent breakage of the clamps, necessitating renewal of the complete yoke.

2.10a Slacken top and ...

2.10b ... bottom yoke pinch bolts to release fork stanchions

3 Fork legs: dismantling

1   Refer to the accompanying illustrations for exact details of the forks being dismantled. Always dismantle fork legs separately and keep the components in separate, clearly-marked containers to avoid exchanging them.
2   If the damper rod Allen screw has not yet been slackened, this should be done now, while spring pressure can be used to prevent the damper rod from rotating.
3   Dismantle the forks legs following the instructions given in the appropriate sub-section below.

CX500, CX500-A, CX500 1978, 1979, CX500 C and CX500 D 1979, 1980

4   As no drain plugs are fitted to the fork lower legs, the damping oil can be removed only after the top plug has been unscrewed. The oil will flow out when the fork is inverted to remove the main fork spring. Note carefully which way up the springs are fitted.
5   Unscrew the chrome plug and lift out the short top spring and spring seat. Invert the fork so that the main spring slides out. Be prepared to catch the damping fluid in a suitable container. After removal of the spring, the fork leg may be pumped in and out to expel the remaining fluid.
6   Clamp the fork lower leg in a vice fitted with soft jaws, or wrap a length of rubber inner tube around the leg to prevent damage. Unscrew the damper rod Allen screw, recessed in the housing which carries the front wheel spindle. Prise the dust excluder from position and slide it up the stanchion, which can now be pulled out of the lower fork leg. Pull the damper rod seat off the rod, invert the stanchion and tip out the damper rod.
7   The oil seal fitted to the top of the lower leg should be removed only if it is to be renewed, because damage will almost certainly be inflicted when it is prised from position. The seal is retained by a spring clip, and removed, after the clip has been withdrawn, by carefully levering it away from the lower leg.

CX500 E-C, CX650 E-D

8   Remove the top plug, noting that it will be under slight spring pressure. Remove the spacer and washer (500 models only), and the main fork spring, then invert the fork leg and expel the fork oil by 'pumping' the fork leg. Clamp the lower leg horizontally in a vice, using rag to protect the paint finish. On no account overtighten the vice or the lower leg will be distorted. Using an Allen key, remove the damper bolt from the underside of the lower leg.
9   Slide the dust seal away from the top of the lower leg followed by the foam seal and plastic washer, and remove the circlip which retains

the seal and top bush. Pull the stanchion sharply outwards until the bush and seal are displaced and the stanchion assembly comes out of the lower leg. Slide the seal, backing ring and top bush off the stanchion, leaving the bottom bush undisturbed unless it is to be renewed.
10 In the case of the left-hand fork leg only, release the wire circlip which retains the spring seat, spring and oil lock valve on the damper rod. Free the second wire circlip, then invert the stanchion and tip out the damper rod. Displace the damper rod seat from the bottom of the lower leg, noting that it is held in position by an O-ring.

All other models

11 Remove the previously slackened fork cap from the top of the stanchion. It should be observed that this cap is under a certain amount of spring tension which will make it exit with a strong force. For this reason unscrew the cap very slowly and remember to keep your face away from the line of the fire. Once removed, the damping oil can be drained and the fork spring(s) shaken from the top of the stanchion. All models except the CX500 D 1981 and the CX650 C have two fork springs, the upper spring being the shorter of the two, separated by a plate washer. The CX500 D however, has only one main spring, while the CX650 C has a short spacer instead of the top spring. Take note when removing the spring(s) which way the closer wound coils of the spring face as a reminder for reassembly.
12 In order to separate the stanchion from the lower leg the Allen bolt, situated in the base of the leg must be removed. This bolt screws into the damper rod. Because the damper rod is able to rotate inside the stanchion it is very often found that when the Allen bolt is turned the damper rod will revolve also and some means of holding the damper rod in place will be necessary. The head of the damper rod is recessed to accommodate a service tool which can be obtained through your dealer. Alternatively the following methods can be utilised to the same effect.
13 Place the lower leg portion of the fork leg between the padded jaws of a vice and tighten the vice just enough to hold the leg in place. Obtain a length of wooden dowelling, slightly taper one end and install the tapered end into the top of the damper rod. With the aid of an assistant, the dowelling protruding from the stanchion can be held firm whilst the Allen bolt is removed. If an assistant is not available, cut the dowelling off just below the top of the stanchion and refit the fork cap.
14 Before the stanchion can be withdrawn, prise off the rubber dust cover to reveal the oil seal and circlip located beneath it. Using a pair of circlip pliers, remove the circlip. The stanchion and lower leg can now be separated by using the lower leg as a slide hammer by which to dislodge the bush, together with the oil seal and backing rings. The damper rod seat can now be shaken out of the lower leg.

3.5a Unscrew fork leg top bolt to release springs ...

3.5b ... noting top spring and washer (where fitted)

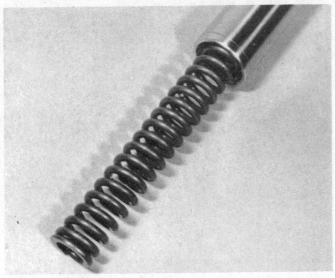

3.5c Note carefully which way up springs are fitted before removing them

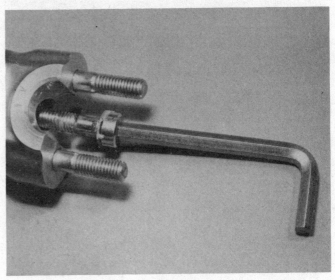

3.6a Use fork spring pressure or holding tool (see paragraph 13) to prevent damper rod rotation as Allen screw is removed

3.6b Slide dust seal upwards off lower leg ...

3.6c ... and pull stanchion out of lower leg – note damper rod seat

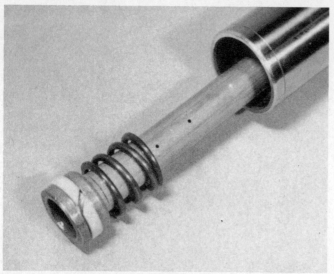

3.6d Invert stanchion to tip out damper rod assembly

3.7 Oil seal is secured by a circlip

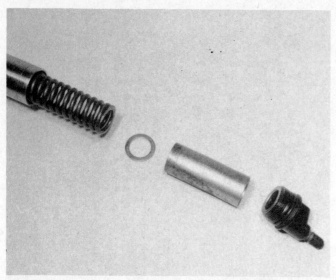

3.8 Some models are fitted with a spacer instead of top spring

3.9a Displace dust seal, followed by foam seal and plastic washer

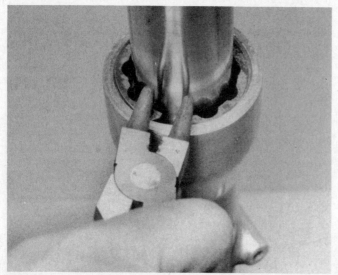

3.9b Remove circlip to release fork oil seal

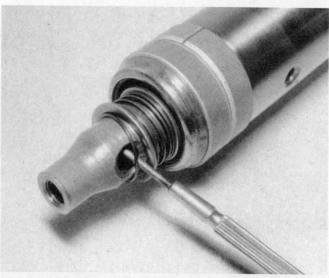

3.10 Displace wire circlip to release anti-dive valve components (where fitted)

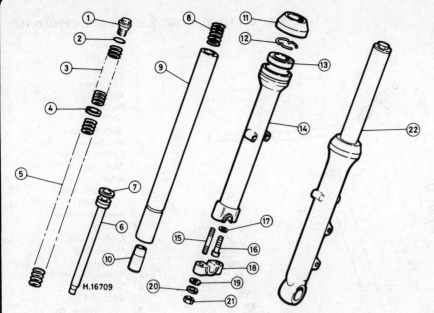

1   Top plug
2   O-ring
3   Top spring
4   Spring seat
5   Main spring
6   Damper rod
7   Piston ring
8   Rebound spring
9   Stanchion
10 Damper rod seat
11 Dust seal
12 Spring clip
13 Oil seal
14 Right-hand lower leg
15 Stud
16 Allen screw
17 Sealing washer
18 Spindle clamp
19 Washer
20 Spring washer
21 Nut
22 Left-hand leg

H.16709

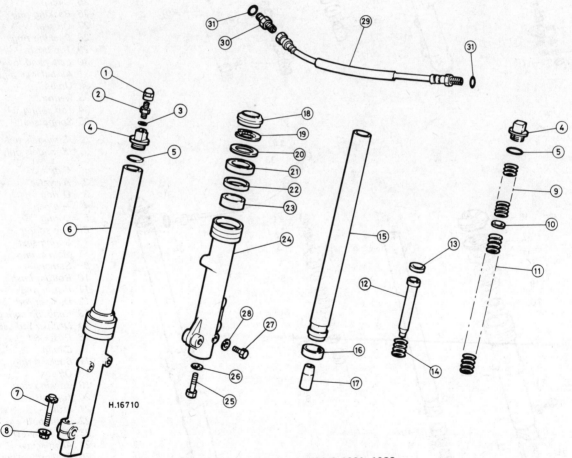

H.16710

Fig. 5.2 Front forks – CX500 C 1981, 1982

1 Cap	9 Top spring	17 Damper rod seat	25 Allen screw
2 Air valve	10 Spring seat	18 Dust seal	26 Sealing washer
3 O-ring	11 Main spring	19 Circlip	27 Drain bolt
4 Top cap	12 Damper rod	20 Backing ring	28 Sealing washer
5 O-ring	13 Piston ring	21 Oil seal	29 Air balance pipe
6 Right-hand leg	14 Rebound spring	22 Backing ring	30 Union
7 Pinch bolt	15 Stanchion	23 Top bush	31 O-ring
8 Nut	16 Bottom bush	24 Left-hand lower leg	

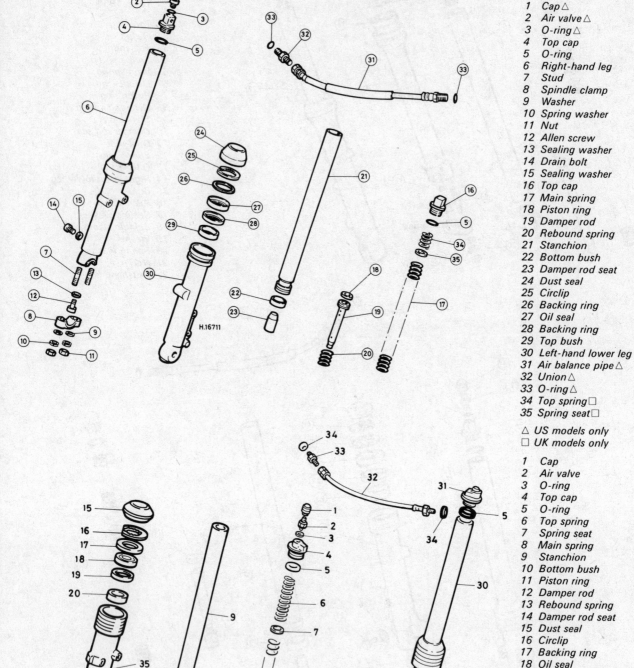

Fig. 5.3 Front forks – CX500-B, CX500 C-B, CX500 D 1981

1 Cap △
2 Air valve △
3 O-ring △
4 Top cap
5 O-ring
6 Right-hand leg
7 Stud
8 Spindle clamp
9 Washer
10 Spring washer
11 Nut
12 Allen screw
13 Sealing washer
14 Drain bolt
15 Sealing washer
16 Top cap
17 Main spring
18 Piston ring
19 Damper rod
20 Rebound spring
21 Stanchion
22 Bottom bush
23 Damper rod seat
24 Dust seal
25 Circlip
26 Backing ring
27 Oil seal
28 Backing ring
29 Top bush
30 Left-hand lower leg
31 Air balance pipe △
32 Union △
33 O-ring △
34 Top spring □
35 Spring seat □

△ US models only
□ UK models only

1 Cap
2 Air valve
3 O-ring
4 Top cap
5 O-ring
6 Top spring
7 Spring seat
8 Main spring
9 Stanchion
10 Bottom bush
11 Piston ring
12 Damper rod
13 Rebound spring
14 Damper rod seat
15 Dust seal
16 Circlip
17 Backing ring
18 Oil seal
19 Backing ring
20 Top bush
21 Stud
22 Drain bolt
23 Sealing washer
24 Allen screw
25 Sealing washer
26 Spindle clamp
27 Washer
28 Spring washer
29 Nut
30 Left-hand leg
31 Top cap
32 Air balance pipe
33 Union
34 O-ring
35 Right-hand lower leg

Fig. 5.4 Front forks – all GL500 models

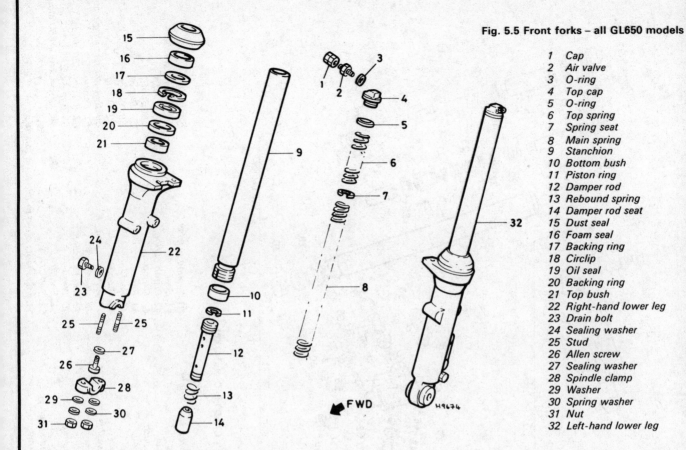

Fig. 5.5 Front forks – all GL650 models

1 Cap
2 Air valve
3 O-ring
4 Top cap
5 O-ring
6 Top spring
7 Spring seat
8 Main spring
9 Stanchion
10 Bottom bush
11 Piston ring
12 Damper rod
13 Rebound spring
14 Damper rod seat
15 Dust seal
16 Foam seal
17 Backing ring
18 Circlip
19 Oil seal
20 Backing ring
21 Top bush
22 Right-hand lower leg
23 Drain bolt
24 Sealing washer
25 Stud
26 Allen screw
27 Sealing washer
28 Spindle clamp
29 Washer
30 Spring washer
31 Nut
32 Left-hand lower leg

FWD

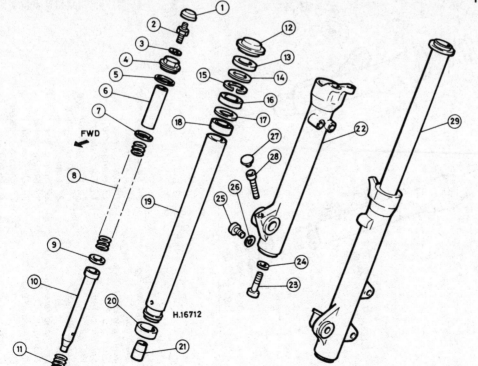

Fig. 5.6 Front forks – CX650 C

1 Cap
2 Air valve
3 Seal
4 Top cap
5 O-ring
6 Spacer
7 Spring seat
8 Spring
9 Piston ring
10 Damper rod
11 Rebound spring
12 Dust seal
13 Foam seal
14 Backing ring
15 Circlip
16 Oil seal
17 Backing ring
18 Top bush
19 Stanchion
20 Bottom bush
21 Damper rod seat
22 Right-hand lower leg
23 Allen screw
24 Sealing washer
25 Drain bolt
26 Sealing washer
27 Plug
28 Allen pinch bolt
29 Left-hand leg

FWD

H.16712

146

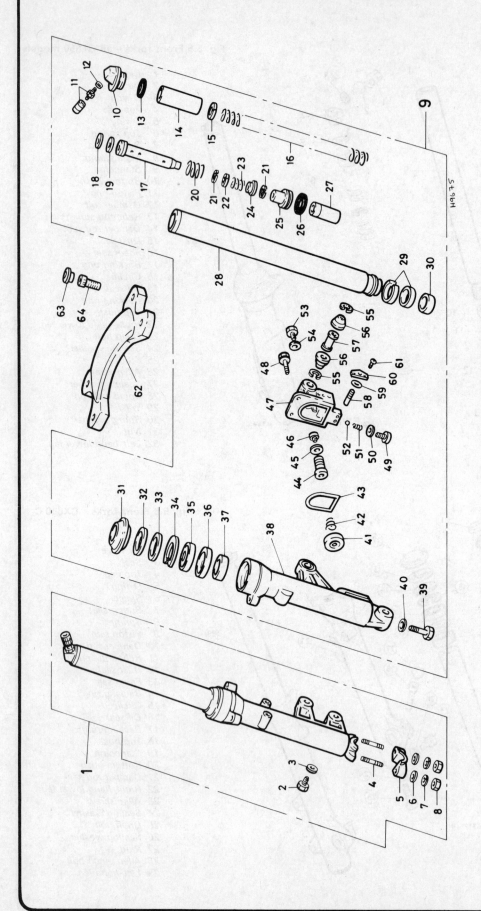

Fig. 5.7 Front forks – CX500 E-C and CX650 E-D

1 Right-hand fork leg
2 Drain bolt
3 Sealing washer
4 Stud
5 Spindle clamp
6 Washer
7 Spring washer
8 Nut
9 Left-hand fork leg
10 Top plug
11 Air valve and cap
12 O-ring
13 O-ring

14 Spacer – 500 model
15 Spring seat
16 Main spring
17 Damper rod
18 Top piston ring*
19 Bottom piston ring
20 Rebound spring
21 Circlip*
22 Spring seat*
23 Spring*
24 Oil lock valve*
25 Damper rod seat*
26 O-ring*

27 Damper rod seat – right-
      hand leg
28 Stanchion
29 Bottom bush
30 Bottom bush
31 Dust seal
32 Foam seal
33 Plastic washer
34 Circlip
35 Oil seal
36 Backing ring
37 Top bush
38 Left-hand lower leg
39 Allen screw

40 Sealing washer
41 Seal
42 Piston spring
43 O-ring
44 Piston
45 Seal
46 Stopper
47 Anti-dive unit
48 Allen bolt
49 Detent bolt
50 Washer
51 Spring
52 Ball

53 Drain bolt
54 Sealing washer
55 Circlip
56 Boot
57 Sleeve
58 Selector
59 O-ring
60 Selector retaining plate
61 Screw
62 Brace – 650 model
63 Cap – 650 model
64 Allen screw – 650 model

* Components fitted to left-hand fork leg only

4 Fork legs: examination and renovation

1   Lay all the components out on a clean work surface and examine for wear and damage as follows.

2   It is not generally possible to straighten forks which have been damaged in an accident, particularly when the correct jigs are not available. It is always best to err on the side of safety and fit new ones, especially since there is no easy means to detect whether the forks have been overstressed or metal fatigued. Fork stanchions can be checked, after removal from the lower legs, by rolling them on a dead flat surface. Any misalignment will be immediately obvious. If the facilities are available, set the stanchion in V-blocks and with the use of a dial gauge check the fork runout along its length. Maximum runout is 0.2 mm (0.01 in).

3   Check the free length of the fork spring(s) to ensure that they are not shorter than the specified service limit. Fork springs will take a permanent set after considerable usage and will require renewal if the fork action becomes spongy. Always renew the springs in both legs at the same time.

4   The parts most likely to wear are the internal surfaces of the lower leg, the outer face of the stanchion and both bushes. Visually inspect the contact face of the fork bushes for chipping and scoring. Honda recommend that they be renewed if the Teflon surface coating has worn to such an extent that the copper bush is showing for more than three-quarters of the entire surface. If the bottom bush is to be renewed, insert a flat screwdriver blade into the slot and spring it apart sufficiently for it to be worked over the end of the stanchion. Always renew all the bushes in both legs at the same time. Check for distortion the backing ring above the top bush; renew if necessary.

5   Where, on early models, the forks do not contain bushes, the lower legs slide directly against the outer hard chrome surface of the stanchions. If wear occurs, indicated by slackness, the fork lower leg will have to be renewed, possibly also the stanchion. Wear of the stanchion is indicated by scuffing the penetration of the hard chrome surface.

6   Renew the oil seals as a matter of course whenever the forks are dismantled. On early models the seals can be levered out of the housing once the retaining circlip has been removed: take great care not to damage the seal housing. Fit the seals with a tubular drift such as a socket spanner which bears only on the seal hard outer edge. Drive the seal in just far enough to expose the circlip groove, then refit the circlip and grease the seal lips.

7   Check the condition of the rubber dust covers. If they are worn or cracked around the stanchion area dust and water will penetrate the cover and eventually damage the oil seal. It is essential that all damaged components are renewed. All seals, O-rings and gaskets should be renewed as a matter of course.

8   If damping action is lost, the piston ring around the damper rod head should be renewed. Clear any obstructions from the small holes in the rod and check the action of the short rebound spring.

9   Certain models are fitted with a plastic washer and a foam grease holder between the oil seal retaining circlip and the dust seal. Since it is possible that these foam grease holders, or seals, can promote oil leaks by damaging the oil seal lips it is recommended that these are discarded whenever the forks are dismantled. If the fork seals begin to weep, check first to ensure that these seals are no longer fitted, and remove them if necessary.

5 Anti-dive unit: examination and renovation – CX500 E-C, CX650 E-D

1   The anti-dive unit takes the form of a rectangular valve casing bolted to the fork lower leg and connected by a short torque link to the brake caliper. As a general rule it is not necessary to disturb the valve assembly except as part of a fork overhaul, when the valve unit should be removed for inspection and cleaning. If it proves necessary to remove the unit at any other time, note that it will first be necessary to disconnect the caliper at the torque link, release fork air pressure and to drain the fork oil. To facilitate this a drain plug is provided on the valve body. Note that the detent bolt, which passes up into the unit from the underside, should not be disturbed at this stage.

2   Remove the four Allen bolts which retain the unit to the lower leg and lift it away. Referring to the accompanying photographs, remove the detent bolt and shake out the spring and ball, placing them in a container for safe keeping. The seal and piston spring will probably have remained in place in the lower leg and can be lifted out for inspection and cleaning. The piston should be withdrawn from the valve body. Remove the two screws which retain the selector retainer plate and lift the selector out of the valve body.

3   Check the O-ring seals on the selector and valve piston, and the piston to body seal and valve body seal for wear or damage, renewing them if necessary. Examine the piston surface for scoring.

4   Clean the internal components before reassembly commences, paying particular attention to the orifices in the selector; if these become obstructed, the anti-dive effect will be seriously affected. Lubricate the seals with automatic transmission fluid (ATF) prior to installation. Reassemble by reversing the dismantling sequence, taking care not to damage the seal faces. Refit the unit, tightening the Allen bolts to the specified torque setting to avoid distortion. Remove and clean the sleeve to which the torque link attaches, lubricating it with silicone grease. Add the recommended quantity of oil to the fork leg and set the fork air pressure correctly before using the machine.

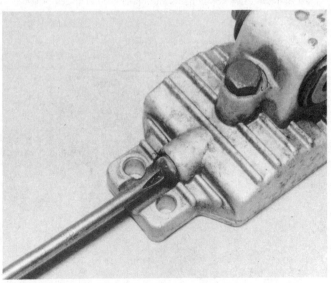

4.4 Spring bush ends apart as shown to remove and refit – check seal backing ring for distortion

5.2a Unscrew detent bolt ...

5.2b ... to release selector detent ball and spring

5.2c Check seal in lower leg – renew if worn or damaged

5.2d Remove valve components and check for blockages or wear

5.2e Remove retaining screws and plate ...

5.2f ... to release selector – renew all O-rings whenever disturbed

6 Fork legs: reassembly

1   Reassembly is essentially a reversal of the dismantling sequence, noting the following points. Conduct the reassembling in clean working conditions and ensure that all components are absolutely clean.

CX500, CX500-A, CX500 1978, 1979, CX500 C and CX500 D 1979, 1980

2   Place the rebound spring over the damper rod end, check that the piston ring is in place and insert the assembly into the stanchion upper end. Use the fork spring to push the damper rod down until it protrudes from the stanchion lower end. Fit the damper rod seat and insert the assembly into the stanchion.

3   Applying thread locking compound to its threads, refit and tighten to a torque setting of 1.8 – 2.5 kgf m (11 – 18 lbf ft) the damper rod Allen screw. Pack grease above the fork seal and slide the dust excluder down into place. Check that the stanchion moves smoothly in and out.

4   Fill each fork leg with exactly 135cc (4.6 US fl oz/4.8 Imp fl oz) of ATF (Automatic Transmission Fluid) then refit the fork springs. When

inserting the fork main spring into each stanchion, note that the spring tapers slightly at one end. The tapered end **must** point downwards when fitted. Fit the thick washer and the top spring, then refit and tighten to the specified torque setting the fork top plug; do not forget to renew its sealing O-ring.

CX500 E-C, CX650 E-D

5   Fit the rebound spring over the damper rod and drop it into the stanchion. In the case of the left-hand fork leg, fit the circlip, spring seat, spring, lock valve and second circlip to the damper rod. Check, and if necessary renew, the O-ring on the damper rod seat and place it over the end of the damper rod. Temporarily refit the fork springs and top bolt to retain the damper, then lower the lower leg over the stanchion assembly.

6   Coat the damper Allen screw threads with non-hardening locking fluid, then secure the bolt, tightening it to 1.5 – 2.5 kgf m (11 – 18 lbf ft). Remove the top plug and fork springs. Coat the fork seal in ATF (automatic transmission fluid) and press the bush, backing ring and seal home over the stanchion and into the lower leg recess, using a length of tubing. Secure the assembly with the circlip and fit the dust seal. Do not forget to refit the plastic washer and, if required, the foam seal.

7   Compress the assembled fork fully and stand it vertically. Add exactly the correct amount of ATF to each fork leg. Note that the quantity varies between the two legs, but that the level is identical. Refit the springs and spring seat, noting that the main spring tapered coils should face downward with the closer pitched coils uppermost and install the top plugs. Refit the fork legs, remembering to tighten fully the top plugs.

All other models

8   Place the rebound spring over the damper rod and slide the assembly into the stanchion. Fit the damper rod seat onto the damper rod. The stanchion can then be replaced in the lower leg; the lower bush can be lubricated with ATF to aid reassembly. Coat the threads of the Allen screw with a locking compound and refit it and the sealing washer into the base of the lower leg. Use one of the methods described in the dismantling sequence for holding the damper rod while the Allen bolt is tightened. Install the upper bush over the stanchion and drive it into place in the lower leg. Honda market a special tool to drive the bush into position which can be obtained from an authorized Honda dealer. Alternatively a tool can be fabricated from a piece of metal tubing. Obtain a length of tube of an internal diameter slightly larger than that of the stanchion and an external diameter slightly less than that of the bush. Ensure that the end of this tube is both square to its sides and free of burrs. Refit the backing ring over the

bush followed by the new oil seal. Before fitting the seal, lubricate it thoroughly in ATF and ensure that it is correctly positioned over the stanchion with the marked face uppermost. The seal can be driven into position with the tool used for fitting the bush. It is in the correct position once the circlip groove can be seen above its upper surface. Refit the backing ring and secure it in place with the circlip. Finally slide the dust cover over the stanchion ensuring that it locates properly in the lower leg. Do not forget to refit the plastic washer and, if required, the foam seal.

9   Refill each leg with the correct amount of ATF as shown in the specifications. The fork spring(s) must be refitted with the closer wound coils facing in the direction noted when dismantling. The main springs are usually refitted with their tapered coils facing downwards, ie closer-pitched coils uppermost. On machines where two springs are fitted remember to refit the washer between them.

10   Inspect the condition of the O-ring around the fork cap and replace if it is damaged. It is essential that an airtight seal is maintained on models with air assisted forks otherwise a loss of air pressure will occur. Refit the fork cap and tighten by hand because it will be much easier to fully tighten the cap when the leg is secured in the machine.

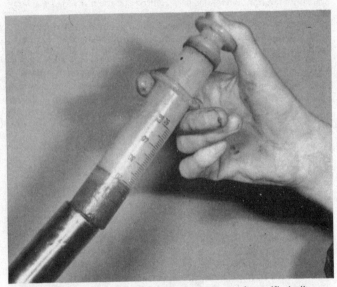

6.4 Fill each leg with exactly the correct amount of specified oil

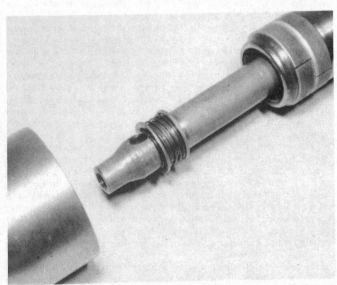

6.5 Where applicable, fit damper rod to stanchion and assemble anti-dive valve

6.6 Fit backing ring as shown before driving new seal into lower leg

7 Front fork legs: refitting

1   Install the fork legs into the steering yokes and tighten the pinch bolts. Models with air assisted suspension must have the line scribed around the circumference of the stanchion aligned with the top face of the upper yoke. On all other models, align the top of the stanchion with the yoke top surface.
2   Finally tighten the fork top plugs to the correct torque setting.
3   On models with linked air forks, connect the air balance pipe connector to the right-hand fork cap and refit the balance pipe. Ensure that all pipe O-rings are in a serviceable condition and that the pipe is not cracked or damaged in any way. The balance pipe must be routed behind the headlamp and immediately in front of the ignition switch body. On all models wth air forks, fill them to the required pressure (see Routine Maintenance). Locate the instrument console on its mounting plate, securing it with the two bolts and reconnect the speedometer and tachometer drive cables.
4   Finally refit the front mudguard, brake caliper(s) and front wheel. When refitting the front mudguard note which way round the water channels must be installed so that water is directed away from the disc and not on to it. On 650 models, do not tighten fully the fork brace bolts until the front wheel has been refitted and the fork legs have been aligned.
5   When the wheel has been refitted, push the machine off its stand, apply the front brake and depress the forks vigorously several times to settle all components and to align them with each other. Place the machine back on its stand and work from the wheel spindle up to the top yoke, tightening securely all nuts, bolts and other fasteners that have been disturbed. Use the recommended torque settings, where specified.

7.4 Note correct fitted position of water channels, where separate

8 Steering head bearings: removal

1   On machines with Interstate fairings, remove the fairing.
2   Remove the front forks. See Section 2.
3   Remove the fuel tank as described in Chapter 3 Section 2, or at the very least protect it with a thick layer of old blanket or similar padding. Remove the clamp bolts and top clamp and lift away the handlebars. Without straining or distorting the brake hose, control cables or wiring move the handlebars to the rear, clear of the steering head area and secure them on the frame top tube.

4   On CX500 models, remove the headlamp unit from the headlamp shell and disconnect the electrical leads at the connectors. Pass the connectors through the rear of the shroud and then disconnect the leads running from the fuse box and the ignition switch. Disconnect the speedometer and tachometer drive cables at the instrument heads by unscrewing the knurled union rings. After freeing all the connections, the headlamp shroud may be lifted away, together with the instruments.
5   On CX500 E-C and CX650 E-D models tilt the headlamp fairing forwards and remove it, then unplug its connector and remove its two mounting bolts to withdraw the headlamp. Unscrew their knurled retaining rings and disconnect the instrument drive cables from the instruments, then disconnect all electrical leads at their connectors. Remove the four retaining nuts and their washers and withdraw the instrument panel. Remove the two retaining nuts and their washers to release the headlamp bracket from the top yoke; the bracket can be lifted away once the top yoke has been removed.
6   On all other models, release the headlamp retaining screw, then lift out and disconnect the headlamp. Separate the wiring connectors and push them clear of the shell, then release the shell mounting bolts and lift the shell away. Disconnect the instrument drive cables from the underside of the instrument panel. Remove the instrument panel retaining bolts and lift away the panel.
7   On all models, slacken the steering stem clamp bolt (where fitted), then unscrew the steering stem nut. The top yoke may now be lifted away and the headlamp bracket(s), where applicable, can be withdrawn.
8   Removing their mounting bolts, withdraw from the bottom yoke all components such as horns, turn signal assemblies and brake hose junction blocks which will prevent its removal.
9   Place an old blanket or some large rags beneath the bottom yoke to catch any steering head balls which drop free. Slacken the steering stem adjuster, holding the bottom yoke in position while it is removed. Carefully lower the yoke clear of the steering head and place it to one side. Collect the steering head balls which can now be degreased prior to examination; note that there are a total of 37 No 8 ($\frac{1}{4}$ in) steel balls, eighteen in the top race and nineteen in the lower.

9 Steering head bearings: examination and renovation

1   Clean and examine the cups and cones of the steering head bearings. They should have a polished appearance and show no signs of indentation. Renew the set if necessary.
2   Clean and examine the ball bearings which should also be polished and show no signs of surface cracks or blemishes. If any require replacement the whole set must be replaced.
3   Eighteen balls are fitted in the top race and nineteen in the lower race; all are No 8 ($\frac{1}{4}$ in) in diameter. This arrangement will leave a gap but an extra ball must not be fitted otherwise the balls will press against each other, accentuating wear and making the steering stiff.
4   The outer races are a drive fit in the steering head lug and may be drifted out, using a suitable long handled drift passed through the centre of the lug. The lower inner race may be levered from position on the steering stem. When driving the new inner races into place, ensure that they remain square to the housing in the lug or the housing may be damaged.
5   Fit a new dust seal over the steering stem not forgetting the steel washer beneath it, then drive the lower cone into position using a long tubular drift. Ensure that it seats squarely and fully. A large socket can be used to drive the bearing cups into the steering head. Coat both the upper cup and lower cone with general purpose grease and use this to hold the bearing balls in place. Offer up the lower yoke and fit the top cone and adjusting nut to retain it. Tighten the nut until it seats, then back it off by $\frac{1}{8}$ of a turn. Check that the steering stem turns easily and smoothly, and that no discernible play is evident, and make any minor adjustments necessary.
6   Temporarily refit the fork legs, holding them in place with the lower yoke pinch bolts. Refit the top yoke, and headlamp bracket(s), locating it over the fork stanchions, then tighten the top nut to 9.0 – 12.0 kgf m (65 – 87 lbf ft). Continue reassembly by reversing the dismantling sequence.

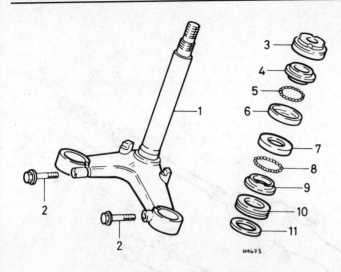

Fig. 5.8 Steering head – typical

1 Steering stem	*7 Bottom cup*
2 Pinch bolts	*8 Bottom bearing balls*
3 Adjusting nut	*9 Bottom cone*
4 Top cone	*10 Seal*
5 Top bearing balls	*11 Washer*
6 Top cup	

10 Steering head bearings: refitting

1   Fit the steering head bearing components to the frame and refit the fork yokes as described in the previous Section. When the front forks and wheel have been refitted, check the steering head bearing adjustment as described in Routine Maintenance and make any alterations necessary. When adjustment is complete, tighten the steering stem nut (and clamp bolt, where fitted) to the specified torque setting.

2   Refitting is a straightforward reversal of the dismantling sequence. Be very careful to ensure that all control cables, brake hoses and wiring are correctly routed and secured by any clamps or ties provided. Re-connect the wiring using the colour-coding of the wires and by reference to the wiring diagrams at the back of this manual; most connections are simplified by the use of coloured plastic block connectors which can only be connected in the correct way.

3   Tighten securely all disturbed nuts, bolts and other fasteners, using the correct torque settings, where given.

4   Before taking the machine out on the road check that all components are correctly refitted and securely fastened, that the steering is correctly adjusted and that the front brakes and suspension are working properly. Check also, at all handlebar positions, that the controls are correctly adjusted and working smoothly.

11 Steering lock: removal and refitting

1   On early models the steering lock is fitted within a lug on the left-hand side of the steering head, and is retained by a Mills pin. If the lock fails, or the keys are lost, the lock cylinder can be removed after drilling out the pin.

2   Later models incorporate the steering lock in the ignition switch, which is located in the warning lamp console on the handlebars. The forks become immovable when the key is turned to the 'lock' position and removed. If the lock is to be renewed, the switch assembly must be disconnected and the retaining bolts which secure it to the underside of the top yoke removed.

12 Frame: examination and renovation

1   The frame is unlikely to require attention unless accident damage has occurred. In some cases, renewal of the frame is the only

satisfactory remedy if the frame is badly out of alignment. Only a few frame specialists have the jigs and mandrels necessary for resetting the frame to the required standard of accuracy, and even then there is no easy means of assessing to what extent the frame may have been overstressed.

2   After the machine has covered a considerable mileage, it is advisable to examine the frame closely for signs of cracking or splitting at the welded joints. Rust corrosion can also cause weakness at these joints. Minor damage can be repaired by welding or brazing, depending on the extent and nature of the damage.

3   Remember that a frame which is out of alignment will cause handling problems and may even promote 'speed wobbles'. If misalignment is suspected, as a result of an accident, it will be necessary to strip the machine completely so that the frame can be checked, and if necessary, renewed.

13 Swinging arm pivot bearings: adjustment

1   The rear fork assembly pivots on two tapered roller bearings, one of which is placed either side of the swinging arm cross-member. Worn swinging arm bearings will cause handling problems, making the rear end twitch and hop, especially when accelerating or shutting off whilst banked over. Play in the bearings may be detected by pulling and pushing horizontally on the rear fork ends. Any play will be magnified by the leverage obtained. Because the bearings are of the tapered roller type, adjustment of play due to wear is possible within certain limits.

2   Adjustment is carried out by means of the left-hand pivot shaft, which is threaded and fitted with a locknut. Remove the rubber boot covering the nut and then loosen the nut. Tighten the pivot shaft, using an Allen key, to the specified torque setting and then tighten the locknut, also to its specified torque setting. If the play has not been eliminated, or if harshness occurs in the swinging arm movement, the bearings are worn to an unacceptable level and must be renewed.

13.2 Swinging arm pivot bearings can be adjusted to remove play

14 Swinging arm: removal

1   Place the machine on its centre stand so that it rests securely on level ground and the rear wheel is well clear of the ground. Remove the silencers after slackening off the clamps at the silencer/balance box unions and removing the silencer support bolts. Remove the rear wheel as described in Chapter 6.

2   On all models with Pro-Link rear suspension, remove the

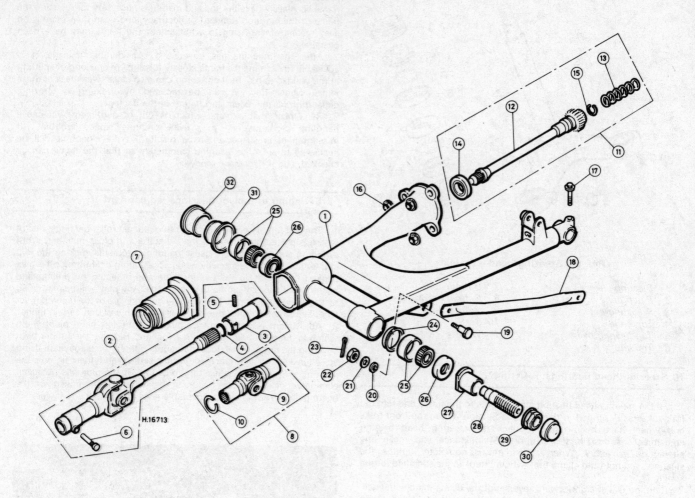

Fig. 5.9 Swinging arm and final drive shaft – all 500 models, CX650 E-D and GL650 models

1   Rear fork/shaft housing
2   Propeller shaft – 500 models,
    GL650 and GL650 I
3   Shaft joint
4   O-ring
5   Spring pin
6   Bolt
7   Gaiter

8   Universal joint – CX650 E-D,
    GL650 D2-E
9   Universal joint
10 Circlip
11 Propeller shaft – CX650 E-D,
    GL650 D2-E
12 Propeller shaft
13 Spring
14 Oil seal

15 Circlip
16 Nut
17 Bolt
18 Brake torque arm
19 Bolt
20 Spring washer
21 Washer
22 Nut
23 Split pin

24 Grease cap
25 Bearing
26 Oil seal
27 Threaded sleeve
28 Pivot shaft
29 Locknut
30 Cap
31 Bearing holder
32 Pivot cap

suspension linkage and the rear suspension unit. See Sections 17 and 18.

3   On CX500 E-C and CX650 E-D models release the brake hose from its clamps, disconnect the torque arm at its front mounting and secure the rear brake assembly out of harm's way.

4   On all models with twin suspension units, disconnect the brake torque arm at the forward mounting by removing the bolt. Free the drain hoses from the guide bracket on the swinging arm cross-member. Remove the right-hand rear suspension unit completely and unscrew the left-hand suspension unit lower mounting bolt, leaving it in position to support temporarily the swinging arm's weight.

5   On all models, remove the nuts securing the final drive case to the swinging arm and withdraw the gear case. On CX650 C, CX650 E-D and GL650 D2-E models the propellor shaft will be withdrawn with the case; on all other models the shaft will stay in place. To remove it, peel back the rubber gaiter at the rear of the gearbox. Unscrew and

withdraw the pinch bolt from the coupling sleeve, so that the driveshaft will be able to slip off the end of the final output shaft when the swinging arm is detached.

6   On all models remove the rear brake pedal if it prevents access to the swinging arm pivot components.

7   On CX650 C models, displace the caps from the pivot bearings and slacken the left-hand shaft locknut. Unscrew the left-hand pivot shaft, followed by the right-hand shaft. Manoeuvre the swinging arm out of the frame.

8   On all other models, displace the cap from the left-hand pivot bearing, slacken the locknut and then unscrew the pivot shaft. Support the weight of the swinging arm and pull out the left-hand rear suspension unit lower mounting bolt (where applicable). Lift the swinging arm over to the left so that the small stub leaves the bearing in the right-hand swinging arm support plate. The swinging arm can now be manoeuvred from position.

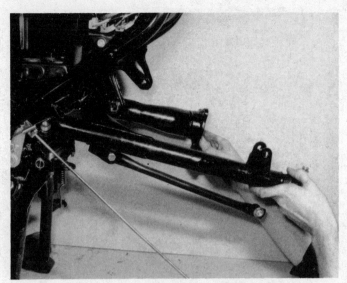

14.8 Remove rear wheel, brake, final drive and suspension components to release swinging arm

1 Rear fork/shaft housing
2 Gaiter
3 Circlip
4 Universal joint
5 Oil seal
6 Spring
7 Propeller shaft
8 Circlip
9 Cap
10 Pivot shaft
11 Locknut
12 Pivot shaft
13 Oil seal
14 Bearing
15 Grease cap
16 Nut
17 Bolt
18 Brake torque arm
19 Bolt
20 Washer
21 Spring washer
22 Nut
23 Split pin

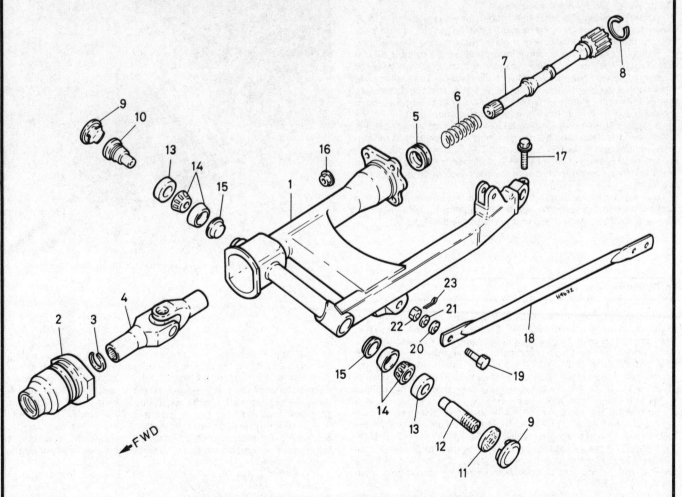

Fig. 5.10 Swinging arm and final drive shaft – CX650 C

15 Swinging arm: examination and renovation

1   On CX650 C models, lever out the oil seals over each bearing and withdraw the bearing inner races. If the bearings are worn or damaged they must be renewed. To withdraw the outer races, drill a suitably-sized hole in the right-hand bearing grease retainer, pass the shaft of a slide hammer through it and attach a washer of the same size as the grease retainer to the shaft. Draw out the right-hand outer race and grease retainer, then repeat the procedure to remove the left-hand bearing components. Renew the oil seals and grease retainers as a matter of course whenever they are disturbed in this way.

2   On all other models, it will be seen that the right-hand swinging arm bearing and oil seal is fitted in a removeable cup in the frame lug. The cup can be drifted from position from.the outside after prising off the plastic cap. The left-hand bearing is fitted in the end of the swinging arm cross-member, similarly protected by an oil seal. Removal of the bearing inner races requires that the oil seals be prised from position; the inner races can then be lifted out. Prising oil seals from position will almost certainly damage the sealing lips, rendering the seals useless. As a result they should be renewed. Both bearing outer races are a tight drive fit and because they are fitted in blind housings they must be drawn out of position. This operation requires the use of a slide hammer fitted with two internally locating legs (see the accompanying photograph). If this tool is not available, it is suggested that the swinging arm and right-hand bearing cup be placed in the hands of a Honda Service Agent for the work to be carried out. Attempting to remove the outer races using levers will prove fruitless and may well damage the components.

3   When removing the left-hand bearing outer race using a slide hammer, better access may be gained by drilling out the sealing plate fitted to the rear of the race. A new plate must be fitted on assembly.

4   New outer races may be drifted into place using a suitable tubular drift. Ensure that the race being driven in is kept square with the housing to prevent tying and damage to the housing. The sealing plate fitted behind the bearings(s) should be so placed that the dished face points outwards. Before inserting the inner races lubricate them thoroughly with a high melting point multi-purpose grease. Fit the inner races and then drift the new oil seals into place using a tubular drift. When the right-hand bearing assembly is complete, drift the cup back into position in the frame. Except on CX650 C models, note that if the left-hand bearing threaded sleeve is disturbed, it must be refitted so that the milled portion locates with the flat shoulder on the inside of the frame.

5   Do not forget to insert the driveshaft or universal joint (as applicable) into the swinging arm before refitting it to the machine. Renew the rubber gaiter if it is split or otherwise damaged.

15.2a Right-hand bearing assembly may be drifted out of frame on most models

15.2b Bearing outer races must be extracted using slide hammer arrangement

16 Swinging arm: refitting

1   On CX650 C models insert the swinging arm into the frame, not forgetting to fit the universal joint on the final output shaft and to refit the rubber gaiter on both gearbox and swinging arm flanges. Align the swinging arm and refit both pivot shafts. Tighten the right-hand shaft to a torque setting of 9.0 – 12.0 kgf m (65 – 87 lbf ft) and refit the bearing cap. Tighten the left-hand shaft to a torque setting of 2.0 kgf m (14.5 lbf ft). Check that the swinging arm moves smoothly throughout its full travel with no sign of free play.

2   Slacken the left-hand pivot shaft fully, then re-tighten it to a torque setting of 1.0 – 1.4 kgf m (7 – 10 lbf ft). Hold the shaft steady while the locknut is tightened to a torque setting of 10.0 – 13.0 kgf m (72 – 94 lbf ft). Note that if the locknut is tightened using the Honda service tool 07908-ME90000 the extra leverage imposed by the tool means that a lower torque setting (see Specifications) must be applied to tighten the nut by the same amount.

3   On all other models manoeuvre the swinging arm into the frame. On CX650 E-D and GL650 D2-E models do not forget to fit the universal joint on the final output shaft before the swinging arm is refitted, or to refit the rubber gaiter on both gearbox and swinging arm flanges. On all models, insert the right-hand pivot stub into the bearing, align the left-hand side and refit the pivot shaft.

4   Tighten the shaft as hard as possible using hand pressure only on an Allen key of the correct size; this will approximate the torque setting necessary to preload the bearing. Check that the swinging arm moves smoothly throughout its full travel with no sign of free play, then slacken the pivot shaft and re-tighten it to the specified torque wrench setting. Hold the shaft steady while the locknut is tightened to its specified torque setting. Again, note that if the Honda service tool 07908-4690001 is used to tighten the locknut (Pro-Link models only), the extra leverage imposed by the tool means that a lower torque setting (see Specifications) must be applied to tighten the nut by the same amount.

5   On CX650 C, CX650 E-D and GL650 D2-E models insert the shaft (if removed) into the final drive gear case. See Chapter 6. Apply a generous coat of the specified grease (see Routine Maintenance) to the shaft splines and insert the assembly into the swinging arm. Rotate the gear case output flange to assist the shaft splines in engaging those of the universal joint. Apply a thin coat of sealant to the mating surfaces to prevent grease seepage and fit the gear case to the rear of the swinging arm. Do not fully tighten the nuts until the rear wheel has been refitted.

6   On all other models pull the driveshaft out of the swinging arm and onto the final output shaft splines; do not forget to lubricate the splines with the recommended grease. When replacing the driveshaft coupling pinch bolt ensure that it locates correctly with the groove in the splined portion of the final output shaft. Tighten the bolt to the specified torque setting. After tightening, check that the splines do not project from the end of the coupling sleeve by more than 10 mm (0.4 in). After refitting the final drive case 45cc (1.5 US fl oz/1.6 Imp fl oz) of lithium based multi-purpose grease should be pumped into the swinging arm via the angled grease nipple fitted to the drive casing (500 models only). Do not fully tighten the retaining nuts until the rear wheel has been refitted. Replace the rubber gaiter on both the gearbox and swinging arm flanges.

7   The remainder of the reassembly procedure is a straightforward reversal of the removal procedure. Tighten all nuts and bolts to the specified torque settings, where given; do not forget to tighten the final drive case nuts once the rear wheel has been refitted. Check that the rear suspension is working smoothly and that the rear brake is operating correctly before taking the machine out on the road.

17 Rear suspension units: removal, examination and refitting

Models with twin suspension units

1   The units are adjustable for load in five positions. Turn the adjuster clockwise with the toolkit 'C' spanner to increase the spring preload. Both units must be at the same setting.

2   Examine the damper units for oil leaks. Bounce the rear of the machine. If it does not come to rest after a couple of oscillations the dampers may be faulty.

3   The dampers are sealed units, and cannot be repaired. If oil leaks are apparent, the oil seal is damaged and the damper requires renewal. It is best to renew both damper units, as a matched pair.

4   Removal of the suspension units is straightforward; each unit is held at the top mounting by a single bolt as is the lower mounting on the left-hand unit. The right-hand unit lower mounting eye fits onto a stud projecting from the final drive casing where it is secured by a nut. The weight of the machine must, of course, be taken on the centre stand before the suspension units are detached.

5   To dismantle either unit the spring must be compressed and the locknut at the top of the damper rod slackened. The upper eye can then be unscrewed to free the spring seat and spring. Compressing the spring may be difficult if a special compressor is not available. Failing this a second person can compress the spring sufficiently for access to the locknut to be made. After removal of the spring check the damper efficiency. It should be possible to feel an equal resistance throughout the travel of the damper rod, in both directions. Check the spring free length.

6   Reverse this procedure to reassemble the suspension unit.

Pro-Link models

7   The rear suspension unit can be removed as described in Section 18, or without fully dismantling the rest of the suspension linkage. Start by removing the silencers and releasing the air hose and drain tube from their clamps. Remove the upper mounting bolt, then support the unit while the lower bolt is freed. The unit can be manoeuvred clear of the frame once the linkage front pivot bolt has been removed and the linkage has been swung backwards out of the way.

8   The suspension unit cannot be rebuilt, but it is possible to renew the oil seal and guide bush provided that the damper rod surface is not scored or pitted. This operation requires the use of an hydraulic press and a selection of special tools, and thus is best entrusted to a Honda Service Agent.

9   Clean the upper mounting sleeve and coat it with molybdenum disulphide grease, then stick the seals in place with grease. Refit the unit by reversing the above sequence, tightening the various bolts to their recommended torque settings. Refer to Routine Maintenance for details of adjusting the unit air pressure.

16.3a Engage right-hand pivot stub with bearing ...

16.3b ... and refit pivot shaft

17.6 Tighten all fasteners to specified torque settings

17.7 Dismantle linkage and remove top mounting bolt to withdraw suspension unit – Pro-Link models

arm/swinging arm pivot bolts. Separate the two arms by removing the pivot bolt and withdraw them from the frame. Note the small drain tube from the suspension unit which is clipped to the linkage on some models.

3   The pivot points of the suspension arm and links take the form of sleeves carried in renewable bushes, the assembly being protected by a seal at each end. These should be degreased and inspected for wear or scoring. The bushes are pressed into their respective bores, and must not be disturbed unless renewal proves necessary, in which case the old bushes can be driven out using a stepped drift. If care is taken, the new bushes can be inserted using a long bolt and nut and suitable spacers and plain washers. Note that the bushes must be drawn squarely into place and should be positioned so that an equal amount protrudes on each side of the casting to which they are fitted.

4   Examine the lips of the grease seals and renew any that are split, scored or distorted. Lubricate the bushes and the seals using molybdenum disulphide grease prior to reassembly.

5   On reassembly, check that all pivot bearing components are thoroughly greased and correctly refitted, then clean and grease the pivot bolts. Fit the rear arm to the swinging arm and the front arm to the frame, ensuring that each is the correct way up and pivots smoothly with no sign of stiffness or of free play. Fit the two together and refit the pivot bolt, then align the suspension unit bottom mounting and refit its bolt. Tighten all retaining nuts to a torque setting of 4.5 – 5.5 kgf m (32.5 – 40 lbf ft). Refit the suspension unit drain tube, where fitted.

18 Rear suspension linkage: removal, renovation and refitting – Pro-Link models

1   Place the machine on its centre stand and remove the silencers. See Chapter 3.

2   Remove their retaining nuts and withdraw the front arm/frame pivot bolt, the suspension unit bottom mounting bolt and the rear

19 Footrests: examination

The footrests are of the folding type, and are bolted to the frame on each side of the machine. Little maintenance is required except regular lubrication of the pivot. The footrests are designed to fold if the machine is dropped, but if more extensive damage is incurred they may be unbolted for renewal.

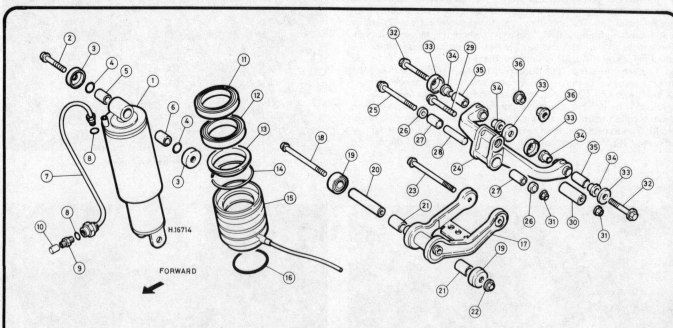

Fig. 5.11 Rear suspension unit and linkage – Pro-link models

1	Suspension unit	10	Cap	19	Seal	28	Sleeve
2	Bolt	11	Guide bush	20	Sleeve	29	Bolt
3	Sealing cap	12	Seal	21	Bush	30	Bush
4	O-ring	13	Backing ring	22	Nut	31	Nut
5	Bush	14	Circlip	23	Bolt	32	Bolt
6	Sleeve	15	Gaiter	24	Rear arm	33	Sealing cap
7	Air hose	16	O-ring	25	Bolt	34	Bush
8	O-ring	17	Front arm	26	Seal	35	Sleeve
9	Air valve	18	Bolt	27	Bush	36	Nut

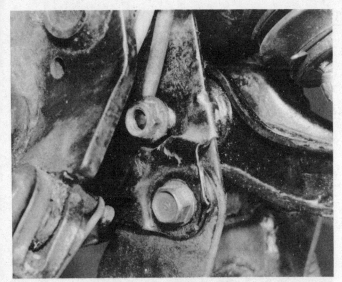

18.2a Remove linkage front arm/frame pivot bolt ...

18.2b ... and rear arm/swinging arm pivot bolts to release linkage

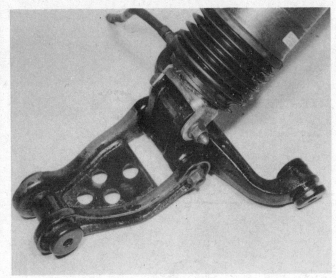

18.2c Suspension unit can be left in place or removed with the linkage

18.4 Renew all seals if damaged or worn – check carefully

18.5a Use specified grease to lubricate all bearing surfaces on reassembly ...

18.5b ... pack sealing caps with grease to prevent entry of dirt or water ...

18.5c ... and grease all pivot bolts before refitting

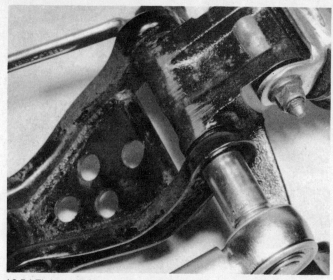

18.5d Tighten all fasteners to specified torque settings

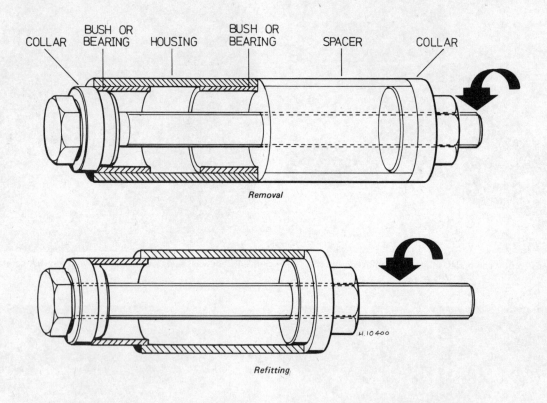

Fig. 5.12 Drawbolt tool for removing and refitting suspension linkage bearings

20 Instrument heads: removal

1 On CX500 models, the speedometer and tachometer heads are mounted in the headlamp shroud from which they must be detached before they can be separated. First, remove the two screws from the headlamp shell which secure the headlamp glass/reflector unit in place. Pull the unit out at the lower edge and then upwards to free it from the top of the shell rim. Disconnect the wires from the rear of the reflector unit and then place the complete assembly to one side. Disconnect the drive cables from the base of each instrument by unscrewing the knurled union ring. The shared bracket on which both heads are mounted is secured in the top of the shroud by two studs which pass through the headlamp shell. Remove the nut from each stud. Carefully ease the assembly upwards and then disconnect the various wires running from the instrument heads. All the wires are colour coded to aid correct reconnection.

2 Each instrument head is retained on the mounting bracket by two studs and nuts. After removal of the nuts, the instruments may be eased from position. The bulb holder in the base of each instrument is a push fit.

3 It is not possible to repair a faulty instrument head. If the instrument

fails completely, or moves jerkily, first check the drive cable. If the mileage recorder of the speedometer ceases to function but the speedometer continues to work or vice versa, the instrument head is faulty.

4   Remember that a working speedometer, accurate at 30 mph, is a statutory requirement in the UK and many other countries.

5   On all other models the instruments form a separate unit and are removed and refitted as described in the relevant paragraphs of Section 8.

21 Screen: removal, examination and renovation – CX500-A, CX500-B

1   Removal of the screen is accomplished by unscrewing the two screws on its outer face. Each screw passes through a spring washer and collar and is secured at the back of the headlamp surround by a plastic nut.

2   To clean the screen, rinse in a mild soapy solution and polish with a chamois leather. On no account use any abrasive compound because this will damage the surface of the screen. Minor scratches can be removed with the appliction of plastic polish but a cracked or broken screen must be replaced.

3   When refitting the screen take care not to overtighten the screws because this could strip the threads of the plastic nuts or place undue stress on the screen.

22 Fairing: removal and refitting – CX500 E-C, CX650 E-D

1   Remove the two screws, one on each side at the rear, which retain the fairing then allow it to pivot forwards. Unhook the stopper cable and slacken the two front pivot bolts. Unhook the fairing from the pivot rubber to remove it.

2   The screen can be renewed if damaged; using a pointed instrument prise off the four retaining clips and withdraw the screen. New clips will be required on refitting.

3   Refer to Section 21 for information on cleaning.

23 Fairing: removal and refitting – GL500 D, GL500 I, GL650 D2 and GL650 I models

1   Place the machine on its centre stand and disconnect the fairing wiring loom at the nine-pin block connector on the left-hand side, under the tank nose.

2   Remove the three screws and their spacers which retain each fairing lower panel (two at the top edge, one at the bottom corner). Withdraw the lower panels. Remove from around the lower front edge of the fairing the four cap nuts or bolts with their spring and plain washers and spacers.

3   With an assistant holding the fairing steady remove the four nuts (two on each side) which secure the fairing to the main mounting bracket. With one person standing on each side of the machine, carefully lift the fairing to the front to clear the handlebars.

4   Reverse the removal procedure to refit the fairing. Tighten securely the various fasteners but do not overtighten them or the fairing material may crack.

5   If the main mounting bracket is to be removed, disconnect their wires, remove their mounting bolts and detach the horns. Remove the two mounting bolts which secure each front bracket to the main bracket and lift them away. Unscrew the two 10 mm nuts securing the main bracket to the cylinder head/engine front mounting bracket studs, then remove the 8 mm bolt which passes through the frame to the rear of the steering head. Carefully withdraw the mounting bracket from the frame, taking care not to scratch the paintwork or to damage any wiring etc. Refitting is the reverse of the removal procedure.

6   Windscreen blades are available in varying heights, and can be adjusted 1 inch (25 mm) above or below the standard height as follows. Slacken first the rear view mirror mounting screws, then the two front panel screws. Adjust the screen to the desired height and tighten the screws in the reverse order ie, panel screws first, then mirror mounting screws (both inner screws, then both outer.) Do not overtighten the screws and be careful to tighten them in the order specified or the screen blade may crack.

7   With the fairing removed, the pockets and air vent duct components can be withdrawn if required; all are retained by a number of self-tapping screws.

24 Seat: removal and refitting

All CX500 and CX500 D models

1   The seat is retained at the rear by two spring loaded catches which are fitted to the frame and which engage with the rear of the seat pan. The front of the seat is located by a tongue projecting from the seat pan which engages with the bridge into which the fuel tank rear mounting bolt passes.

2   To remove the seat depress the catches and lift it at the rear before moving it backwards. Refitting is equally straightforward.

All other models

3   On all GL models the trunk can be replaced by an optional pillion seat which is hooked under a frame bracket at the front and secured by the helmet lock latches at the rear. Unhook the helmet locks to release the seat. On refitting engage the front mounting and press down firmly on the grabrail to lock the catches.

4   The rider's seat on GL models, and the dual seat fitted to all other models is retained by two bolts at the rear. Once these have been unscrewed the seat is removed and refitting as described above.

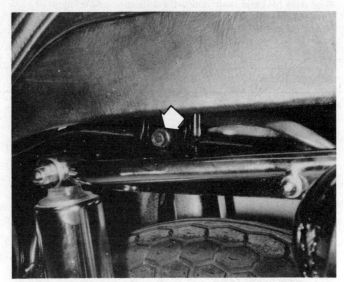

24.4 Seat is retained by a single bolt on each side (arrowed) on most models

25 Rear trunk: removal and refitting – GL models

The rear trunk is removed and refitted exactly as described in Section 24 paragraph 3.

26 Panniers: removal and refitting – GL500 I, GL650 D2-E and GL650 I models

1   To remove the panniers, turn the ignition key clockwise to unlock each helmet holder lock under the rear of the trunk. Push the 'press' button in the centre of the buckle mounting attached to the pillion footrest, then pull the pannier upwards and backwards to release it.

2   To refit, hang the hooks on each pannier's rear face on the frame-mounted support rail beneath the trunk. Slide the pannier forwards until its buckle plate locks into the frame-mounted tongue plate, then press down the helmet holder/latch until it clicks to lock the pannier on the machine.

Chapter 6 Wheels, brakes and tyres

Contents

Specifications

Note: *unless otherwise stated, information applies to all models*

Wheels

Type:
  500 models, CX650 E-D .... Comstar
  650 models except CX650 E-D .... Cast alloy
Rim maximum runout – radial and axial .... 2.0 mm (0.0787 in)
Spindle maximum runout .... 0.2 mm (0.0079 in)

Brakes

	Front	Rear
Type:		
All US CX500, CX500 C, CX500 D models, GL500 1981, 1982, CX650 C	Single hydraulic disc	Drum
GL500 I 1981, 1982, GL650, GL650 I, all UK models except those below	Twin hydraulic disc	Drum
CX500 E-C, CX650 E-D	Twin hydraulic disc	Single hydraulic disc
Single front disc master cylinder:	**Standard**	**Service limit**
Cylinder ID	15.870 – 15.913 mm (0.6248 – 0.6265 in)	15.925 mm (0.6270 in)
Piston OD – CX500 1978	15.827 – 15.854 mm (0.6231 – 0.6242 in)	15.810 mm (0.6224 in)
Piston OD – all other models	15.827 – 15.854 mm (0.6231 – 0.6242 in)	15.815 mm (0.6226 in)
Twin front disc master cylinder, rear master cylinder:		
Cylinder ID	14.000 – 14.043 mm (0.5512 – 0.5529 in)	14.055 mm (0.5534 in)
Piston OD – CX500 (UK)	13.957 – 13.984 mm (0.5495 – 0.5506 in)	13.940 mm (0.5488 in)
Piston OD – all other models	13.957 – 13.984 mm (0.5495 – 0.5506 in)	13.945 mm (0.5490 in)
Caliper – CX500, CX500-A, CX500-B, CX500 C-B:		
Bore ID	42.850 – 42.900 mm (1.6870 – 1.6890 in)	42.915 mm (1.6896 in)
Piston OD	42.772 – 42.822 mm (1.6839 – 1.6859 in)	42.765 mm (1.6837 in)
Caliper – CX500 1978, 1979, CX500 C and CX500 D 1979, 1980:		
Bore ID	38.180 – 38.230 mm (1.5032 – 1.5051 in)	38.240 mm (1.5055 in)
Piston OD	38.098 – 38.148 mm (1.4999 – 1.5019 in)	38.090 mm (1.4996 in)

Brakes (continued)

	Standard	Service limit
Caliper – twin piston type, front and rear:		
Bore ID – CX650 C	32.030 – 32.080 mm (1.2610 – 1.2630 in)	32.090 mm (1.2634 in)
Bore ID – all other models	30.230 – 30.280 mm (1.1902 – 1.1921 in)	30.290 mm (1.1925 in)
Piston OD – CX650 C	31.998 – 32.048 mm (1.2598 – 1.2617 in)	31.940 mm (1.2575 in)
Piston OD – all other models	30.148 – 30.198 mm (1.1869 – 1.1889 in)	30.140 mm (1.1866 in)
Disc thickness:		
Single front disc – CX650 C	4.9 – 5.1 mm (0.1929 – 0.2008 in)	4.0 mm (0.1575 in)
Single front disc – CX500 1978, 1979, CX500 C 1979, 1980, 1981, all CX500 D models	6.8 – 7.2 mm (0.2677 – 0.2835 in)	6.0 mm (0.2362 in)
Single front disc – CX500 C 1982, GL500 1981, 1982	6.9 – 7.1 mm (0.2717 – 0.2795 in)	6.0 mm (0.2362 in)
Twin front discs – all models	4.9 – 5.1 mm (0.1929 – 0.2008 in)	4.0 mm (0.1575 in)
Rear disc – CX500 E-C, CX650 E-D	6.9 – 7.1 mm (0.2717 – 0.2795 in)	6.0 mm (0.2362 in)
Disc maximum runout	0.3 mm (0.0118 in)	
Drum rear brake:		
Drum ID	160.0 – 160.3 mm (6.2992 – 6.3110 in)	161.0 mm (6.3386 in)
Friction material thickness	4.9 – 5.0 mm (0.1929 – 0.1969 in)	2.0 mm (0.0787 in)

Tyres

	Front	Rear
All CX500 models	3.25S-19 4PR	3.75S-18 4PR
All CX500 C, CX500 D and GL500 models	3.50S-19 4PR	130/90-16 67S
CX500 E-C	100/90-18 56S	120/80-18 62S
CX650 C	100/90-19 57H	140/90-15 70H
CX650 E-D	100/90-18 56H	120/80-18 62H
All GL650 models	3.50H-19 4PR	130/90-16 67H
Minimum tread depth	1.5 mm (0.0591 in)	2.0 mm (0.0787 in)

Tyre pressures – tyres cold

	Front	Rear
CX500 (UK), CX500 1978, 1979:		
Solo	25 psi (1.75 kg/cm²)	28 psi (2.00 kg/cm²)
Pillion	25 psi (1.75 kg/cm²)	36 psi (2.50 kg/cm²)
CX500-A, CX500-B, CX500 E-C, all GL500 models:		
UK models solo, US models up to 90 kg (198 lb)*	28 psi (2.00 kg/cm²)	28 psi (2.00 kg/cm²)
UK models pillion, US models over 90 kg (198 lb)*	28 psi (2.00 kg/cm²)	36 psi (2.50 kg/cm²)
All CX500 C and CX500 D models:		
UK model solo, US models up to 90 kg (198 lb)*	28 psi (2.00 kg/cm²)	28 psi (2.00 kg/cm²)
UK model pillion, US models over 90 kg (198 lb)*	28 psi (2.00 kg/cm²)	32 psi (2.25 kg/cm²)
All 650 models:		
UK models solo, US models up to 90 kg (198 lb)*	32 psi (2.25 kg/cm²)	32 psi (2.25 kg/cm²)
UK models pillion, US models over 90 kg (198 lb)*	32 psi (2.25 kg/cm²)	40 psi (2.80 kg/cm²)

*Loads represent total weight of rider, passenger, accessories and luggage

Torque wrench settings

Component	kgf m	lbf ft
Front wheel spindle or spindle nut	5.5 – 6.5	40.0 – 47.0
Front wheel spindle clamp nuts	1.8 – 2.5	13.0 – 18.0
Front wheel spindle clamp bolt	1.5 – 2.5	11.0 – 18.0
Rear wheel spindle nut:		
CX650 C, CX650 E-D	6.0 – 8.0	43.0 – 58.0
All other Pro-Link models	5.0 – 8.0	36.0 – 58.0
All other models	5.5 – 6.5	40.0 – 47.0
Rear wheel spindle clamp bolt	2.0 – 3.0	14.5 – 22.0
Front brake disc mountings:		
CX650 C	3.5 – 4.0	25.0 – 29.0
All other models	2.7 – 3.3	19.5 – 24.0
Rear brake disc mountings	1.0 – 1.2	7.0 – 9.0
Drive flange/rear wheel mounting bolts:		
CX650 C	5.0 – 6.0	36.0 – 43.0
All other models	4.0 – 5.0	29.0 – 36.0
Brake caliper/fork lower leg mounting bolts:		
CX500 E-C, CX650 E-D – left-hand caliper top bolt	3.5 – 4.5	25.0 – 32.5
CX500 E-C, CX650 E-D – left-hand caliper bottom bolt (torque link/anti-dive case)	2.0 – 2.4	14.5 – 17.0
CX650 C, all GL500 models	3.0 – 4.5	22.0 – 32.5
All other models	3.0 – 4.0	22.0 – 29.0

Torque wrench settings (continued)
Component

	kgf m	lbf ft
Single piston brake caliper axle bolts	1.5 – 2.0	11.0 – 14.5
Twin piston brake caliper bolts:		
Top (or front) – pivot	2.5 – 3.0	18.0 – 22.0
Bottom (or rear)	2.0 – 2.5	14.5 – 18.0
Bleed nipple	0.4 – 0.7	3.0 – 5.0
Brake hose junction block mounting bolts	1.0 – 1.4	7.0 – 10.0
Brake hose union bolts	2.5 – 3.5	18.0 – 25.0
Rear brake torque arm mountings:		
CX500 E-C, CX650 E-D	1.8 – 2.8	13.0 – 20.0
All GL models, CX650 C	1.5 – 2.5	11.0 – 18.0
All CX500, CX500 C and CX500 D models	1.5 – 2.3	11.0 – 16.5
Brake pedal pinch bolt	1.0 – 1.5	7.0 – 11.0
Brake pedal adjuster locknut	0.6 – 0.9	4.0 – 6.5
Rear master cylinder mounting bolts	2.4 – 2.9	17.0 – 21.0
Rear master cylinder pushrod locknut	1.5 – 2.0	11.0 – 14.5
Final drive gear case cover bolts:		
8 mm	2.3 – 2.8	16.5 – 20.0
10 mm	3.5 – 4.5	25.0 – 32.5
Final drive gear pinion – 650 models only:		
Retainer	10.0 – 12.0	72.0 – 87.0
Holder nut	10.0 – 12.0	72.0 – 87.0

1 General description

On 500 models and the CX650 E-D the wheels are of the Comstar type, in which steel spoke plates are bolted to a light alloy hub and riveted to a light alloy rim; although their appearance is to the contrary these wheels should not be dismantled and must be renewed if damaged. CX650 C and all GL650 models are fitted with cast-alloy wheels. All models are fitted with tubeless tyres.

The various types of brakes fitted are listed in the Specifications section. Early models use a single piston caliper design, but all models from 1981 on (except CX500 C and CX500 D models) are fitted with a twin piston caliper. The drum rear brake fitted to most models is a rod-operated single leading shoe design.

This Chapter describes only the removal and refitting of the wheels, brakes and tyres and the overhaul of wheel bearings and brake components; for details of regular checks on the wheels, wheel bearings, brakes and tyres, for brake adjustment and pad renewal, refer to Routine Maintenance.

2 Front wheel: removal and refitting

All CX500 models, CX500 C-B, CX500 C 1979, 1980, all CX500 D models
1 With the front wheel supported well clear of the ground, remove the cross head screw which retains the speedometer cable to its drive gearbox, on the right-hand side of the hub. Pull the cable out and replace the screw, to prevent loss.
2 On models with twin front discs, remove the two bolts holding one of the caliper support brackets to the fork leg and lift the caliper and bracket assembly off the disc. Support the weight of the caliper with a length of string or wire attached to the frame or engine.
3 Displace the split pin from the end of the wheel spindle and remove the castellated nut. Slacken the two nuts at the base of the right-hand fork leg to free the spindle from the clamp. Support the weight of the wheel and push out the spindle. The wheel can now be lowered and lifted forward, out of the forks.
4 Do not operate the front brake lever while the wheel is removed since fluid pressure may displace the pistons and cause leakage. Additionally, the distance between the pads will be reduced, making refitting of the brake discs more difficult.
5 Refit the wheel by reversing the dismantling procedure. Do not omit the spacer which is a push fit in the oil seal on the right-hand side of the wheel or the speedometer gearbox which is a push fit on the left-hand side. Ensure that the speedometer drive dogs engage with the notches in the gearbox drive sleeve. Lift the wheel into position and insert the wheel spindle so that the spindle head is flush with the outer face of the clamp and fork leg. Tighten the clamp front nut first to a torque setting of 1.8 – 2.5 kgf m (13 – 18 lbf ft) and then tighten the rear nut to the same figure. Do not try and eliminate the gap between

the clamp face and fork lower leg face. If the clamp is removed at any time, it must be fitted so that the higher mating surface is at the front (also indicated by a cast arrow mark). Rotate the speedometer so that the cable boss is approximately horizontal and then fit and tighten the wheel spindle nut. Fit a new split pin to prevent the spindle nut unscrewing in service.
6 Spin the wheel to ensure that it revolves freely and check the brake operation. Check that all nuts and bolts are fully tightened. If the clearance between the discs and pads is incorrect, pump the front brake lever several times to adjust.

CX500 C 1981, 1982, CX650 C
7 Position the machine on its main stand and ensure that the front wheel is raised clear of the ground.
8 On 500 models, release the small spring clip locating the speedometer cable in its drive gearbox and withdraw the cable, making sure that it is supported well clear of the wheel. On the 650 model remove its retaining screw and pull the cable out of its housing.
9 On the 650 model prise off its cap and remove the Allen-headed clamp bolt from the right-hand lower leg; on 500 models, remove the spindle clamp bolt and nut. Support the weight of the wheel while the spindle is slackened and carefully removed. The speedometer drive gearbox and the right-hand spacer will drop free as the wheel is removed from the fork legs.
10 Do not operate the brake lever whilst the wheel is removed, otherwise there is a danger of the brake pads moving inwards making refitting of the brake disc difficult. It is a good idea to position a wooden wedge between the pads, to prevent movement, if the lever is inadvertently operated.
11 To reassemble the wheel lightly grease the outer face of the oil seal and fit the spacer. Locate the speedometer drive gearbox ensuring that the drive dogs engage with the notches in the gearbox sleeve. Lift the wheel into position and install the wheel spindle, tightening it to the specified torque setting; the speedometer drive gearbox must be positioned correctly in order for the drive cable to be easily connected. Install the spindle clamp bolt (and nut) and tighten it to 1.5 – 2.5 kgf m (11 – 18 lbf ft).
12 Spin the wheel to ensure that it revolves freely and check operation of the brakes.

All GL650 models, CX500 E-C, CX650 E-D
13 Place the machine on its centre stand and place wooden blocks or similar under the crankcase to raise the front wheel clear of the ground.
14 Remove the right-hand brake caliper, supporting it clear of the wheel and forks. Place a wooden wedge between the brake pads to prevent them from being expelled if the brake lever is accidentally operated. Disconnect the speedometer drive cable at the wheel end by releasing its retaining screw. Slacken the clamp nuts at the bottom of the right-hand lower leg, then unscrew and withdraw the wheel spindle, lowering the wheel clear of the forks.
15 To refit the wheel, fit the speedometer drive gearbox, making sure

that it engages correctly in the slots in the wheel hub. Lift the wheel into position and screw the spindle loosely home. Check that the speedometer drive gearbox locates correctly against the lug on the lower leg. Tighten the spindle to 5.5 – 6.5 kgf m (40 – 47 lbf ft). If the clamp was removed, refit it so that the higher mating surface is at the front (also indicated by a cast arrow mark). Tighten first the front clamp nut, then the rear nut; both nuts should be tightened only lightly at this stage. Refit the caliper, tightening the mounting bolts to 3.0 – 4.0 kgf m (22 – 29 lbf ft). Using a 0.7 mm (0.028 in) feeler gauge, check the clearance between both faces of the brake discs and the caliper bracket. If the gauge does not fit easily between the two, push or pull the fork leg until the clearance is correct. Tighten the front clamp nut, followed by the rear clamp nut to 1.5 – 2.5 kgf m (11 – 18 lbf ft) to secure it. Operate the brake lever several times, then re-check the clearances. Note that if the clearances are inadequate, the brake will tend to drag.

All GL500 models

16 Place the machine on its centre stand and place wooden blocks or similar under the crankcase to raise the front wheel clear of the ground.
17 On models with twin front discs, remove the left-hand brake caliper, supporting it clear of the wheel and forks. Place a wooden wedge between the brake pads to prevent them from being expelled if the brake lever is accidentally operated. Disconnect the speedometer drive cable at the wheel end by releasing its retaining screw. Remove the four clamp nuts and the two clamps and lower the wheel clear of the forks.
18 To refit the wheel, fit the speedometer drive gearbox making sure that it engages correctly in the slots in the wheel hub. Position the wheel below the fork legs, then lower the machine until the fork ends rest on the spindle ends. Check that the speedometer drive gearbox locates correctly against the lug on the lower leg. Fit the clamps with the arrow marks (higher mating surface) facing forward and fit the retaining nuts finger tight, starting with the two front nuts. Refit the caliper, tightening the mounting bolts to the prescribed torque figure. Tighten the right-hand clamp nuts only, starting with the front nut.
19 Using a 0.7 mm (0.028 in) feeler gauge, check the clearance between both faces of the brake discs and the caliper bracket. If the gauge does not fit easily between the two, push or pull the fork leg until the clearance is correct. Once the clearance is correct, tighten the remaining clamp nuts, again starting with the front nut. Operate the brake lever several times, then re-check the clearances. Note that if the clearances are inadequate, the brake will tend to drag.

2.1 Remove countersunk screw to release speedometer drive cable

2.3 Slacken clamp nuts and remove spindle nut to release wheel spindle

2.5a Dogs on speedometer gearbox drive ring must engage with slots in gearbox

2.5b Tighten spindle nut to specified torque setting and secure with a new split pin

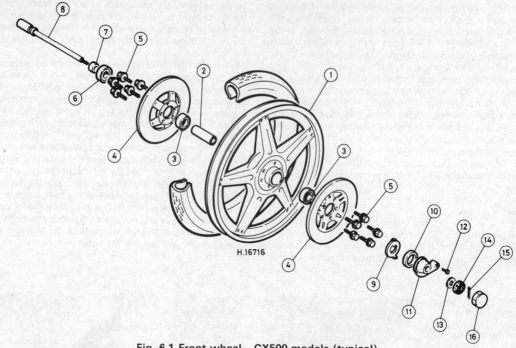

H.16716

Fig. 6.1 Front wheel – CX500 models (typical)

1	Wheel	5	Bolt	9	Gearbox retainer	13	Washer
2	Spacer	6	Seal	10	Seal	14	Nut
3	Bearing	7	Collar	11	Speedometer drive gearbox	15	Split pin
4	Brake disc	8	Spindle	12	Screw	16	Cap

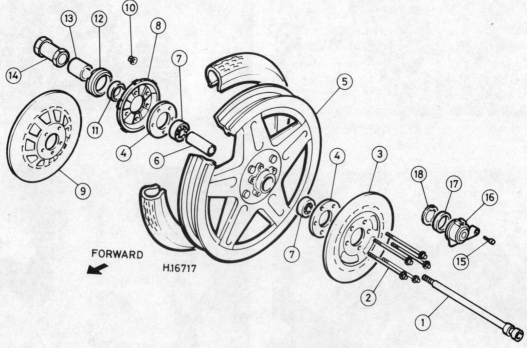

FORWARD

H.16717

Fig. 6.2 Front wheel – GL500 models (typical)

1	Spindle	6	Spacer	10	Nut	15	Screw
2	Bolt	7	Bearing	11	Oil seal	16	Speedometer drive gearbox
3	Left-hand brake disc	8	Flange – single disc system	12	Threaded retainer	17	Oil seal
4	Damping shim	9	Right-hand brake disc	13	Spacer	18	Gearbox retainer
5	Wheel			14	Nut		

3 Rear wheel: removal and refitting

Drum rear brake

1 Place the machine on the centre stand so that the rear wheel is raised clear of the ground. On all GL models remove both panniers (if fitted) and the tail trunk or pillion seat. The rear mudguard rear section is retained by two bolts on GL500 models and by four bolts on GL500 D, GL500 I and all 650 models. Remove the bolts and swing up the rear section on its pivot.

2 Detach the torque arm from the brake back plate after removing the nut secured by a split pin. Unscrew the brake rod adjuster nut fully and depress the brake pedal so that the rod leaves the trunnion in the brake operating arm. Refit the nut to secure the rod spring.

3 Remove the wheel spindle nut after displacing the split pin, and then slacken the spindle pinch bolt. Withdraw the wheel spindle from the left-hand side. Do not lose the spacer. Pull the wheel across towards the left and support its weight as it leaves the final drive casing.

4 Refit the wheel by reversing the dismantling procedure. Apply a coat of the specified grease to the splines of the final drive components before refitting the wheel. To ensure correct alignment tighten the wheel spindle nut first, then the spindle clamp bolt; tighten both bolts to the specified torque settings. Ensure that the torque arm is secure and that the securing split pins are fitted. Likewise do not omit the wheel spindle nut securing pin. Adjust the rear brake as described in Routine Maintenance.

Disc rear brake

5 Place the machine on its centre stand so that the rear wheel is clear of the ground.

6 Unscrew the spindle nut and slacken the clamp bolt, then pull out the wheel spindle. Lift up the caliper and tie it to the frame as high up as possible without straining the brake hose.

7 With the aid of an assistant, tilt the machine to the right, pull the wheel off the splines and manoeuvre it clear of the frame. Take care not to damage the rear mudguard.

8 On refitting apply a coat of the specified grease to the splines of the final drive components before refitting the wheel. Lower the caliper into place, while the wheel is held on the splines, and guide the disc between the brake pads. Insert the spindle and hold it with a spanner applied to its left-hand end while it is tightened to the specified torque setting. Tighten the spindle clamp bolt, also to its specified torque setting.

9 Apply the brake pedal repeatedly until the pads are back in firm contact with the disc and normal brake pressure is restored.

3.3a Slacken spindle clamp bolt, withdraw spindle ...

3.3b ... and remove wheel from the machine

3.4a Always use new split pins to secure torque arm ...

3.4b ... and wheel spindle nuts. Tighten to specified torque settings

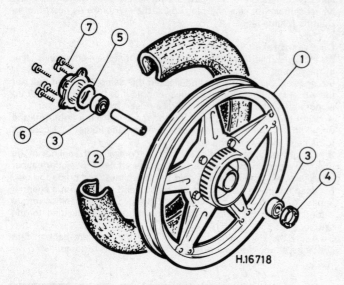

Fig. 6.3 Rear wheel – typical

1 Wheel	5 O-ring
2 Spacer	6 Rear bevel drive
3 Bearing	7 Bolt
4 Bearing retainer	

4 Wheel bearings: removal, examination and refitting

1 Remove from the machine the wheel to be overhauled and withdraw the hub spacer and, where applicable, the brake backplate; refer to Section 2 or 3 of this Chapter.

2 On the front wheel of GL500 models, unscrew the spindle sleeve nut and withdraw the spindle and spacer. The speedometer drive ring can be lifted out of the hub after the retaining oil seal on that side has been levered from its housing.

3 On all GL500 models the front hub right-hand bearing is secured by a threaded retainer, and all Comstar type rear wheels have a threaded retainer fitted to secure the left-hand bearing. These retainers are screwed into the hub and staked at four points to prevent them from unscrewing. A retainer is removed and refitted using a special tool, Part Number 07710-0010401 in conjunction with an adaptor, Part Number 07710-0010200. If these are not available, and a peg spanner of suitable size cannot be found, obtain a steel strip (an old tyre lever would be ideal) and drill two holes in it to correspond with two diagonally opposite holes in the retainer. Pass a small bolt through each hole and secure them with nuts to complete a fabricated peg spanner of the type shown in the accompanying illustration.

4 In all cases, the oil seal into which the hub spacer is inserted may be either levered out or driven out with the appropriate bearing.

5 The bearings are easiest to remove if an internally-expanding bearing puller is used. If this is not available an expanding bolt such as a Rawlbolt could be tightened on to the inner race of one bearing so that it can be driven out by passing a drift through the hub. The simplest method is given below but will require some skill and care if it is to be successful. On GL650 machines the rear wheel right-hand bearing must be removed first, but in all other cases either bearing can be removed first.

6 Position the wheel on a work surface with its hub well supported by wooden blocks so that enough clearance is left beneath the wheel to drive out the first bearing. Ensure the blocks are placed as close to the bearing as possible, to lessen the risk of distortion of the hub casting whilst the bearings are being removed or refitted.

7 Place the end of a long-handled drift through the hub and against the upper face of the bearing inner race and tap the bearing downwards out of the wheel hub. The spacer located between the two bearings may be moved sideways slightly in order to allow the drift to be positioned against the face of the bearing. Move the drift around the

face of the bearing whilst drifting it out of position, so that the bearing leaves the hub squarely.

8 With the one bearing removed, the wheel may be lifted and the spacer withdrawn from the hub. Invert the wheel and remove the second bearing, using a similar procedure. The dust seal should be closely inspected for any indication of damage, hardening or perishing and renewed if necessary. It is advisable to renew all seals as a matter of course if the bearings are found to be defective.

9 Remove all the old grease from the hub and bearings, giving the latter a final wash in petrol. Check the bearings for signs of play or roughness when they are turned. If there is any doubt about the condition of a bearing, it should be renewed.

10 If the original bearings are to be refitted, they should be repacked with high-melting point grease before being fitted into the hub. New bearings must also be packed with grease. Ensure that the bearing recesses in the hub are clean and both bearings and recess mating surfaces lightly greased to aid fitting. Check the condition of the hub recesses for evidence of abnormal wear which may have been caused by the outer race of a bearing spinning. If evidence of this is found, and the bearing is a loose fit in the hub, it is best to seek advice from a Honda Service Agent or a competent motorcycle engineer. Alternatively, a proprietary product such as Loctite Bearing Fit may be used to retain the bearing outer race; this will mean, however, that the bearing housing must be cleaned and degreased before the locking compound can be used.

11 With the wheel hub and bearing thus prepared, fit the bearings and central spacer as follows. With the hub again well supported by the wooden blocks, drift the first of two bearings into position. Use a soft-faced hammer in conjunction with a socket or length of metal tube which has an overall diameter which is slightly less than that of the outer race of the bearing, but which does not bear at any point on the bearing sealed surface or inner race. Tap the bearing into place against the locating shoulder machined in the hub, remembering that the sealed surface of the bearing must always face outwards. With the first bearing in place, invert the wheel, insert the central spacer and pack the hub centre no more than $2/3$ full with high-melting point grease. Fit the second bearing, using the same procedure. Take great care to ensure that each of the bearings enters its housing correctly, that is, square to the housing, otherwise the housing surface may be broached.

12 Insert the speedometer drive ring into the hub so that its protruding tangs fit into the slots in the hub. Oil seals should be greased before being drifted into place using the method described above. Where applicable, renew the retainer if its threads are damaged, apply grease to the threads and screw it into the hub until it is securely tightened. Use a hammer and a punch to stake it to the hub at four diametrically opposite points around its outer edge to prevent any risk of its slackening.

4.1 Withdraw wheel from machine and remove hub spacer (where fitted)

4.3 Bearings are secured by a threaded retainer in some cases

4.11a Tap first bearing into place, then invert wheel ...

4.11b ... fit central spacer and pack cavity with grease ...

4.11c ... and fit second bearing. Note that sealed surface always faces outwards

4.12 Always renew oil seals. Grease lightly before refitting

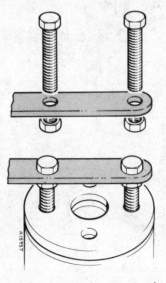
Fig. 6.4 Bearing retainer removal tool

5   Final drive case and shaft: removal, examination and refitting

1   These components can be removed and refitted as described in Sections 14 and 16 of Chapter 5. The final drive case can be removed with the swinging arm in place on the machine, with only the removal of the rear wheel and, where applicable, the suspension unit bottom mounting being necessary.

2   On CX650C, CX650 E-D and GL650 D2-E models the shaft is extracted with the final drive case. To separate the two, hold the case firmly and gently pull the shaft out while moving it around in a circular motion to help dislodge the oil seal. Note the coil spring. Always renew the oil seal and circlip whenever it is disturbed in this way. On reassembly, apply a coat of the recommended grease (see Routine Maintenance) to the shaft splines, check that the new wire circlip is in place on its rear end, and press the shaft into the final drive case input splines. Fit the new oil seal over the shaft end with a film of grease to protect its sealing lips and tap the seal evenly into the case until it is flush with its housing. On these models, note that the universal joint cannot be removed until the swinging arm has been withdrawn.

3   On all other models the shaft is pulled out of the front of the swinging arm after it has been removed from the machine.

4   The driven flange in the rear hub is retained by five bolts; note the O-ring behind the flange if it is disturbed. If the flange splines are ever found to be worn or damaged, it should be renewed. On refitting, renew the O-ring and tighten the bolts to the specified torque setting. Apply the specified grease (see Routine Maintenance) to the splines whenever the wheel is removed.

5   It is strongly advised that the final drive case incorporating the final drive crown and pinion gears be returned to a Honda Service Agent if and when servicing or overhaul is required. Dismantling the gear unit requires the use of special tools which are not generally obtainable by the public. Additionally, reassembly of the unit requires working to very close tolerances and the preload adjustment of bearings and gears.

6   Check the driveshaft, after removal, for slop in the splined joints and for wear in the universal joint. Renew the matching components, where necessary. On all models except the CX650 C, CX650 E-D and GL650 D2-E the internally splined driveshaft coupling on the rear end of the final driveshaft may be separated from the main shaft after driving out the roll pin. Note the O-ring which seals the two components.

5.1 Remove rear wheel and retaining nuts to release final drive gear case

5.3a Driveshaft is pulled forwards out of swinging arm on most models

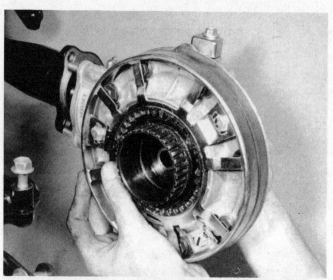

5.3b Check shaft gaiter – renew if split or damaged

5.4 Driven flange is bolted to rear wheel hub. Renew if splines are worn

5.6a Check universal joint bearings for wear or damage, renew if necessary

5.6b Where fitted, drive out roll pin to separate rear coupling sleeve from shaft

6 Brake disc: examination, removal and refitting

1   A brake disc can be checked for wear and for warpage whilst the wheel is still in the machine. Using a micrometer, measure the thickness of the disc at the point of greatest wear. If the measurement is much less than the recommended service limit the disc should be renewed. Check the warpage (runout) of the disc by setting up a suitable pointer close to the outer periphery of the disc and spinning the front wheel slowly. If the total warpage is more than 0.30 mm (0.012 in), the disc should be renewed. A warped disc, apart from reducing the braking efficiency, is likely to cause juddering during braking and will cause the brake to bind when it is not in use.

2   The brake disc should also be checked for bad scoring on its contact area with the brake pads. If any of the above mentioned faults are found, then the disc should be removed from the wheel for renewal or for repair by skimming. Such repairs should be entrusted only to a reputable engineering firm. A local motorcycle dealer may be able to assist in having the work carried out.

3   Remove the wheel from the machine as described in Section 2 or 3 of this Chapter. Removing their retaining nuts (where fitted), unscrew the disc mounting bolts and withdraw the disc. On later models note the damping shim fitted between each disc and the hub.

4   Refitting is the reverse of the above, but when tightening the retaining nuts, do so in an even and diagonal sequence to avoid stress on the disc. Secure to the recommended torque setting.

7 Master cylinder: examination and renovation

1   If the regular checks described in Routine Maintenance reveal the presence of a fluid leak, or if a loss of brake pressure is encountered at any time, the handlebar-mounted master cylinder assembly must be removed and dismantled for checking.

Front master cylinder

2   Disconnect the stop lamp switch wires at the switch, or behind the headlamp, as applicable. The switch need not be disturbed unless the master cylinder is to be renewed. Place a clean container beneath the caliper unit and run a clear plastic tube from the caliper bleed nipple to the container. Unscrew the bleed nipple by one full turn and drain the system by operating the brake lever repeatedly until no more fluid can be seen issuing from the nipple.

3   Position a pad of clean cloth rag beneath the point where the brake hose joins the master cylinder to prevent brake fluid from dripping. Detach the rubber cover from the head of the union bolt and remove the bolt. Once any excess fluid has drained from the union connection, wrap the end of the hose in a rag or polythene and then attach it to a point on the handlebars.

4   Remove the brake lever by unscrewing its shouldered pivot screw and locknut, remove the reservoir cover and diaphragm, then remove the two clamp bolts and withdraw the master cylinder assembly.

5   Use the flat of a small screwdriver to prise out the rubber dust seal boot. This will expose a retaining circlip which must be removed using a pair of circlip pliers which have long, straight jaws. With the circlip removed, the piston and cup assembly can be pulled out. Be very careful to note the exact order in which these components are fitted.

6   Note that if a vice is used to hold the master cylinder at any time during dismantling and reassembly, its jaws must be padded with soft alloy or wooden covers and the master cylinder must be wrapped in soft cloth to prevent any risk of the assembly being marked or distorted.

7   Place all the master cylinder component parts in a clean container and wash each part thoroughly in new brake fluid. Lay the parts out on a sheet of clean paper and examine each one as follows.

8   Inspect the unit body for signs of stress failure around both the brake lever pivot lugs and the handlebar mounting points. Carry out a similar inspection around the hose union boss. Examine the cylinder bore for signs of scoring or pitting. If any of these faults are found, then the unit body must be renewed. Where the reservoir is of translucent plastic, it can be detached and the sealing O-ring can be renewed, if necessary.

9   Inspect the surface of the piston for signs of scoring or pitting and renew it if necessary. It is advisable to discard all the components of the piston assembly as a matter of course as the replacement cost is relatively small and does not warrant re-use of components vital to safety. If measuring equipment is available, compare the dimensions of the master cylinder bore and the piston with those given in the Specifications Section of this Chapter and renew any component that is worn beyond the set wear limit. Inspect the threads of the brake hose union (bolt) for any signs of failure and renew the bolt if in the slightest doubt. Renew each of the gasket washers located one either side of the hose union.

10 Check before reassembly that any traces of contamination remaining within the reservoir body have been removed. Inspect the diaphragm to see that it is not perished or split. It must be noted at this point that any reassembly work must be undertaken in ultra-clean conditions. Particles of dirt entering the component will only serve to score the working points of the cylinder and thereby cause early failure of the system.

11 When reassembling and fitting the master cylinder, follow the removal and dismantling procedures in the reverse order whilst paying particular attention to the following points. Make sure that the piston components are fitted the correct way round and in the correct order.

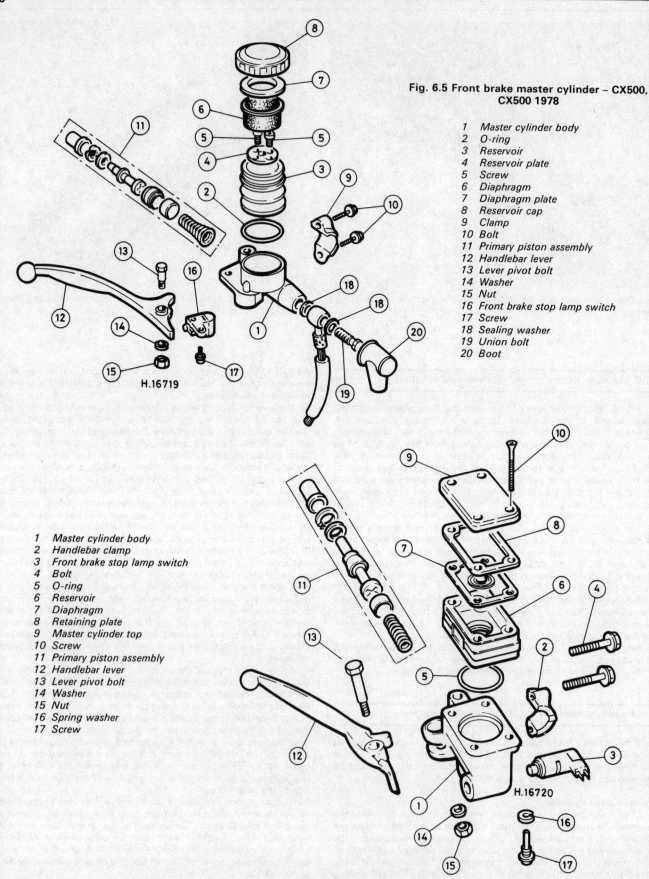

Fig. 6.5 Front brake master cylinder – CX500, CX500 1978

1   Master cylinder body
2   O-ring
3   Reservoir
4   Reservoir plate
5   Screw
6   Diaphragm
7   Diaphragm plate
8   Reservoir cap
9   Clamp
10 Bolt
11 Primary piston assembly
12 Handlebar lever
13 Lever pivot bolt
14 Washer
15 Nut
16 Front brake stop lamp switch
17 Screw
18 Sealing washer
19 Union bolt
20 Boot

H.16719

1   Master cylinder body
2   Handlebar clamp
3   Front brake stop lamp switch
4   Bolt
5   O-ring
6   Reservoir
7   Diaphragm
8   Retaining plate
9   Master cylinder top
10 Screw
11 Primary piston assembly
12 Handlebar lever
13 Lever pivot bolt
14 Washer
15 Nut
16 Spring washer
17 Screw

H.16720

Fig. 6.6 Front brake master cylinder – all later models (typical)

Immerse all of these components in new brake fluid prior to reassembly and refer to the figure accompanying this text when in doubt as to their fitted positions. When refitting the master cylinder assembly to the handlebar, position the assembly so that the reservoir will be exactly horizontal when the machine is in use. Tighten the clamp top bolt first, and then the bottom bolt. Connect the brake hose to the master cylinder, ensuring that a new sealing washer is placed on each side of the hose union, and tightening the hose union (bolt) to the specified torque setting. Finally, replace the rubber union cover.

12 Bleed the brake system after refilling the reservoir with new hydraulic fluid, then check for leakage of fluid whilst applying the brake lever. Push the machine forward and bring it to a halt by applying the brake. Do this several times to ensure that the brake is operating correctly before taking the machine for a test run. During the run, use the brakes as often as possible and on completion, recheck for signs of fluid loss.

Rear master cylinder – CX500 E-C, CX650 E-D

13 Remove the right-hand sidepanel. To avoid splashing brake fluid over any plastic or painted components, drain the system as described in paragraph 2 above before disconnecting the fluid reservoir hose from the master cylinder top union; place plenty of clean rag around the components to catch any spilt fluid.

14 Straighten and remove the retaining split pin, then withdraw the clevis pin from the end of the pushrod. Unscrew the two Allen bolts retaining it to the pillion footrest plate and withdraw the master cylinder.

15 With the unit on the work surface, remove the rubber cover from its lower end, and use a pair of circlip pliers to displace the circlip which retains the pushrod. Withdraw the pushrod, noting that the piston assembly behind it may be ejected under spring pressure; wrap the cylinder in clean rag to prevent the piston from flying out. Withdraw the primary cup and spring; it may be necessary to apply a jet of compressed air to the fluid passage to force them out.

16 Clean and check the unit following the instructions given in paragraphs 7 – 10 above.

17 Reassembly is a straightforward reversal of the removal sequence, noting those points made in paragraphs 11 – 12 above which do not specifically apply to the front unit. Tighten the mounting bolts to the specified torque setting.

8 Brake hose: examination and renovation

1 A flexible brake hose is used as a means of transmitting hydraulic pressure to the caliper unit once the front brake lever is applied.

2 When the brake assembly is being overhauled, or at any time during a routine maintenance or cleaning procedure, check the condition of the hose for signs of leakage, damage, deterioration or scuffing against any cycle components. Any such damage will mean that the hose must be renewed immediately. The union connections at either end of the hose must also be in good condition, with no stripped threads or damaged sealing washers. Do not tighten these union bolts over the recommended torque setting as they are easily sheared if overtightened.

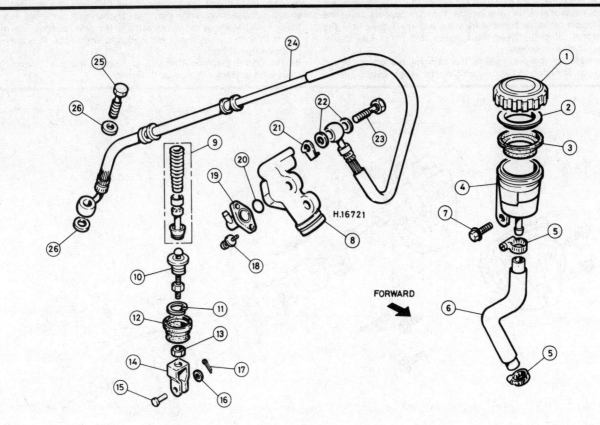

H.16721

FORWARD

Fig. 6.7 Rear brake master cylinder – CX500 E-C, CX650 E-D

1 Reservoir cap	8 Master cylinder	15 Clevis pin	22 Sealing washer – 1 off 650,
2 Retaining plate	9 Piston assembly	16 Washer	2 off 500
3 Diaphragm	10 Pushrod	17 Split pin	23 Union bolt
4 Reservoir	11 Circlip	18 Screw	24 Brake hose
5 Hose clamp	12 Rubber cover	19 Union	25 Union bolt
6 Hose	13 Locknut	20 O-ring	26 Sealing washer
7 Bolt	14 Link	21 Plate	

9 Brake caliper: examination and renovation

1 Start by removing the pads as described in Routine Maintenance. If working on a twin-disc system, dismantle the calipers separately and store their components in separate, clearly-marked containers to avoid the accidental interchange of part-worn components.

Single piston caliper – all CX500 and CX500 D models, CX500 C 1979, 1980, 1981

2 Drain the system of fluid as described in Section 7, paragraph 2.
3 Having separated the caliper from its mounting bracket, withdraw the two slider pins and their rubber dust covers. Displace the circlip which retains the piston dust seal and then carefully prise it out using a small screwdriver, taking care not to scratch the surface of the cylinder bore. The piston can be displaced most easily by applying an air jet to the hydraulic fluid feed orifice. Be prepared to catch the piston as it falls free. Displace the annular piston seal from the cylinder bore groove.
4 Clean the caliper components thoroughly in trichlorethylene or in hydraulic brake fluid. **Caution:** Never use petrol for cleaning hydraulic brake parts otherwise the rubber components will be damaged. Discard all the rubber components as a matter of course. The replacement cost is relatively small and does not warrant re-use of components vital to safety. Check the piston and caliper cylinder bore for scoring, rusting or pitting. If the necessary measuring equipment is available, compare the dimensions of the caliper bore and piston with those specified. If any component is found to be excessively worn or if any of the above defects are evident it is unlikely that a good fluid seal can be maintained and for this reason the components must be renewed. Inspect the slider spindles for wear and check their fit in the support bracket. Slack between the spindles and bores may cause brake judder if wear is severe.
5 To assemble the caliper, reverse the removal procedure. When assembling pay attention to the following points. Apply caliper grease

(high heat resistant) to the caliper spindles. Apply a generous amount of brake fluid to the inner surface of the cylinder and to the periphery of the piston, then reassemble. Do not reassemble the piston with it inclined or twisted. When installing the piston push it slowly into the cylinder while taking care not to damage the piston seal. Apply brake pad grease around the periphery of the moving pad. Bleed the brake after refilling the reservoir with new hydraulic fluid, then check for leakage while applying the brake lever tightly. Repeat the entire procedure for the second brake caliper. After a test run, check the pads and brake disc.
6 Note that any work on the hydraulic system must be undertaken under ultra-clean conditions. Particles of dirt will score the working parts and cause early failure.

Twin piston caliper – all later models

7 The general procedure for dealing with twin piston front calipers is similar to that described above noting the following points. After draining the hydraulic system remove the caliper mounting and pivot bolts and lift the caliper clear of the mounting bracket and disc. Remove the pads and anti-rattle shim as described in Routine Maintenance. The pistons can be displaced from their bores using compressed air as described above. Note that a strip of wood or similar should be placed in the caliper to prevent one of the pistons from emerging before the other; once both are nearly clear of the bores they can be pulled clear manually. In the absence of compressed air, temporarily reconnect the brake hose and use hydraulic pressure to push the pistons out. With either method, wrap some rag around the caliper to catch any spilt fluid and take care to avoid trapped fingers.
8 Examination and reassembly of the caliper is as described above with the obvious exception that there are two sets of pistons and bores to be dealt with. It is recommended that the pistons are marked internally with a spirit-based felt marker to ensure that they are refitted in their original bores.
9 On CX500 E-C and CX650 E-D models, the rear caliper can be dealt with in the same way as the front.

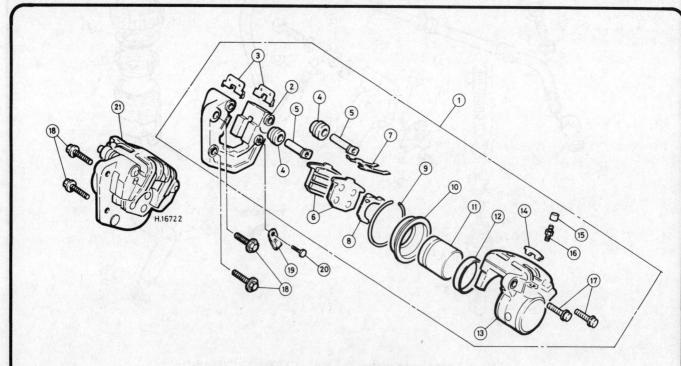

Fig. 6.8 Single piston brake caliper

1	Caliper assembly	7	Pad spring	12	Piston seal	17	Bolt
2	Front bracket	8	Anti-chatter shim	13	Caliper	18	Bolt
3	Bracket retainer	9	Clip	14	Inspection window cap	19	Clamp
4	Dust cover	10	Piston boot	15	Cap	20	Bolt
5	Pin	11	Piston	16	Bleed nipple	21	Caliper
6	Brake pads						

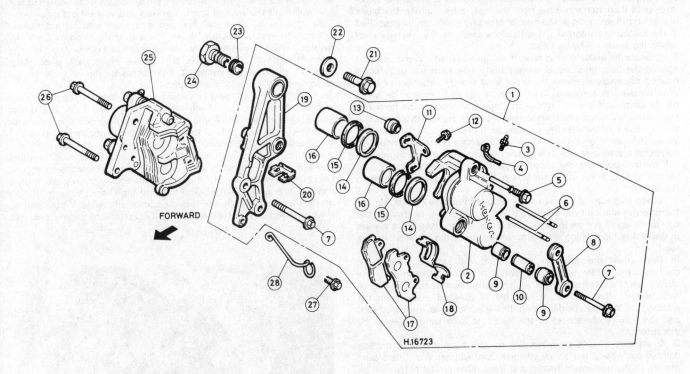

Fig. 6.9 Twin piston brake caliper – CX500 E-C, CX650 E-D (other models similar)

1 Left-hand caliper assembly
2 Caliper
3 Bleed nipple
4 Cap
5 Caliper pivot bolt
6 Pad retaining pin
7 Caliper mounting bracket/fork
   lower leg mounting bolt
8 Torque link*
9 Dust seal
10 Sleeve

11 Pin retaining plate
12 Bolt
13 Dust seal
14 Piston fluid seal
15 Piston dust seal
16 Piston
17 Brake pads
18 Pad spring
19 Mounting bracket
20 Bracket retainer

21 Caliper mounting bracket/fork
   lower leg mounting bolt
22 Washer*
23 O-ring*
24 Pivot*
25 Right-hand caliper
26 Caliper mounting bracket/fork
   lower leg mounting bolt
27 Bolt
28 Cable guide

* CX500 E-C and CX650 E-D only

9.3 Piston dust seal is retained by a circlip on single-piston calipers

10 Bleeding the hydraulic brake system

1   The method of bleeding a brake system of air and the procedure described below apply equally to either a front brake or rear brake of the hydraulically actuated type.

2   If the brake action becomes spongy, or if any part of the hydraulic system is dismantled (such as when a hose is replaced) it is necessary to bleed the system in order to remove all traces of air. The procedure for bleeding the hydraulic system is best carried out by two people.

3   Check the fluid level in the reservoir and top up with new fluid of the specified type if required. Keep the reservoir at least half full during the bleeding procedure; if the level is allowed to fall too far air will enter the system requiring that the procedure be started again from scratch. Refit the cap onto the reservoir to prevent the ingress of dust or the ejection of a spout of fluid.

4   Remove the dust cap from the caliper bleed nipple and clean the area with a rag. Place a clean glass jar below the caliper and connect a pipe from the bleed nipple to the jar. A clear plastic tube should be used so that air bubbles can be more easily seen. Where two calipers are fitted a pipe should be fitted to each nipple, and the two nipples operated simultaneously. Place some clean hydraulic fluid in the glass jar so that the pipe(s) are immersed below the fluid surface throughout this operation.

5   If parts of the system have been renewed, and thus the system must be filled, open the bleed nipple about one turn and pump the brake lever until fluid starts to issue from the clear tube. Tighten the bleed nipple and then continue the normal bleeding operation as described in the following paragraphs. Keep a close check on the reservoir level whilst the system is being filled.

6   Operate the brake lever as far as it will go and hold it in this position against the fluid pressure. If spongy brake operation has occurred it may be necessary to pump the brake lever rapidly a number of times until pressure is achieved. With pressure applied, loosen the bleed nipple about half a turn. Tighten the nipple as soon as the lever has reached its full travel and then release the lever. Repeat this operation until no more air bubbles are expelled with the fluid into the glass jar. When this condition is reached, the air bleeding operation should be complete, resulting in a firm feel to the brake operation. If sponginess is still evident continue the bleeding operation; it may be that an air bubble trapped at the top of the system has yet to work down through the caliper.

7   When all traces of air have been removed from the system, top up the reservoir and refit the diaphragm and cap or cover, as appropriate. Check the entire system for leaks, and check also that the brake system in general is functioning efficiently before using the machine on the road.

8   Brake fluid drained from the system will almost certainly be contaminated, either by foreign matter or more commonly by the absorption of water from the air. All hydraulic fluids are to some degree hygroscopic, that is, they are capable of drawing water from the atmosphere, and thereby degrading their specifications. In view of this, and the relative cheapness of the fluid, old fluid should always be discarded.

9   Great care should be taken not to spill hydraulic fluid on any painted cycle parts; it is a very effective paint stripper. Also, the plastic glasses in the instrument heads, and most other plastic parts, will be damaged by contact with this fluid.

11 Rear drum brake: examination and renovation

1   After removal of the rear wheel, the brake backplate complete with brake shoes can be removed.

2   Examine the brake linings for oil, dirt or grease. Surface dirt can be removed with a stiff brush but oil soaked linings should be renewed. High spots can be carefully eased down with emery cloth.

3   Examine the condition of the brake linings and if they have worn to less than 2.0 mm (0.08 in) in thickness they should be renewed. The brake linings are bonded to the brake shoes and thus separate linings are not available.

4   To remove the shoes displace the split pin from each fulcrum post and lift off the link plate. Remove the pinch bolt from the operating arm and pull it off the splined shaft, followed by the wear indicator plate. Note the punch mark on the arm and shaft end to aid correct positioning on reassembly. Push the camshaft through from the outside whilst simultaneously easing the brake shoe ends off the fulcrum posts at the opposite side of the brake back plate. After removal, displace the camshaft from between the shoe ends and separate the shoes from the springs.

5   The brake drum should be checked for scoring. This happens if the brake shoe linings have been allowed to get too thin. The drums should be quite smooth. Remove all traces of lining dust and wipe with a clean rag soaked in petrol to remove all traces of grease and oil.

6   Reassemble the brake back plate assembly by reversing the dismantling procedure. Check the return spring for wear or other damage at the hook ends and for stretching. Renew the springs, if necessary. Grease the operating camshaft and fulcrum posts with a high melting point grease before refitting the shoes. Note the O-ring on the camshaft, which prevents the escape of grease. When refitting the camshaft and arm, realign the marks to restore the original position.

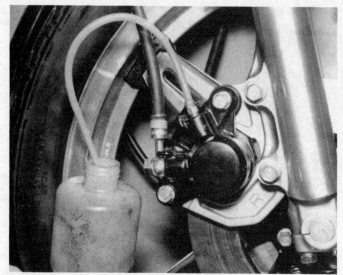

10.4 Bleeding hydraulic brake system

11.4 Note alignment punch marks before dismantling

11.6 Secure brake shoes with new split pins on reassembly. Grease all pivots lightly

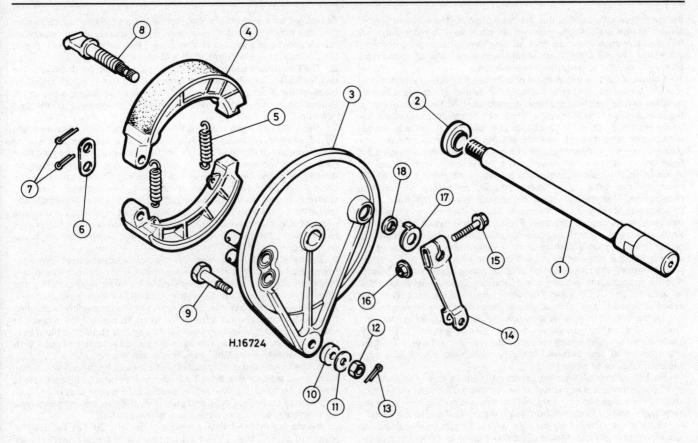

Fig. 6.10 Rear drum brake

1	Wheel spindle	6	Anchor plate	11	Washer	15	Bolt
2	Collar	7	Split pin	12	Nut	16	Nut
3	Brake backplate	8	Brake cam	13	Split pin	17	Wear indicator
4	Brake shoe	9	Bolt	14	Brake arm	18	Dust seal
5	Return spring	10	Rubber washer				

12 Tyres: removal and refitting

1   It is strongly recommended that should a repair to a tubeless tyre be necessary, the wheel is removed from the machine and taken to a tyre fitting specialist who is willing to do the job or taken to an official dealer. This is because the force required to break the seal between the wheel rim and tyre bead is considerable and considered to be beyond the capabilities of an individual working with normal tyre removing tools. Any abortive attempt to break the rim to bead seal may also cause damage to the wheel rim, resulting in an expensive wheel replacement. If, however, a suitable bead releasing tool is available, and experience has already been gained in its use, tyre removal and refitting can be accomplished as follows.

2   Remove the wheel from the machine by following the instructions for wheel removal as described in the relevant Section of this Chapter. Deflate the tyre by removing the valve insert and when it is fully deflated, push the bead of the tyre away from the wheel rim on both sides so that the bead enters the centre well of the rim. As noted, this operation will almost certainly require the use of a bead releasing tool.

3   Insert a tyre lever close to the valve and lever the edge of the tyre over the outside of the wheel rim. Very little force should be necessary: if resistance is encountered it is probably due to the fact that the tyre beads have not entered the well of the wheel rim all the way round the tyre. Should the initial problem persist, lubrication of the tyre bead and the inside edge and lip of the rim will facilitate removal. Use a recommended lubricant, a diluted solution of washing-up liquid or french chalk. Lubrication is usually recommended as an aid to tyre fitting but its use is equally desirable during removal. The risk of lever damage to wheel rims can be minimised by the use of proprietary plastic rim protectors placed over the rim flange at the point where the

tyre levers are inserted. Suitable rim projectors may be fabricated very easily from short lengths (4 – 6 inches) of thick-walled nylon petrol pipe which have been split down one side using a sharp knife. The use of rim protectors should be adopted whenever levers are used and, therefore, when the risk of damage is likely.

4   Once the tyre has been edged over the wheel rim, it is easy to work around the wheel rim so that the tyre is completely free on one side.

5   Working from the other side of the wheel, ease the other edge of the tyre over the outside of the wheel rim, which is furthest away. Continue to work around the rim until the tyre is freed completely from the rim.

6   Refer to the following Section for details relating to puncture repair and the renewal of tyres. See also the remarks relating to the tyre valves in Section 14.

7   Refitting of the tyre is virtually a reversal of removal procedure. If the tyre has a balance mark (usually a spot of coloured paint) indicating its lightest point, as on the tyres fitted as original equipment, this must be positioned alongside the valve. Similarly any arrow indicating direction of rotation must face the right way.

8   Starting at the point furthest from the valve, push the tyre bead over the edge of the wheel rim until it is located in the central well. Continue to work around the tyre in this fashion until the whole of one side of the tyre is on the rim. It may be necessary to use a tyre lever during the final stages. Here again, the use of a lubricant will aid fitting. It is recommended strongly that when refitting the tyre only a recommended lubricant is used because such lubricants also have sealing properties. Do not be over generous in the application of lubricant or tyre creep may occur.

9   Fitting the upper bead is similar to fitting the lower bead. Start by pushing the bead over the rim and into the well at a point diametrically opposite the tyre valve. Continue working round the tyre, each side of

the starting point, ensuring that the bead opposite the working area is always in the well. Apply lubricant as necessary. Avoid using tyre levers unless absolutely essential, to help reduce damage to the soft wheel rim. The use of the levers should be required only when the final portion of bead is to be pushed over the rim.

10 Lubricate the tyre beads again prior to inflating the tyre, and check that the wheel rim is evenly positioned in relation to the tyre beads. Inflation of the tyre may well prove impossible without the use of a high pressure air hose. The tyre will retain air completely only when the beads are firmly against the rim edges at all points and it may be found when using a foot pump that air escapes at the same rate as it is pumped in. This problem may also be encountered when using an air hose on new tyres which have been compressed in storage and by virtue of their profile hold the beads away from the rim edges. To overcome this difficulty, a tourniquet may be placed around the circumference of the tyre, over the central area of the tread. The compression of the tread in this area will cause the beads to be pushed outwards in the desired direction. The type of tourniquet most widely used consists of a length of hose closed at both ends with a suitable clamp fitted to enable both ends to be connected. An ordinary tyre valve is fitted at one end of the tube so that after the hose has been secured around the tyre it may be inflated, giving a constricting effect. Another possible method of seating beads to obtain initial inflation is to press the tyre into the angle between a wall and the floor. With the airline attached to the valve additional pressure is then applied to the tyre by the hand and shin, as shown in the accompanying illustration. The application of pressure at four points around the tyre's circumference whilst simultaneously applying the airhose will often effect an initial seal between the tyre beads and wheel rim, thus allowing inflation to occur.

11 Having successfully accomplished inflation, increase the pressure to 40 psi and check that the tyre is evenly disposed on the wheel rim. This may be judged by checking that the thin positioning line found on each tyre wall is equidistant from the rim around the total circumference on the tyre. If this is not the case, deflate the tyre, apply additional lubrication and reinflate. Minor adjustments to the tyre position may be made by bouncing the wheel on the ground.

12 Always run the tyre at the recommended pressures and never under or over-inflate. The correct pressures for various weights and configurations are given in the Specifications Section of this Chapter.

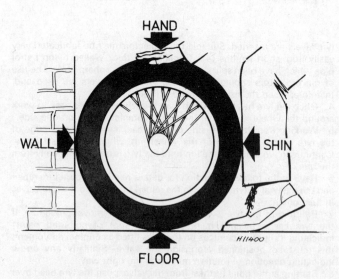

Fig. 6.11 Method of seating the beads on tubeless tyres

13 Puncture repair and tyre renewal

1 If a puncture occurs, the tyre should be removed for inspection for damage before any attempt is made at remedial action. The temporary repair of a punctured tyre by inserting a plug from the outside should not be attempted. Although this type of temporary repair is used widely on cars, the manufacturers strongly recommend that no such repair is

carried out on a motorcycle tyre. Not only does the tyre have a thinner carcass, which does not give sufficient support to the plug, the consequences of a sudden deflation are often sufficiently serious that the risk of such an occurrence should be avoided at all costs.

2 The tyre should be inspected both inside and out for damage to the carcass. Unfortunately the inner lining of the tyre – which takes the place of the inner tube – may easily obscure any damage and some experience is required in making a correct assessment of the tyre condition.

3 There are two main types of tyre repair which are considered safe for adoption in repairing tubeless motorcycle tyres. The first type of repair consists of inserting a mushroom-headed plug into the hole from the inside of the tyre. The hole is prepared for insertion of the plug by reaming and the application of an adhesive. The second repair is carried out by buffing the inner lining in the damaged area and applying a cold or vulcanised patch. Because both inspection and repair, if they are to be carried out safely, require experience in this type of work, it is recommended that the tyre be placed in the hands of a repairer with the necessary skills, rather than repaired in the home workshop.

4 In the event of an emergency, the only recommended 'get-you-home' repair is to fit a standard inner tube of the correct size. If this course of action is adopted, care should be taken to ensure that the cause of the puncture has been removed before the inner tube is fitted. It will be found that the valve hole in the rim is considerably larger than the diameter of the inner tube valve stem. To prevent the ingress of road dirt, and to help support the valve, a spacer should be fitted over the valve. A conversion spacer for most Honda models equipped with Comstar wheels is available from Honda dealers.

5 In the event of the unavailability of tubeless tyres, ordinary tubed tyres fitted with inner tubes of the correct size may be fitted. Refer to the manufacturer or a tyre fitting specialist to ensure that only a tyre and tube of equivalent type and suitability is fitted, and also to advise on the fitting of a valve nut/spacer to the rim hole.

6 Honda recommend that a repaired tyre should not be used at speeds of over 50 mph (80 km/h) for the first 24 hours after the repair has taken place, and that a repaired tyre should **never** be used at speeds of over 80 mph (120 km/h).

14 Tyre valves: description and renewal – Comstar wheel type

1 It will be appreciated from the preceding Sections, that the adoption of tubeless tyres has made it necessary to modify the valve arrangement, as there is no longer an inner tube which can carry the valve core. The problem has been overcome by using a moulded rubber valve body which locates in the wheel rim hole. The valve body is pear-shaped, and has a groove around its widest point which engages with the rim forming an airtight seal.

2 The valve is fitted from the rim well, and it follows that it can only be renewed whilst the tyre itself is removed from the wheel. Once the valve has been fitted, it is almost impossible to remove it without damage, and so the simplest method is to cut it as close as possible to the rim well. The two halves of the old valve can then be removed.

3 The new valve is fitted by inserting the threaded end of the valve body through the rim hole, and pulling it through until the groove engages in the rim. In practice, a considerable amount of pressure is required to pull the valve into position, and most tyre fitters have a special tool which screws onto the valve end to enable purchase to be obtained. It is advantageous to apply a little tyre bead lubricant to the valve to ease its insertion. Check that the valve is seated evenly and securely.

4 The incidence of valve body failure is relatively small, and leakage only occurs when the rubber valve case ages and begins to perish. As a precautionary measure, it is advisable to fit a new valve when a new tyre is fitted. This will preclude any risk of the valve failing in service. When purchasing a new valve, it should be noted that a number of different types are available. The correct type for use in the Comstar wheel is a Schrader 413, Bridgeport 183M or equivalent.

5 The valve core is of the same type as that used with tubed tyres, and screws into the valve body. The core can be removed with a small slotted tool which is normally incorporated in plunger type pressure gauges. Some valve dust caps incorporate a projection for removing valve cores. Although tubeless tyre valves seldom give trouble, it is

Tyre changing sequence - tubeless tyres

Deflate tyre. After releasing beads, push tyre bead into well of rim at point opposite valve. Insert lever adjacent to valve and work bead over edge of rim.

Use two levers to work bead over edge of rim. Note use of rim protectors.

When first bead is clear; remove tyre as shown.

Before fitting, ensure that tyre is suitable for wheel. Take note of any sidewall markings such as direction of rotation arrows.

Work first bead over the rim flange.

Use a tyre lever to work the second bead over rim flange.

possible for a leak to develop if a small particle of grit lodges on the sealing face. Occasionally, an elusive slow puncture can be traced to a leaking valve core, and this should be checked before a genuine puncture is suspected.

6   The valve dust caps are a significant part of the tyre valve assembly. Not only do they prevent the ingress of road dirt into the valve, but also act as a secondary seal which will reduce the risk of sudden deflation if a valve core should fail.

7   Note that while this Section specifically concerns Comstar wheel valves, those valves fitted to the CX650 C and GL650 models cast wheels are treated in exactly the same manner.

15 Wheel balancing

1   It is customary on all high performance machines to balance the wheels complete with tyre and tube. The out of balance forces which exist are eliminated and the handling of the machine is improved in consequence. A wheel which is badly out of balance produces through the steering a most unpleasant hammering effect at high speeds.

2   Some tyres have a balance mark on the sidewall, usually in the form of a coloured spot. This mark must be in line with the tyre valve, when the tyre is fitted. Even then the wheel may require the addition of balance weights, to offset the weight of the tyre valve itself.

3   If the wheel is raised clear of the ground and is spun, it will probably come to rest with the tyre valve or the heaviest part downward and will always come to rest in the same position. Balance weights must be added to a point diametrically opposite this heavy spot until the wheel will come to rest in ANY position after it is spun.

4   Weights are available in 20 and 30 gram sizes for all models; note that a 10 gram weight is also available for the CX500 E-D and CX650 E-D models. On all wheel types the weights are clamped to the rim central flange.

5   It is necessary to disconnect the speedometer cable and to remove the brake caliper(s) to enable the front wheel to spin with absolute freedom; in some cases it may be necessary to slacken the front wheel spindle nut. When balancing the rear wheel, it must be removed from the machine and arranged on a purpose-built stand to release it from the drag of the brake and final drive.

Chapter 7 Electrical system

Contents

Specifications

Note: *unless otherwise stated, information applies to all models*

Electrical system
Voltage	12
Earth (ground)	Negative (-)
Battery capacity	12v 14Ah

Alternator
Type	Three-phase
Output:	
All CX500, CX500 D and CX500 C models except	
CX500 C 1982	170W @ 5000 rpm
All other models	252W @ 5000 rpm (18A min/14.0V)

Voltage regulator
Type	Electronic integrated circuit, non-adjustable
Regulated voltage	14 – 15 volts

Starter motor
Brush spring pressure	495 – 605 gm (17.5 – 21.3 oz)
Service limit	400 gm (14.1 oz)
Brush length	11.0 – 12.5 mm (0.43 – 0.49 in)
Service limit	5.5 mm (0.22 in)

Bulbs

	UK models	US models
Headlamp:		
CX500 1978, 1979, CX500 D 1979	N/App	12V, 50/40W*
CX500 C 1979, 1980, CX500 D 1980	N/App	12V, 65/50W*
All other models	12V, 60/55W	12V, 60/55W
Parking lamp	12V, 4W	N/App
Front turn signals	12V, 21W	12V, 23/8W (32/3cp)
Rear turn signals	12V, 21W	12V, 23W (32cp)
Number/license plate lamp – where fitted	12V, 10W	12V, 10W (4cp)
Stop/tail lamp	12V, 21/5W	12V, 23/8W (32/3cp)
Instrument illuminating and warning lamps	12V, 3.4W	12V, 3.4W (2 cp)

*Indicates a sealed-beam unit

Fuses

	All CX500 models, CX500 C 1979, 1980 CX500 D 1979, 1980	CX500 C, CX500 D 1981	CX500 E-C, GL500 1981, GL500 I 1981
Main	20A	30A	30A
Headlamp	10A	10A	10A
Instruments, tail lamp	10A	10A	10A

	GL500 D-C, CX500 C 1982, GL500 and GL500 I 1982	CX650 C, CX650 E-D	All GL650 models
Main	30A	30A	30A
Headlamp	10A	15A	15A
Oil and neutral warning lamps	10A	15A	15A
Turn signals, stop lamp, horn	10A	15A	15A
Instruments, tail lamp	10A	15A	15A
Accessory	5A	N/App	5A
Interstate fairing fuses:			
GL500 I 1981	10A, 5A		
GL500 I 1982, GL500 D-C, GL650 I, GL650 D2-E	1A, 2A, 5A		

1 General description

The 12 volt electrical system is powered by a three-phase alternating current generator (alternator). The permanent magnet rotor is mounted on the extreme rear end of the crankshaft where it is driven at engine speed and the multi-coil stator is mounted in the engine rear cover.

The alternating current (ac) produced by the alternator is passed through a combined integrated circuit (IC) regulator/rectifier unit. The

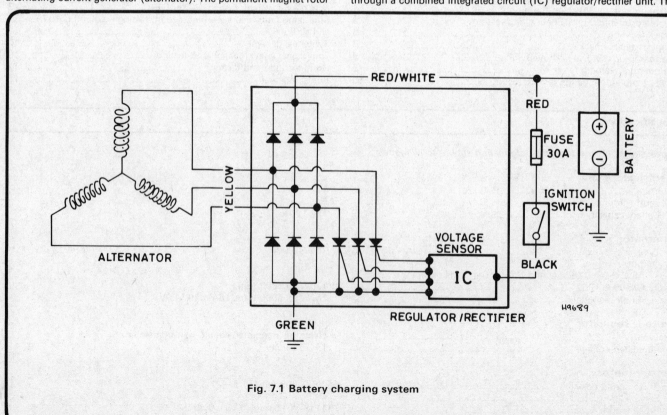

Fig. 7.1 Battery charging system

ac curent passed through the rectifying components of the unit is converted to direct current (dc) to enable the lights and ancillary electrical equipment to be powered and the battery to be charged. The regulator side of the unit ensures that the voltage does not exceed a predetermined level of 14 – 15 volts.

2 Electrical system: general information and preliminary checks

1 In the event of an electrical system fault, always check the physical condition of the wiring and connectors before attempting any of the test procedures described here and in subsequent Sections. Look for chafed, trapped or broken electrical leads and repair or renew these as necessary. Leads which have broken internally are not easily spotted, but may be checked using a multimeter or a simple battery and bulb circuit as a continuity tester. This arrangement is shown in the accompanying illustration. The various multi-pin connectors are

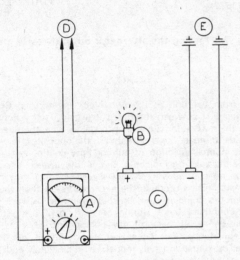

Fig. 7.2 Simple testing arrangement for checking the wiring

A Multimeter	D Positive probe
B Bulb	E Negative probe
C Battery	

generally trouble-free but may corrode if exposed to water. Clean them carefully, scraping off any surface deposits, and pack with silicone grease during assembly to avoid recurrent problems. The same technique can be applied to the handlebar switches.

2 A sound, fully charged battery is essential to the normal operation of the system. There is no point in attempting to locate a fault if the battery is partly discharged or worn out. Check battery condition and recharge or renew the battery before proceeding further.

3 Many of the test procedures described in this Chapter require that voltages or resistances be checked. This necessitates the use of some form of test equipment such as a simple and inexpensive multimeter of the type sold by electronics or motor accessory shops.

4 If you doubt your ability to check the electrical system entrust the work to a Honda Service Agent. In any event have your findings double checked before consigning expensive components to the scrap bin.

3 Battery: examination and maintenance

1 Details of the regular checks needed to maintain the battery in good condition are given in Routine Maintenance, together with

instructions on removal and refitting and general battery care. Batteries can be dangerous if mishandled; read carefully the 'Safety First' section at the front of this Manual before starting work, and always wear overalls or old clothing in case of accidental acid spillage. If acid is ever allowed to splash into your eyes or on to your skin, flush it away with copious quantities of fresh water and seek medical advice immediately.

2 When new, the battery is filled with an electrolyte of dilute sulphuric acid having a specific gravity of 1.280 at 20°C (68°F). Subsequent evaporation, which occurs in normal use, can be compensated for by topping up with distilled or demineralised water only. Never use tap water as a substitute and do not add fresh electrolyte unless spillage has occurred.

3 The state of charge of a battery can be checked using an hydrometer.

4 The normal charge rate for a battery is $1/10$ of its rated capacity, thus for a 14 ampere hour unit charging should take place at 1.4 amp. Exceeding this figure could cause the battery to overheat, buckling the plates and rendering it useless. Few owners will have access to an expensive current controlled charger, so if a normal domestic charger is used check that after a possible initial peak, the charge rate falls to a safe level. If the battery becomes hot during charging **stop.** Further charging will cause damage. Note that cell caps should be loosened and vents unobstructed during charging to avoid a build-up of pressure and risk of explosion.

5 After charging top up with distilled water as required, then check the specific gravity and battery voltage. Specific gravity should be above 1.270 and a sound, fully charged battery should produce 15 – 16 volts. If the recharged battery discharges rapidly if left disconnected it is likely that an internal short caused by physical damage or sulphation has occurred. A new battery will be required. A sound item will tend to lose its charge at about 1% per day.

4 Charging system: checking the output

1 Remove the left-hand side cover from the frame so that access can be made to the battery, which must be fully charged for the test to be accurate. Connect a 0 – 20 volt dc voltmeter (or a multi-meter switched to a similar range) across the battery terminals. Place an ammeter in line with the starter solenoid negative (–) terminal and the Red/White lead that runs from the terminal on early models; on later models with the 30A ribbon-type fuse (see Specifications), remove the fuse and connect the ammeter across the fuse terminals. See the accompanying diagrams. Disconnect the Black wire from the voltage regulator/rectifier unit. This will isolate the regulator, taking it out of circuit.

2 Start the engine and allow it to warm up. Increase the engine speed from tick-over. At 1300 – 1600 rpm a definite charge should be evident. Increase the engine speed smoothly to 5000 rpm, keeping it at this speed only long enough for a reading of the ammeter and voltmeter to be taken. The specified output at this speed is 5 amperes minimum and 14.5 volts on early models, 8A/14.0 volts on later models.

3 If the output is erratic or noticeably below the specified amount, either the alternator or the rectifier may be at fault. The rectifier may be tested as described in Section 5. The alternator stator should be tested as follows, using a multi-meter set to the resistance function. Disconnect the block connector which interconnects the three Yellow wires running from the alternator. The block connector is clipped to the rear mudguard, below the dualseat. Using the multi-meter check that continuity exists between all three wires when tested in pairs. Check also that no lead has continuity with earth. If the results of the check do not correspond with those specified, there is evidence of short-circuits or open-circuits in the stator windings or the leads.

4 A visual check of the stator coils and leads can be carried out only after the engine has been removed and the rear cover detached. Removal of the stator is described in Chapter 1. To check for breakage in the individual leads, use a multi-meter set to the resistance function. Due to the high cost of a replacement stator, which is available only as a complete unit, including the ignition fixed pulser on CDI ignition systems only, it is wise to have the assembly double-checked by a Honda Service Agent or auto-electrician, before consigning it to the scrap bin.

4.3 Alternator stator is tested at connectors on rear mudguard

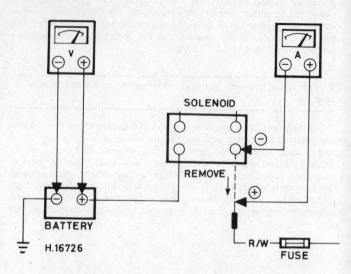

Fig. 7.3 Testing the alternator output – early models

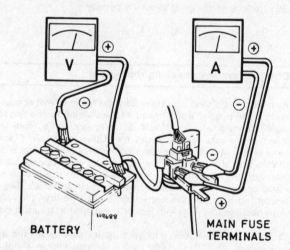

Fig. 7.4 Testing the alternator output – later models

5 Regulator/rectifier unit: location and testing

1   If performance of the charging system is suspect, but the alternator is found to be in good condition, it is probable that one side of the combined regulator/rectifier unit is malfunctioning. Exactly which side is malfunctioning is of academic interest only because the sealed unit cannot be repaired, but must be renewed.

2   Voltage regulator performance may be checked using a voltmeter connected across the battery. Connect the positive voltmeter lead to the positive (+) battery terminal and the negative voltmeter lead to the negative (–) battery terminal. Start the engine and increase the speed until a 14 volt output is registered. Increase the engine speed to approximately 5,000 rpm. If the regulator is functioning correctly, the voltage will not rise to above 15 volts. If this voltage is exceeded, the unit is malfunctioning.

3   A static test on the rectifier side of the unit may be carried out using a multi-meter set to the resistance function. Disconnect the leads

running from the unit at the two block connectors. Connect the multi-meter first between the green wire and each yellow wire and between the red/white wire and each yellow wire, then transpose the meter leads to repeat these tests, but in the opposite direction.

4   In the normal direction of current flow continuity (ie very little resistance) should be indicated, but in the reverse direction there should be no continuity (very heavy resistance). If any one of the twelve tests fails to produce the expected result then that diode is faulty and the complete unit must be renewed. Always have your findings confirmed by a Honda Service Agent using the correct equipment before purchasing a new component.

5   The regulator/rectifier is a sealed, heavily finned unit mounted to the rear of the battery on early models, or immediately underneath the battery carrier on later models.

6 Fuses: location

1   Refer to Specifications and/or the relevant wiring diagram for details of the fuses fitted.

2   The main fuse is a 20A tubular type on early models, retained in a white plastic holder which is clipped next to the starter relay. On later models a 30A ribbon type fuse is fitted in a black plastic holder mounted in the same place. To renew these fuses unplug the connector block, open the cover and slacken the two retaining screws. Always have a spare on the machine as these are likely to be difficult to find; in cases of real emergency a short length of 30A fuse wire should serve as a temporary substitute.

3   Further fuses protect the various circuits, these being housed below a cover on the handlebar clamp for easy access. Fuses are fitted to protect the electrical system in the event of a short circuit or sudden surge; they are, in effect, an intentional 'weak link', which will blow, in preference to the circuit burning out.

4   Before replacing a fuse that has blown, check that no obvious short circuit has occurred, otherwise the replacement fuse will blow immediately it is inserted. It is always wise to check the electrical circuit thoroughly, to trace the fault and eliminate it.

5   When a tubular fuse blows while the machine is running and no spare is available, a 'get you home' remedy is to remove the blown fuse and wrap it in silver paper before replacing it in the fuseholder. The silver paper will restore the electrical continuity by bridging the broken fuse wire. This expedient should **never** be used if there is evidence of a short circuit or other major electrical fault, otherwise more serious damage will be caused. Replace the 'doctored' fuse at the earliest possible opportunity, to restore full circuit protection. It follows that spare fuses that are used should be replaced as soon as possible to prevent the above situation from arising.

5.5a Location of regulator/rectifier unit – early models ...

5.5b ... and later models

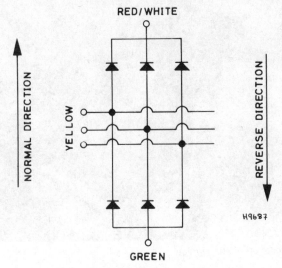

Fig. 7.5 Rectifier test

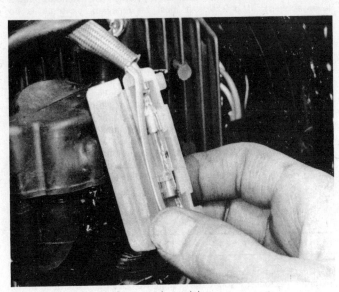

6.2a Location of main fuse – early models ...

6.2b ... and later models

7 Starter system: checks

1 In the event of a starter malfunction, always check first that the battery is fully charged. A partly discharged battery may be able to provide enough power for the lighting circuit, but not the very heavy current required for starting the engine.

2 Remove the left-hand side panel and note the location of the starter relay. This is mounted on the rear of the battery tray and can be identified by the two heavy duty cables connected to two of its four terminals. Switch on the ignition and press the starter button. If the relay is operating a distinct click will be heard as the internal solenoid closes the starter lead contact. A silent relay can be assumed to be defunct.

3 Disconnect the heavy duty starter lead at the motor terminal and connect a 12 volt test bulb between it and a sound earth point. Operate the starter switch again. If the bulb lights, the motor is being supplied with power and should be removed for overhaul.

4 To test the relay itself, disconnect all cables and wires and check that there is continuity across the relay battery and starter motor cable terminals when a fully-charged 12 volt battery is connected to the relay

switch terminals. If this is not the case, the relay is faulty and must be renewed.

5   If the relay is working properly, but not receiving any power, check back through the circuit; there may be a fault in the clutch interlock switch, the neutral interlock switch or in the diode between them. Test each as described below.

Neutral indicator switch

6   A small switch fitted to the engine rear cover operates a warning lamp in the instrument panel to indicate that neutral has been selected. More importantly, it is interconnected with the starter solenoid and will only allow the engine to be started if the gearbox is in neutral, unless the clutch is disengaged. It can be checked by setting a multimeter on the resistance scale and connecting one probe to the switch terminal and the other to earth. The meter should indicate continuity when neutral is selected and infinite resistance when in any gear. If the switch is being tested when removed there should be continuity between the top and bottom terminals and none between the top terminal and the body. If it is faulty, the switch must be renewed.

Clutch interlock switch

7   A small plunger-type switch is incorporated in the clutch lever, serving to prevent operation of the starter circuit when any gear has been selected, unless the clutch lever is held in. Check that there is continuity across the switch terminals only when it is extended (clutch lever applied). If defective it must be renewed, as there is no satisfactory means of repair. The switch can be removed after releasing the clutch cable and lever blade.

Interlock circuit diode

8   A small diode unit is fitted in the interlock circuit to ensure that the clutch switch can override the neutral switch, ie so that the machine can be started in gear when the clutch lever is pulled in. The diode is a small component mounted in the handlebar fuse case (see accompanying photograph) on all models up to 1981; after that it is a small black rectangular block with two spade terminals that is located just above the battery.

9   To test the diode, check that there is current flowing one way only, in the direction indicated by the arrow on the unit or in the relevant wiring diagram. If current flows in both directions, or in neither, the unit is faulty and must be renewed.

8   Starter motor: removal and overhaul

1   Check that the ignition switch is off and disconnect the battery negative (–) lead to isolate the system. Remove the starter lead at the motor terminal. Release the two motor retaining bolts and remove the motor by pulling it clear of the casing.

2   Remove the two long screws which retain the motor end covers and lift away the brush cover. Ease the brush plate away from the motor casing until the brushes spring clear of the commutator. Disengage the field coil brushes from the brush plate to free it.

3   Remove the end cover and reduction gear assembly then slide the armature out of the motor casing. Note that shims are fitted to both ends of the armature. These must be kept square and refitted in their original positions.

4   Measure the length of the brushes using a vernier caliper, renewing them as a set if any have worn beyond the 6.5 mm (0.26 in) service limit. If a spring balance or some weights are available, check that the brush springs are able to exert at least 680 gm (24.0 oz) pressure.

5   Clean the commutator segments with methylated spirit and inspect each one for scoring or discolouration. If any pair of segments is discoloured, a shorted armature winding is indicated. The manufacturer supplies no information regarding skimmed and re-cutting the armature in the event of serious scoring or burning, and so by implication suggests that a new armature is the only solution. It is suggested, however, that the advice of a vehicle electrical specialist is sought first; professional help may work out a lot cheaper.

6   Honda advise against cleaning the commutator segments with abrasive paper, presumably because of the risk of abrasive particles becoming embedded in the soft segments. It is suggested, therefore, that an ink eraser be used to burnish the segments and remove any surface oxide deposits before installing the brushes.

7   Using a multimeter set on the resistance scale, check the continuity between pairs of segments, noting that anything other than a very low resistance indicates a partially or completely open circuit. Next check the armature insulation by checking for continuity between the armature core and each segment. Anything other than infinite resistance indicates an internal failure.

8   Check the field coil windings by measuring the resistance between each brush and the terminal stud. A high resistance indicates an internal break in the windings. Repeat the test, this time between the brushes and the casing. No reading should be indicated, anything less than infinite resistance suggesting an insulation failure.

9   Reassemble the motor by reversing the dismantling sequence, noting that the reduction gears will benefit from a coating of molybdenum disulphide grease. Make sure that the shims are fitted correctly. Fit the brushes into their holders and ease the brush plate over the commutator, using a screwdriver to work the brushes into place. Note that the motor end covers are located by a slot which engages over a pin on the motor body.

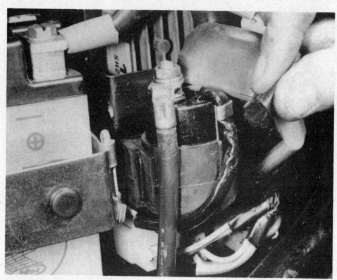

7.2 Location of starter relay

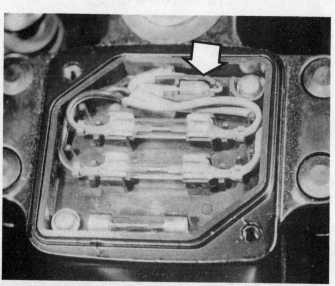

7.8a Starter interlock circuit diode (arrowed) is mounted with auxiliary fuses on early models ...

7.8b ... and separately on later models (arrowed)

9 Oil pressure warning lamp: testing

1   An oil pressure warning lamp is incorporated in the lubrication system to give immediate warning of excessively low oil pressure.
2   The oil pressure switch is screwed into the top of the engine front cover. The switch is connected to a warning light on the lighting panel on the handlebars. The light should be on when the ignition is switched on but will usually go out almost as soon as the engine is started.
3   If the oil warning lamp comes on whilst the machine is being ridden, the engine should be switched off immediately, otherwise there is a risk of severe engine damage due to lubrication failure.
4   If the lamp is thought to be faulty, check that the bulb is in good condition. If so, with the engine stopped disconnect the switch terminal, switch on the ignition and briefly earth the switch lead to the crankcase; the lamp should light. If not, the switch is faulty and must be renewed.
5   On fitting a new switch apply a liquid sealant to its threads and screw it in until two threads are left, then tighten it to a torque setting of 1.8 – 2.3 kgf m (13 – 16.5 lbf ft). Do not overtighten the switch or the engine front cover will be cracked.

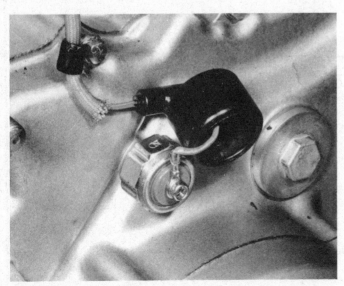

9.2 Oil pressure switch is screwed into engine front cover – do not overtighten

10 Water temperature gauge circuit: testing

1   The circuit consists of the sender unit mounted in the cylinder head water jacket and the gauge assembly mounted in the instrument panel.
2   The sender unit is removed and refitted as described in Chapter 2, while the gauge assembly is removed and refitted as described in Chapter 5. Apply a sealant such as Three Bond No. 1212 or equivalent to the sender unit threads on refitting.
3   To test the system, first ensure that the battery is fully charged by checking that all other electrical circuits work properly, then disconnect its wire from the sender unit and check that the ignition is switched on. The temperature gauge needle should point to 'C'. Earth the sender unit wire on the cylinder head, whereupon the needle should immediately swing over to 'H'. Do not earth the wire for more than 5 seconds or the gauge may be damaged.
4   If the needle moves as described, the sender unit is proven faulty and must be renewed, although a fuller check of its performance is given in paragraph 5 below. If the needle's movement is still faulty, or if it does not move at all, the gauge or wiring is faulty and must be checked until the fault is eliminated. Proceed as described below from paragraph 8 onwards.
5   To test fully the sender unit remove it from the machine; it can be checked by measuring its resistance at various temperatures. The task must be carried out in a well-ventilated area and great care must be taken to avoid the risk of personal injury. The equipment necessary is a heatproof container, a small gas-powered camping stove or similar, a thermometer capable of reading up to 120°C (250°F) and an ohmmeter or a multimeter set to the appropriate resistance scale.
6   Fill the container with oil and suspend the sender unit on some wire so that the probe end is immersed in it. Connect one of the meter leads to the unit body and the other to its terminal. Suspend the thermometer so that the bulb is close to the sender probe.
7   Start to heat the oil, and make a note of the resistance reading at each temperature shown in the table below. If the unit does not give readings which approximate quite closely to those shown, it must be renewed.

60°C (140°F)	104.0 ohms
85°C (185°F)	43.9 ohms
110°C (232°F)	20.3 ohms
120°C (250°F)	16.1 ohms

8   To test the gauge assembly, dismantle the instrument panel as described in Chapter 5 until the gauge terminals are exposed. Remove the three nuts and washers and disconnect the wires from their terminals, then test the wiring as follows.
9   On all models except the CX650 E-D and CX650 C the gauge unit supply is regulated to 7 volts by a voltage regulator (see Section 11); **do not** apply full battery voltage to the gauge during testing. Referring to the accompanying illustration, connect the sender unit, gauge and regulator to a fully-charged 12 volt battery as shown. Heat the sender unit as described above; at 50°C (122°F) the gauge needle should point to 'C', at 80°C (176°F) the needle should be just entering the white band and at 125°C (257°F) the needle should point to 'H'. If the gauge does not react as specified it is faulty and must be renewed.
10 On CX650 C and CX650 E-D models, switch on the ignition and use a dc voltmeter to check that full battery voltage is available, connecting the meter between the black/brown wire (CX650 C), or green/black (CX650 E-D) wire terminal and a suitable earth point on the frame. If battery voltage is not available the fault must be in the feed wire, the fuse, or in the ignition switch; check these carefully until the fault is located. Next use an ohmmeter or a multimeter set to the resistance scale to check the green/blue wire; first check that it is not broken by connecting the meter probe to the terminals at each end (gauge assembly and sender unit) of the wire and checking for continuity, then with the wire disconnected at both ends, connect the meter probes between one of the wire terminals and a good earth point on the frame. No continuity (ie infinite resistance) should be measured; if continuity exists, the wire is trapped or pinched somewhere along its length thus creating a short circuit.
10 If the wiring and sender unit are tested and found to be in good condition, the fault must be in the gauge assembly, which can only be renewed as repairs are not possible.

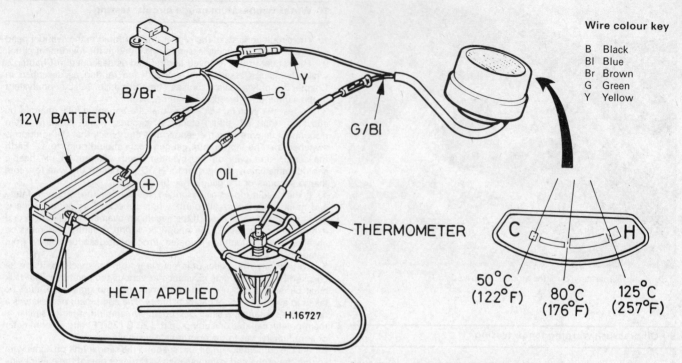

Wire colour key
B   Black
Bl   Blue
Br   Brown
G   Green
Y   Yellow

Fig. 7.6 Testing the temperature gauge unit – except CX650 C, CX650 E-D

11 Auxiliary voltage regulator: testing – except CX650 C, CX650 E-D

1   This unit is fitted to all models except the CX650 C and CX650 E-D to reduce to 7 volts the power supply to the water temperature gauge.

2   To test the unit connect it to a fully-charged 12 volt battery as shown in the accompanying illustration. Using a dc voltmeter or equivalent, check that the regulator output is exactly 7 volts.

3   If the output varies significantly from that specified, the unit is faulty and must be renewed.

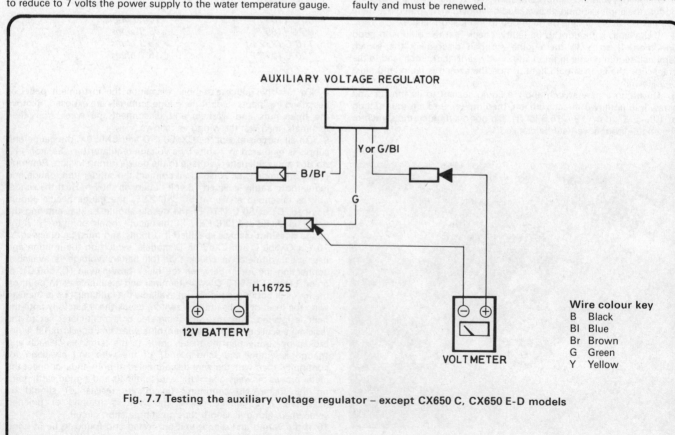

Wire colour key
B   Black
Bl   Blue
Br   Brown
G   Green
Y   Yellow

Fig. 7.7 Testing the auxiliary voltage regulator – except CX650 C, CX650 E-D models

11.1 Location of auxiliary voltage regulator on instrument panel – CX500 E-C

12 Electric fan thermostatic switch: testing – 650 models only

1   The quickest way of testing the switch performance is to remove the radiator filler cap (with the engine cold, only) so that a thermometer can be used to check the coolant temperature as the engine is started and allowed to warm up.

2   The switch should cut in the fan motor within the specified temperature range (see Chapter 2 Specifications) as the coolant warms up. The fan should remain on until the coolant temperature has fallen to the specified range, when the switch should cut out the fan. If the switch performance is reliable, but differs significantly from that specified, the owner must decide whether the switch is to be renewed.

3   If the switch is suspected of being faulty, disconnect the switch wires and bridge them with a short length of wire. Switch on the ignition. The fan motor should start immediately.

4   If the fan does start, either the switch is faulty or the coolant is not reaching full temperature; check this second possible cause as described in paragraphs 1 and 2 above. If the coolant temperature is high enough, then the switch is proved faulty and must be renewed.

5   If the fan still does not start, test for full battery voltage across the black (positive) and green (negative) wires leading to the main loom from the fan motor lead block connector when the ignition is switched on. If there is no power supply, work back through the wiring checking for loose, corroded or broken connectors, broken or shorted wires, loose terminals or faulty fuses.

13 Fuel gauge system: testing – CX500 E-C

1   The fuel gauge is controlled by a float-operated variable resistance, known as a sensor or sender unit, mounted inside the fuel tank. If the accuracy of the gauge is suspect, the sender resistances at various fuel levels should be checked as described below. Note that the gauge unit power supply is reduced to 7 volts by the regulator; **do not** apply full battery voltage.

2   Switch off the ignition and remove the key for safety. Place the machine on its centre stand on level ground and drain completely the fuel tank, taking the normal precautions to avoid any fire risk. Locate and separate the two-pin connector (Yellow/white and Green leads) below the front edge of the fuel tank. Set a multimeter to the 0-100 ohms scale and connect the test probes to the Yellow/white and Green leads on the sender unit side of the connector.

3   Measure the resistance of the sender unit at the reserve, half and full positions by adding the amount of fuel shown below.

Position	Litres	Imp gallon	Resistance
Reserve	6.0	1.32	95 – 100 ohms
Half full	10.5	2.31	31.5 ohms
Full	16	3.52	4 – 10 ohms

If the readings obtained correspond with those shown above, the sender unit can be considered serviceable; this means that the fault must lie in the instrument head. Check this by noting the gauge readings at each position.

4   If the sensor unit readings differed significantly from those shown, drain the fuel tank and remove it. Remove the four nuts which retain the sender unit to the base of the tank and lift it away, taking care not to twist or bend the float arm. Check that the float arm moves smoothly up and down with no signs of sticking. When fitting the sender, make sure that the O-ring seal is in good condition and check for leakage before refitting the tank.

5   Although the CX650 E-D model is fitted with a similar circuit, no information is available by which it can be tested. Note that on this model the gauge unit operates on full battery voltage.

14 Stop lamp bulb failure warning circuit: testing – CX500 C 1979, CX650 C

1   A stop lamp failure warning lamp is combined with the oil pressure warning lamp on the 500 model, and fitted separately on the 650 model.

2   If the lamp in the instrument console lights up whilst the brakes are applied this indicates failure of the combined stop and tail lamp bulb. In the event of this happening, stop the machine and check the operation of the bulb. It is most important that this check is made to obviate the possibility of there being a fault with the oil pressure system.

3   Refer to Section 9 for information relating to the oil pressure warning lamp.

4   If there is any doubt as to the correct operation of the system any faults can be eliminated as follows. Check the correct operation of the stop lamp bulb and the warning lamp bulb. Remove any corrosion that has built up in the bulbholders. By following the wiring diagram for this model trace all the relevant wires, checking for continuity. Make sure all wiring connectors are free from corrosion and are fitted correctly. If the unit continues to malfunction it must be assumed that the fault lies within the unit, which is a sealed component and must be renewed.

15 Switches: general

1   Generally speaking, the handlebar switches give little trouble but if necessary they can be dismantled by separating the halves which form a split clamp around the handlebars. Note that the machine cannot be started until the ignition cut-out on the right-hand end of the handlebars is turned to the central 'ON' position.

2   Always disconnect the battery before removing any of the switches, to prevent the possibility of a short circuit. Most troubles are caused by dirty contacts but in the event of the breakage of some internal part, it will be necessary to renew the complete switch, although in such a case there is little to be lost by attempting to dismantle the switch for repair.

3   Because the internal components of each switch are very small, and therefore difficult to dismantle and reassemble, it is suggested a special electrical contact cleaner be used to clean corroded contacts. This can be sprayed into each switch, without the need for dismantling.

4   In the event of a suspected malfunction the various switch contacts can be checked for continuity in a similar manner to that described in the preceding Sections. Details of the switch connections and the appropriate wiring will be found in the wiring diagrams that follow. Note that the electrical system should be isolated by disconnecting the battery lead to avoid short circuits.

5   In the case of the stop lamp switches, note that these should show continuity when fully extended, ie with the lever or pedal applied. Small switches such as these can only be renewed if faulty.

16 Turn signal relay: location and testing

1   The turn signal relay is a sealed cylindrical metal unit rubber mounted to protect it from the effects of vibration. On CX500 models it is mounted in the headlamp nacelle, on all later models it is immediately to the rear of the battery, or in the toolbox.

2   If the turn signal lamps cease to function correctly, there may be any one of several possible faults responsible which should be checked before the relay is suspected. First check that the lamps are correctly mounted and that all the earth connections are clean and tight. Check that the bulbs are of the correct wattage and that corrosion has not developed on the bulbs or in their holders. Any such corrosion must be thoroughly cleaned off to ensure proper bulb contact. Also check that the turn signal is functioning correctly and that the wiring is in good order. Finally ensure that the battery is fully charged.

3   Faults in any one or more of the above items will produce symptoms for which the turn signal relay may be blamed unfairly. If the fault persists even after the preliminary checks have been made, the relay must be at fault. Unfortunately the only practical method of testing the relay is to substitute a known good one. If the fault is then cured, the relay is proven faulty and must be renewed.

16.1 Location of turn signal relay (arrowed) – CX500 models

17 Horn: locating and testing

1   The horn is mounted on the bottom yoke or on the fairing mounting bracket via a flexible metal strip retained by a single bolt. Note that some models are fitted with twin horns. No maintenance is required other than regular cleaning to remove road dirt and occasional spraying with WD40 or a similar water dispersant spray to minimise internal corrosion.

2   If the horn fails to work, first check that the battery is fully charged. If full power is available, a simple test will reveal whether the current is reaching the horn. Disconnect the horn wires and substitute a 12 volt bulb. Switch on the ignition and press the horn button. If the bulb fails to light, check the horn button and wiring as described elsewhere in this Chapter. If the bulb does light, the horn circuit is proved good and the horn itself must be checked.

3   With the horn wires still disconnected, connect a fully charged 12 volt battery directly to the horn. If it does not sound, a sharp tap on the outside may serve to free the internal contacts. If this fails, the horn must be renewed as repairs are not possible.

4   Different types of horn may be be fitted; if a screw and locknut is

provided on the outside of the horn, the internal contacts may be adjusted to compensate for wear and to cure a weak or intermittent horn note. Slacken the locknut and rotate slowly the screw until the clearest and loudest note is obtained, then retighten the locknut. If no means of adjustment is provided on the horn fitted, it must be renewed.

18 Bulbs: renewal

Headlamp

1   On models with Interstate fairings, slacken the headlamp adjusting knob grub screw, withdraw the knob and unscrew the nut and washer behind it. Push gently on the knob while placing a hand over the headlamp to stop it from dropping clear and withdraw the headlamp assembly. Reverse to refit.

2   On CX500 E-C and CX650 E-D models remove the fairing top retaining screws and swing the fairing forwards. Unhook the cable stopper and withdraw the fairing. To remove the headlamp bulb it may be necessary to remove the two mounting bolts and to move the assembly forwards. Unplug the connector from the rear of the bulbs.

3   On all other models, remove the screws around the circumference of the chrome rim to free the headlamp assembly from the shell or nacelle.

4   Depending on the model, the headlamp bulb is either fitted into a removable bulb holder or is of the sealed beam type, in which case complete unit must be renewed if a fault develops.

5   On models fitted with a separate headlamp bulb, prise off the rubber boot which protects the bulb holder. The bulb is secured by two sprung arms which hinge from one side of the holder. Pinch the arms together to free them and then lift the bulb out. The bulb is of the tungsten-halogen type with a quartz envelope. It is important that the envelope is not touched with the fingers, because any greasy or acidic deposits will etch the quartz leading to the early failure of the bulb. If the envelope is touched inadvertently, it should be cleaned with a solvent such as methylated spirits. The parking lamp bulb holder is a press fit in the reflector, the bulb being of the bayonet fitting type. To remove the bulb, press it inwards and turn it anti-clockwise so that the bayonet pins disengage.

6   If a sealed beam headlamp becomes defective it is necessary to renew the complete glass/reflector unit as the bulb element is not detachable. The unit is held by screws passing through tabs on the headlamp rim. It will also be necessary to disconnect the beam adjustment which comprises two screws and tension springs which pass into captive nuts. Make an approximate note of the screw position to aid resetting the beam adjustment. No parking lamp bulb is fitted to the sealed beam unit.

Turn signal and stop/tail lamp

7   On CX500 and CX500 D models, two twin filament stop/tail lamp bulbs are used, each of which is fitted to a separate detachable holder. To gain access to the holders, the dualseat and tool box must be removed. Both holders pass through the rear wall of the seat fairing.

8   To remove a holder it should be twisted slightly so that the locating lugs disengage from the aperture, and then withdrawn. Both bulbs are of the bayonet type, with offset pins to prevent incorrect positioning.

9   On all other models the lamp lens must be removed to reach the bulb; also on all turn signal lamp assemblies. Except on the front turn signal lamp assemblies on later models where the lenses are clipped in place and can be dislodged by inserting a suitable coin in the slot provided and twisting; all lamp lenses are retained by screws. Take care not to tear the lens seal on removal and do not overtighten the screws on refitting or the lens may crack.

10 All turn signal lamp bulbs and the stop/tail lamp bulb are of the bayonet type and can be released by pushing in, turning anti-clockwise and pulling from the holder. The stop/tail lamp is fitted with a twin-filament bulb which has offset pins to prevent accidental reversal in its holder; the same applies to the front turn signal lamp bulbs of US models where the less powerful filament is separately switched to allow the lamps to function as daylight running lights. Note that it is essential that only the correct wattage bulbs are fitted, or the turn signal relay may break down.

Number/license plate lamp – CX500, CX500 D, CX650 C

11 The bulb on the CX650 C model is reached as described above.

12 On CX500 and CX500 D models the bulb fits into a holder directly

between and below the two stop/tail lamp bulb holders. The holder is a push fit in the seat fairing wall and is secured by a strip of metal held by a single screw.

Instrument illuminating and warning lamps

13 On CX500 models, to reach these bulbs the panel housing the instrument heads must be detached from the headlamp nacelle. Start by removing the headlamp rim/reflector unit which is held to the nacelle by two screws. The complete instrument assembly is secured by two studs and nuts in the roof of the headlamp nacelle. After removal of the nuts, detach the instrument drive cables by unscrewing the union ring below each instrument, and then lift the assembly up away from the nacelle as far as possible within the constraints of the various wiring leads. Disconnect the leads at the block connector and individual snap connectors to free the instrument/warning panel unit.

14 The two instrument heads and the warning lamp holders are mounted on a single bridge plate held to the instrument panel by four cross head screws. To gain access to the warning lamp bulbs, remove the screws and displace the panel face. The bulbs are all of the bayonet fitting type and may be removed with ease. Access to the instrument illumination lamps can be made after separating either instrument from the bridge plate. Each instrument is secured by two nuts on studs. The bulb holders are a push fit in the instrument bases and the bulbs are of the bayonet fitting type.

15 On CX500 E-C and CX650 E-D swing forward the headlamp fairing as described above; the bases of the rubber bulb holders are then easily reached so that they can be pulled out.

16 On all other models each instrument is illuminated by two bulbs which can be inspected by removing the chrome cover at the base of the instrument. Start by removing the relevant drive cable and remove the two domed nuts which retain the instrument to its mounting bracket. Note the positions of the washer, grommet and spacer when refitting. After removal of the chrome cover the rubber bulb holders can be prised from position in the base of the instrument.

17 To gain access to the warning lamp panel bulbs remove the four self-tapping screws and lift the cover away. Twist the bulb to disengage it from its holder and remove for inspection.

18.2 Headlamp bulb renewal, CX500 E-C – swing fairing forwards and remove, unscrew mounting bolts to release assembly ...

18.3a ... on models without fairings, remove retaining screws to release headlamp assembly ...

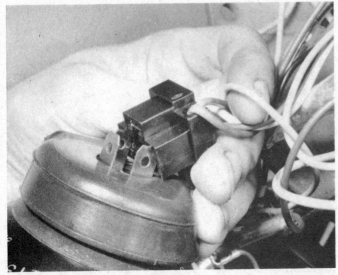

18.3b ... and detach connector to remove

18.5a Peel off rubber cover and displace spring retainer arms ...

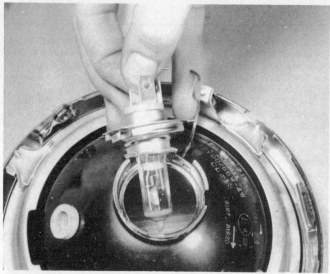

18.5b ... to release headlamp bulb (where fitted). **Do not** touch the glass

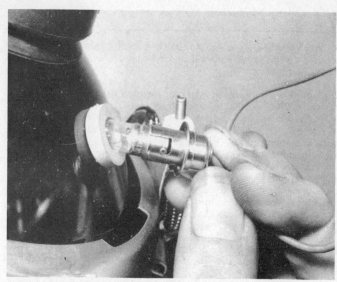

18.5c Parking lamp bulb (UK only) is a bayonet fitting in bulb holder

18.7 Stop/tail lamp bulb renewal – CX500, CX500 D models, twist bulb holder to release ...

18.9a ... on all other models, remove tail lamp lens to reach bulbs

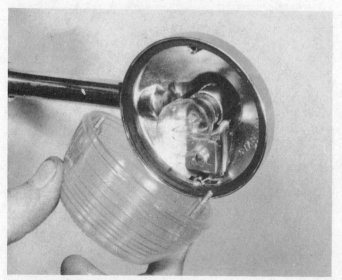

18.9b Turn signal lamp lenses are also retained by two screws on most models

18.10 All bulbs are of the bayonet type, twin filament bulbs will have offset pins to ensure correct refitting

18.12a Number/license plate lamp bulb renewal, CX500, CX500 D models-remove locating plate ...

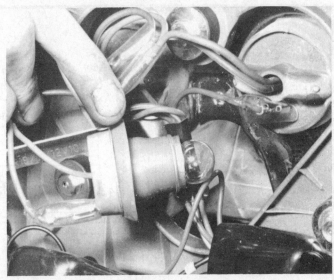

18.12b ... and withdraw bulb holder – bulb is a bayonet type

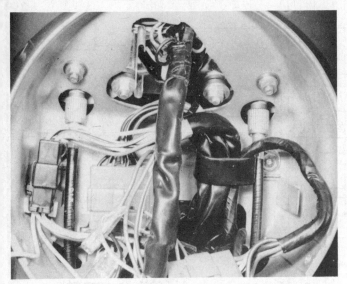

18.13a Instrument panel bulb renewal CX500 models – remove drive cables and two adjacent nuts ...

18.13b ... to allow panel to be raised

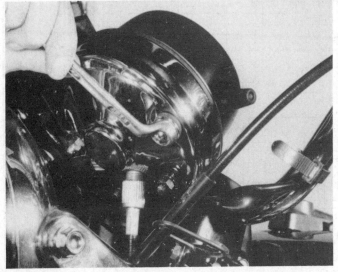

18.16 On most other models detach instrument bottom cover to reach bulb holders ...

18.17 ... or remove central cover to reach warning lamp bulbs

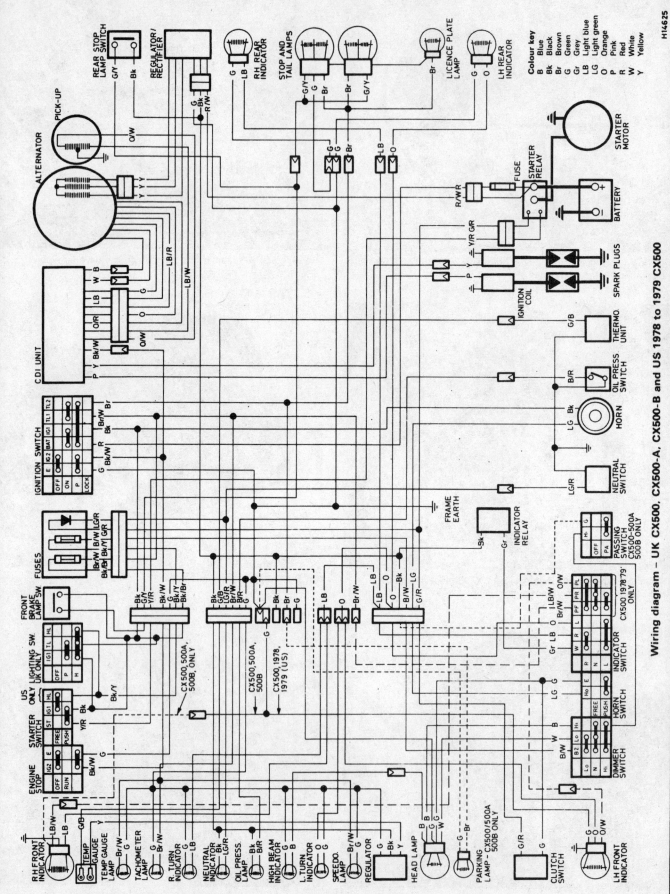

Wiring diagram – UK CX500, CX500-A, CX500-B and US 1978 to 1979 CX500

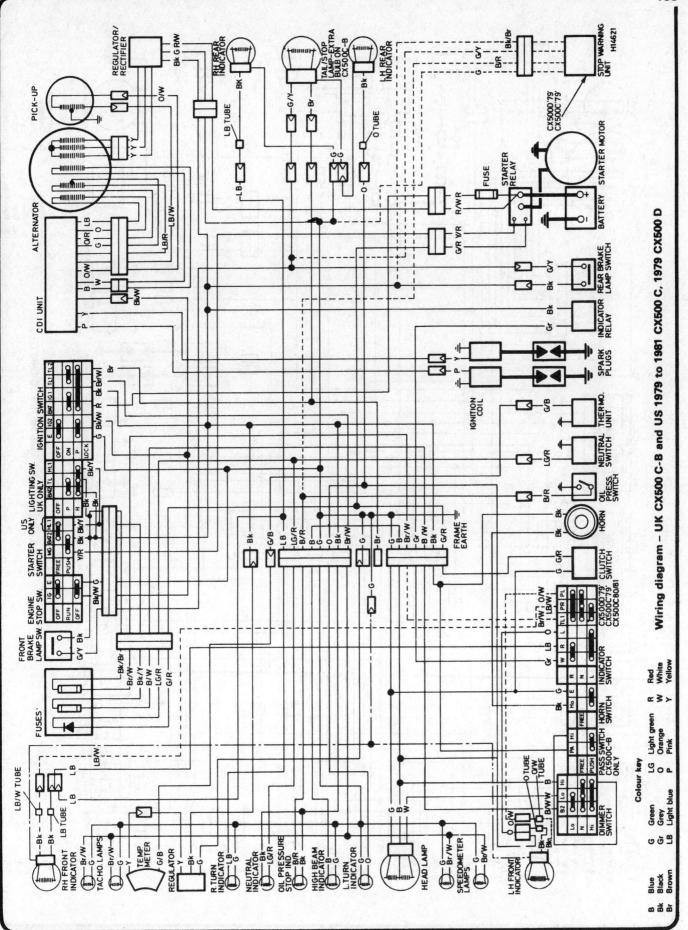

Wiring diagram – UK CX500 C-B and US 1979 to 1981 CX500 C, 1979 CX500 D

Colour key

B	Blue	G	Green	R	Red
Bk	Black	Gr	Grey	W	White
Br	Brown	LB	Light blue	Y	Yellow
		LG	Light green		
		O	Orange		
		P	Pink		

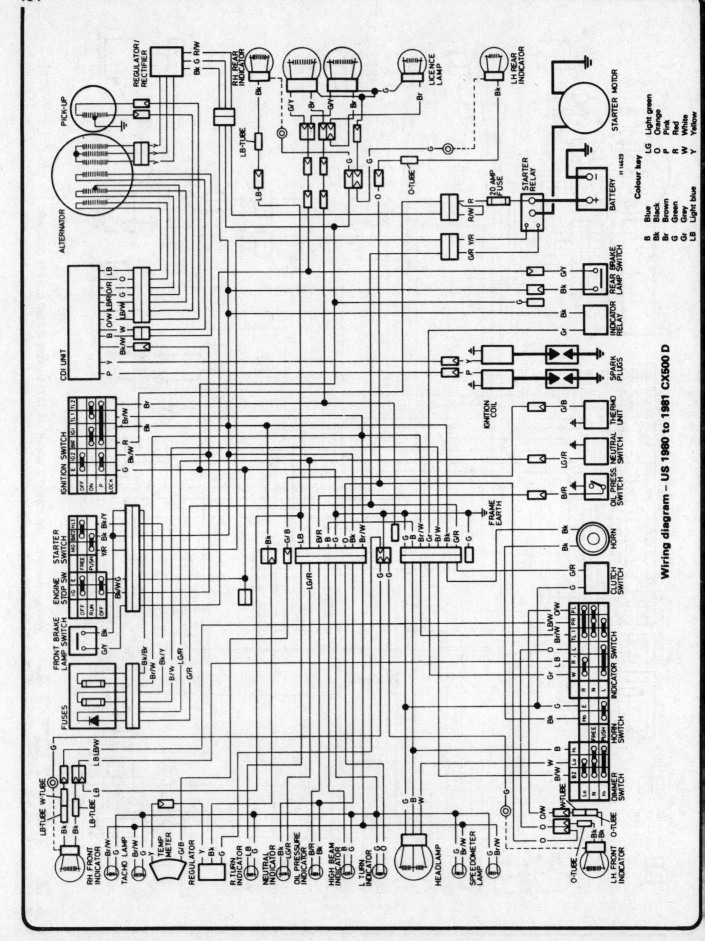

Wiring diagram – US 1980 to 1981 CX500 D

Colour key

B	Blue
Bk	Black
Br	Brown
G	Green
Gr	Grey
LB	Light blue

LG	Light green
O	Orange
P	Pink
R	Red
W	White
Y	Yellow

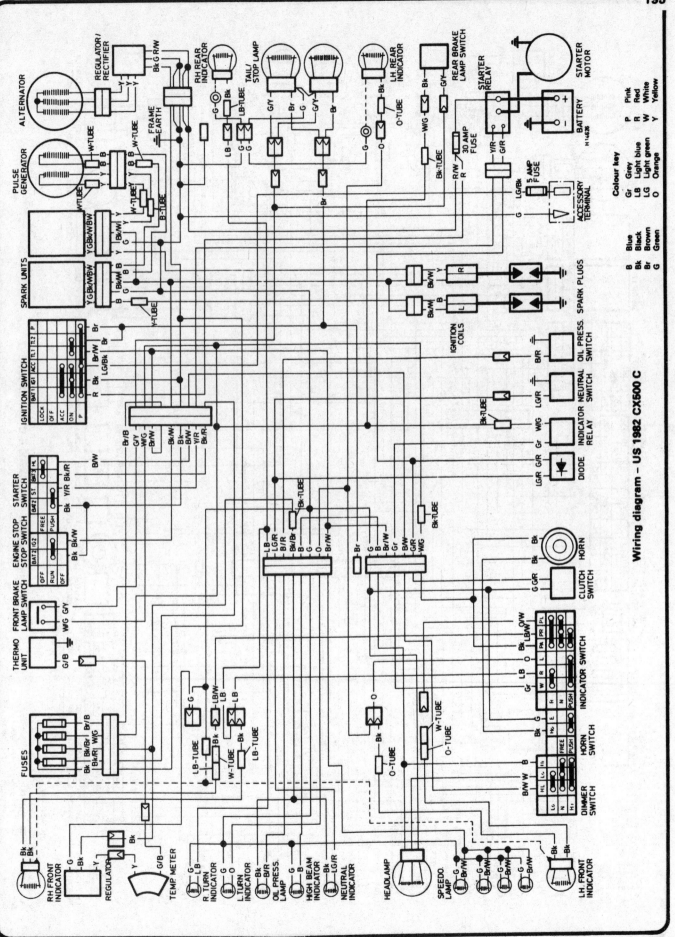

Wiring diagram – US 1982 CX500 C

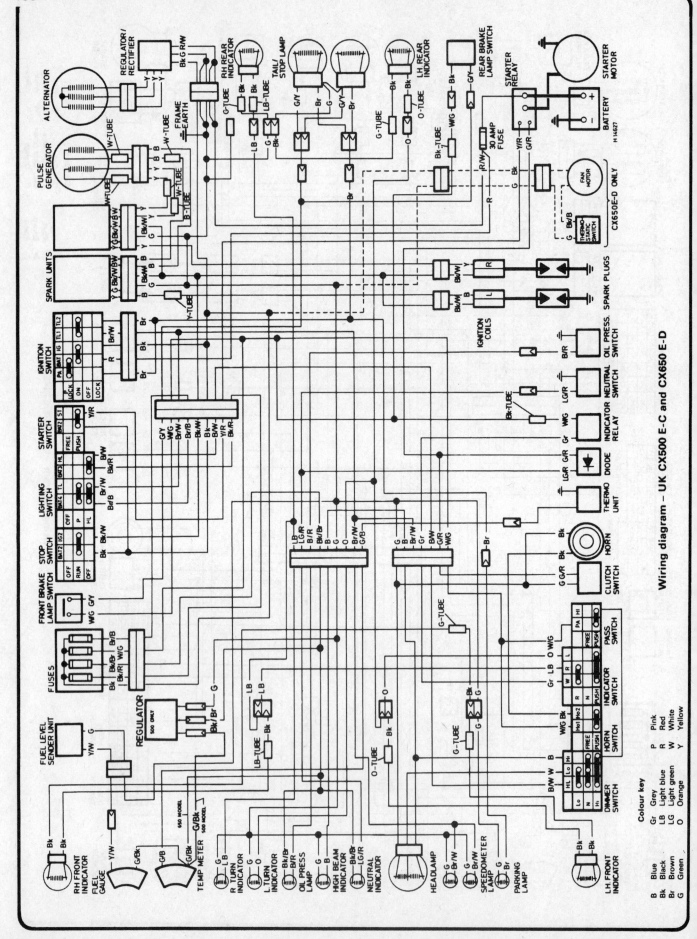

Wiring diagram – UK CX500 E-C and CX650 E-D

Colour key

B	Blue	P	Pink
Bk	Black	R	Red
Br	Brown	W	White
G	Green	Y	Yellow
Gr	Grey		
LB	Light blue		
LG	Light green		
O	Orange		

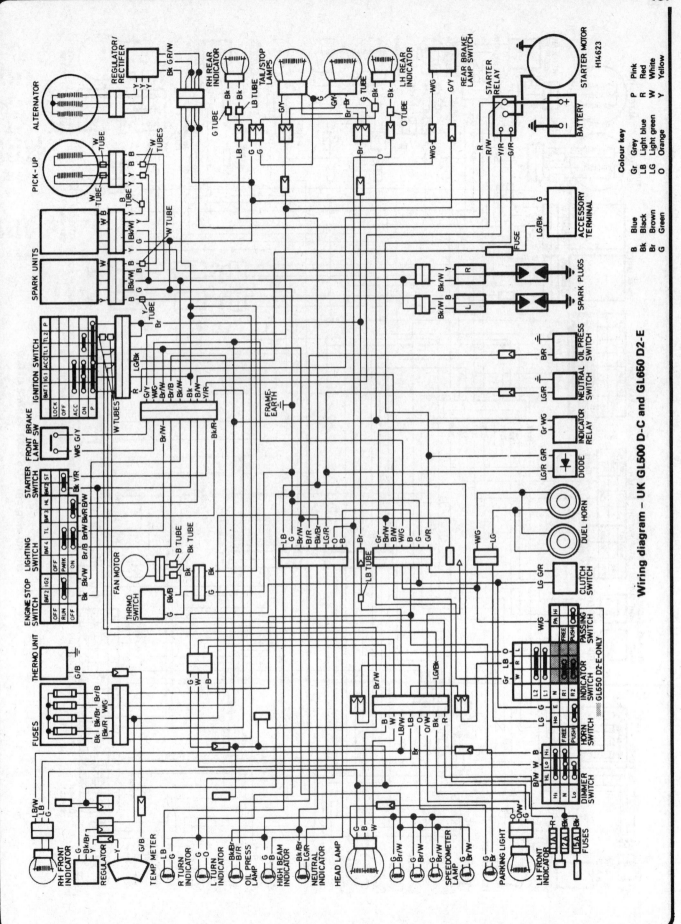

Wiring diagram – UK GL500 D-C and GL650 D2-E

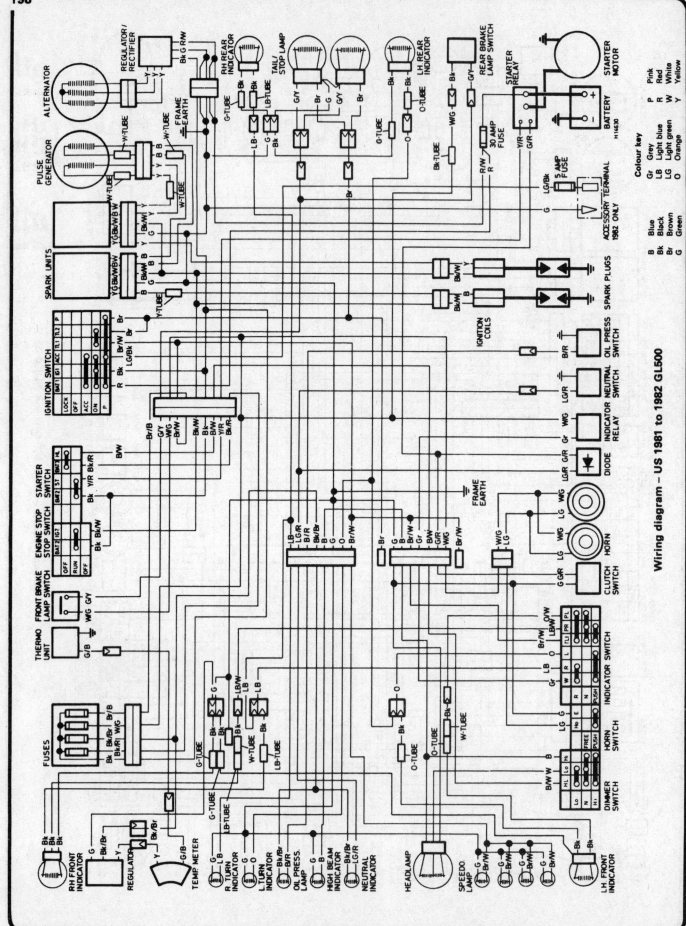

Wiring diagram – US 1981 to 1982 GL500

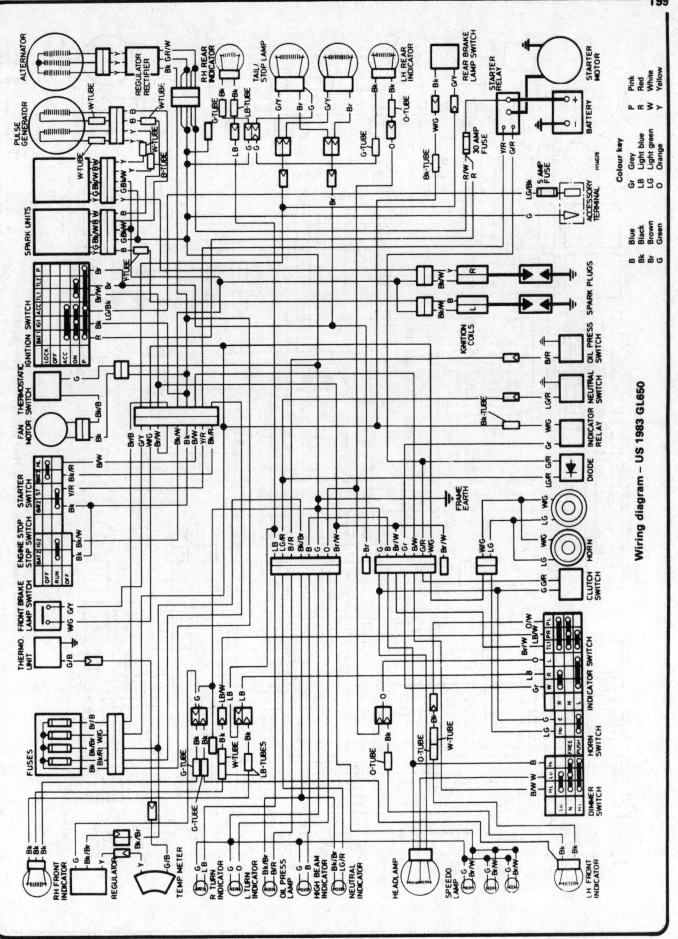

Wiring diagram – US 1983 GL650

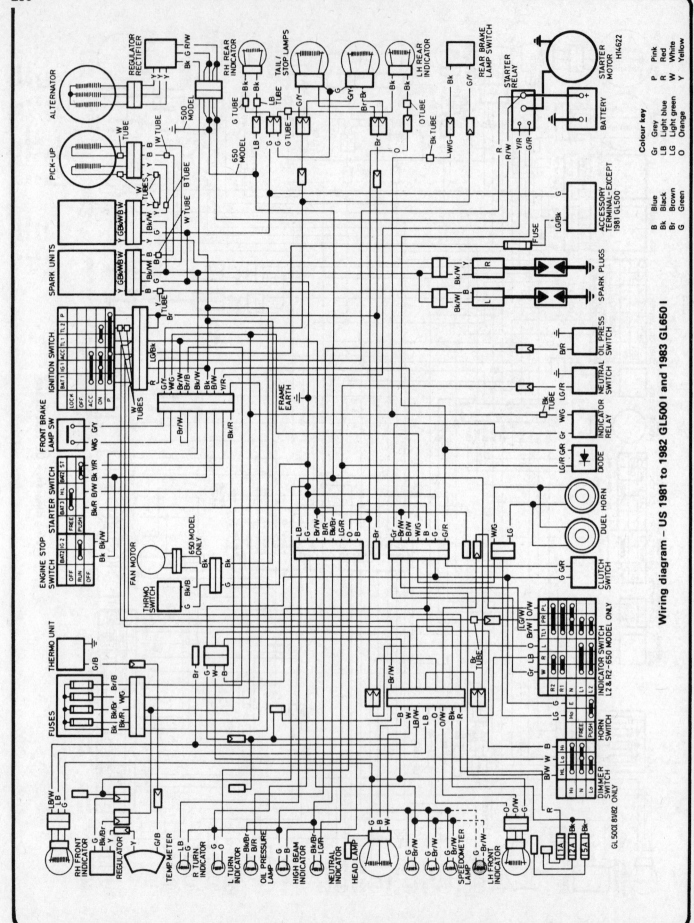

Wiring diagram – US 1981 to 1982 GL500 I and 1983 GL650 I

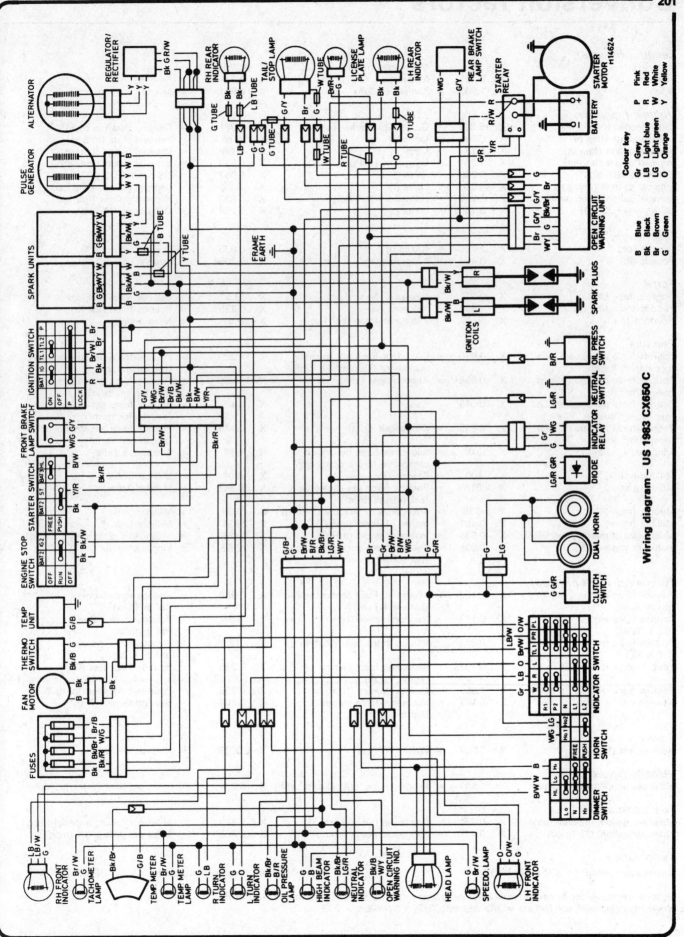

Wiring diagram – US 1983 CX650 C

Conversion factors

Length (distance)

Inches (in)	X	25.4	=	Millimetres (mm)	X	0.0394	= Inches (in)
Feet (ft)	X	0.305	=	Metres (m)	X	3.281	= Feet (ft)
Miles	X	1.609	=	Kilometres (km)	X	0.621	= Miles

Volume (capacity)

Cubic inches (cu in; in³)	X	16.387	=	Cubic centimetres (cc; cm³)	X	0.061	= Cubic inches (cu in; in³)
Imperial pints (Imp pt)	X	0.568	=	Litres (l)	X	1.76	= Imperial pints (Imp pt)
Imperial quarts (Imp qt)	X	1.137	=	Litres (l)	X	0.88	= Imperial quarts (Imp qt)
Imperial quarts (Imp qt)	X	1.201	=	US quarts (US qt)	X	0.833	= Imperial quarts (Imp qt)
US quarts (US qt)	X	0.946	=	Litres (l)	X	1.057	= US quarts (US qt)
Imperial gallons (Imp gal)	X	4.546	=	Litres (l)	X	0.22	= Imperial gallons (Imp gal)
Imperial gallons (Imp gal)	X	1.201	=	US gallons (US gal)	X	0.833	= Imperial gallons (Imp gal)
US gallons (US gal)	X	3.785	=	Litres (l)	X	0.264	= US gallons (US gal)

Mass (weight)

Ounces (oz)	X	28.35	=	Grams (g)	X	0.035	= Ounces (oz)
Pounds (lb)	X	0.454	=	Kilograms (kg)	X	2.205	= Pounds (lb)

Force

Ounces-force (ozf; oz)	X	0.278	=	Newtons (N)	X	3.6	= Ounces-force (ozf; oz)
Pounds-force (lbf; lb)	X	4.448	=	Newtons (N)	X	0.225	= Pounds-force (lbf; lb)
Newtons (N)	X	0.1	=	Kilograms-force (kgf; kg)	X	9.81	= Newtons (N)

Pressure

Pounds-force per square inch (psi; lbf/in²; lb/in²)	X	0.070	=	Kilograms-force per square centimetre (kgf/cm²; kg/cm²)	X	14.223	= Pounds-force per square inch (psi; lbf/in²; lb/in²)
Pounds-force per square inch (psi; lbf/in²; lb/in²)	X	0.068	=	Atmospheres (atm)	X	14.696	= Pounds-force per square inch (psi; lbf/in²; lb/in²)
Pounds-force per square inch (psi; lbf/in²; lb/in²)	X	0.069	=	Bars	X	14.5	= Pounds-force per square inch (psi; lbf/in²; lb/in²)
Pounds-force per square inch (psi; lbf/in²; lb/in²)	X	6.895	=	Kilopascals (kPa)	X	0.145	= Pounds-force per square inch (psi; lbf/in²; lb/in²)
Kilopascals (kPa)	X	0.01	=	Kilograms-force per square centimetre (kgf/cm²; kg/cm²)	X	98.1	= Kilopascals (kPa)
Millibar (mbar)	X	100	=	Pascals (Pa)	X	0.01	= Millibar (mbar)
Millibar (mbar)	X	0.0145	=	Pounds-force per square inch (psi; lbf/in²; lb/in²)	X	68.947	= Millibar (mbar)
Millibar (mbar)	X	0.75	=	Millimetres of mercury (mmHg)	X	1.333	= Millibar (mbar)
Millibar (mbar)	X	0.401	=	Inches of water (inH₂O)	X	2.491	= Millibar (mbar)
Millimetres of mercury (mmHg)	X	0.535	=	Inches of water (inH₂O)	X	1.868	= Millimetres of mercury (mmHg)
Inches of water (inH₂O)	X	0.036	=	Pounds-force per square inch (psi; lbf/in²; lb/in²)	X	27.68	= Inches of water (inH₂O)

Torque (moment of force)

Pounds-force inches (lbf in; lb in)	X	1.152	=	Kilograms-force centimetre (kgf cm; kg cm)	X	0.868	= Pounds-force inches (lbf in; lb in)
Pounds-force inches (lbf in; lb in)	X	0.113	=	Newton metres (Nm)	X	8.85	= Pounds-force inches (lbf in; lb in)
Pounds-force inches (lbf in; lb in)	X	0.083	=	Pounds-force feet (lbf ft; lb ft)	X	12	= Pounds-force inches (lbf in; lb in)
Pounds-force feet (lbf ft; lb ft)	X	0.138	=	Kilograms-force metres (kgf m; kg m)	X	7.233	= Pounds-force feet (lbf ft; lb ft)
Pounds-force feet (lbf ft; lb ft)	X	1.356	=	Newton metres (Nm)	X	0.738	= Pounds-force feet (lbf ft; lb ft)
Newton metres (Nm)	X	0.102	=	Kilograms-force metres (kgf m; kg m)	X	9.804	= Newton metres (Nm)

Power

Horsepower (hp)	X	745.7	=	Watts (W)	X	0.0013	= Horsepower (hp)

Velocity (speed)

Miles per hour (miles/hr; mph)	X	1.609	=	Kilometres per hour (km/hr; kph)	X	0.621	= Miles per hour (miles/hr; mph)

Fuel consumption*

Miles per gallon, Imperial (mpg)	X	0.354	=	Kilometres per litre (km/l)	X	2.825	= Miles per gallon, Imperial (mpg)
Miles per gallon, US (mpg)	X	0.425	=	Kilometres per litre (km/l)	X	2.352	= Miles per gallon, US (mpg)

Temperature

Degrees Fahrenheit = (°C x 1.8) + 32   Degrees Celsius (Degrees Centigrade; °C) = (°F - 32) x 0.56

*It is common practice to convert from miles per gallon (mpg) to litres/100 kilometres (l/100km),
where mpg (Imperial) x l/100 km = 282 and mpg (US) x l/100 km = 235

English/American terminology

Because this book has been written in England, British English component names, phrases and spellings have been used throughout. American English usage is quite often different and whereas normally no confusion should occur, a list of equivalent terminology is given below.

English	American	English	American
Air filter	Air cleaner	Number plate	License plate
Alignment (headlamp)	Aim	Output or layshaft	Countershaft
Allen screw/key	Socket screw/wrench	Panniers	Side cases
Anticlockwise	Counterclockwise	Paraffin	Kerosene
Bottom/top gear	Low/high gear	Petrol	Gasoline
Bottom/top yoke	Bottom/top triple clamp	Petrol/fuel tank	Gas tank
Bush	Bushing	Pinking	Pinging
Carburettor	Carburetor	Rear suspension unit	Rear shock absorber
Catch	Latch	Rocker cover	Valve cover
Circlip	Snap ring	Selector	Shifter
Clutch drum	Clutch housing	Self-locking pliers	Vise-grips
Dip switch	Dimmer switch	Side or parking lamp	Parking or auxiliary light
Disulphide	Disulfide	Side or prop stand	Kick stand
Dynamo	DC generator	Silencer	Muffler
Earth	Ground	Spanner	Wrench
End float	End play	Split pin	Cotter pin
Engineer's blue	Machinist's dye	Stanchion	Tube
Exhaust pipe	Header	Sulphuric	Sulfuric
Fault diagnosis	Trouble shooting	Sump	Oil pan
Float chamber	Float bowl	Swinging arm	Swingarm
Footrest	Footpeg	Tab washer	Lock washer
Fuel/petrol tap	Petcock	Top box	Trunk
Gaiter	Boot	Torch	Flashlight
Gearbox	Transmission	Two/four stroke	Two/four cycle
Gearchange	Shift	Tyre	Tire
Gudgeon pin	Wrist/piston pin	Valve collar	Valve retainer
Indicator	Turn signal	Valve collets	Valve cotters
Inlet	Intake	Vice	Vise
Input shaft or mainshaft	Mainshaft	Wheel spindle	Axle
Kickstart	Kickstarter	White spirit	Stoddard solvent
Lower leg	Slider	Windscreen	Windshield
Mudguard	Fender		

Index